IN

3

Elementary Statistical Quality Control

STATISTICS: Textbooks and Monographs

A SERIES EDITED BY

D. B. OWEN, Coordinating Editor
Department of Statistics
Southern Methodist University
Dallas, Texas

PAUL D. MINTON
Virginia Commonwealth University
Richmond, Virginia

JOHN W. PRATT
Harvard University
Boston, Massachusetts

OTHER VOLUMES IN PREPARATION

Elementary Statistical Quality Control

IRVING W. BURR

Professor Emeritus
Statistics Department
Purdue University
West Lafayette, Indiana

MARCEL DEKKER, INC.   New York and Basel

Library of Congress Cataloging in Publication Data

Burr, Irving Wingate [Date]
  Elementary statistical quality control.

  (Statistics, textbooks and monographs ; v. 25)
  Includes bibliographies and index.
  1. Quality control--Statistical methods.
I. Title.
TS156.B85   620'.004'5   78-11368
ISBN 0-8247-6686-5

MARCEL DEKKER, INC.

270 Madison Avenue, New York, New York 10016

Current printing (last digit):
10 9 8 7 6 5 4 3 2

PRINTED IN THE UNITED STATES OF AMERICA

To my wife, Elsie

NOTE FROM THE SERIES EDITOR

This book is one of a very small number of books on quality control which is both statistically sound and written at a level where an individual without a background in statistics can easily understand it. It is strongly recommended for inspectors and other users of quality control methods who have not had the opportunity to study mathematical statistics, but who still want an authoritative and easily read source describing the methods they are asked to use. Industrial engineers will find it especially useful, as will individuals studying business administration.

The author is one of the pioneers in quality control in the United States. He is one of only a few active founding members of the American Society for Quality Control and was editor of that society's journal, *Industrial Quality Control*, from 1961 to 1965. He received the Brumbaugh Award in 1950. This award is given each year to the author of that paper published in *Industrial Quality Control* which an American Society for Quality Control committee judges has made the largest single contribution to the development of industrial applications of quality control. In 1958 he was the Shewhart medalist, an award presented to an individual who has been deemed by a committee to have made the most outstanding contribution to the science and techniques of quality control or to have demonstrated outstanding leadership in the field of modern quality control. Burr's work has been recognized across the spectrum of professional societies from the American Society for Quality Control to the American Statistical Association, both of which elected him a fellow.

Irving W. Burr joined the mathematics faculty of Purdue University in 1941 where he stayed until his retirement in 1974. During his tenure at Purdue he taught many short and part-time courses in quality control. Many of the present leaders in the field of quality control took courses from Professor Burr, both the short courses and those offered in Purdue's regular curriculum. During this period Professor Burr also consulted extensively with industrial and governmental organizations concerning quality control problems. He is also the author of three other well-received texts, *Engineering Statistics and Quality Control* in 1953; *Applied Statistical Methods* in 1974; and *Statistical Quality Control Methods* in 1976.

In view of Professor Burr's accomplishments and my own reading of the text, I feel honored to be able to recommend this book to everyone who has an interest in quality control.

D. B. Owen
Department of Statistics
Southern Methodist University
Dallas, Texas

FOREWORD

The emergence of engineering technology as a significant portion of the engineering profession has created the need for specifically designed textbooks. The technologist is primarily concerned with the application of engineering principles. Development of the theory is important, but a thorough understanding of the uses of the knowledge is essential. A book which recognizes this is rare indeed. It is also refreshing to discover a text that is appropriate for a statistical quality control course in an engineering technology curriculum.

This book serves the needs of a one-quarter, one-semester, or two-quarter engineering technology quality control course. The first chapters lay a good foundation and provide a tie-in between statistics and quality control. They provide either a convenient starting point or an excellent review, depending on the statistics background of the reader.

An appropriate mixture of background, theory, and applications on the traditional quality control topics of control charts and acceptance sampling highlights the meat of this book. Sufficient advanced topics, such as sequential sampling, reliability engineering, and regression analysis are included to challenge the more inquisitive technologist and fill out a two-quarter course. The treatment of reliability engineering is a very satisfactory coverage of a complex topic. Many texts have difficulty keeping discussions of reliability at the level the technologist can handle. This treatment has succeeded in making a very complicated topic understandable without falling into the trap of oversimplification.

A key to the usefulness of any applications-oriented textbook, especially to the engineering technologist, is the quantity and quality of the practice problems provided. This book succeeds on both counts. Each chapter includes a large number of problems. These are provided in sufficient quantity to be valuable in learning proper solution procedures. Problems are also provided that resemble the actual quality control problems that the engineering technologist might face on the job.

This book should be in the library of any engineering technologist who is interested in quality control. It fills a pressing need in the development of the engineering technology profession.

Lawrence S. Aft
Southern Technical Institute
Marietta, Georgia

PREFACE

The objective of this book is to present the basic methods of statistical quality control in as simple, natural, and straightforward
a manner as possible. No previous knowledge of statistics has been
assumed; only elementary algebra has been used. The aim is thus to
make the book accessible to the maximum number of people. It can
be used for industrial courses and self-study, as well as in trade
schools, junior colleges, and even high schools. Those who master this book enhance their employability in industry, not only in
quality control and inspection departments, but also in many other
positions.

Methods of statistical quality control began to be developed
in the 1920s, especially in the Bell Telephone Laboratories. A big
impetus in their usage was given by production for World War II,
aided by a series of intensive 8-day courses in the subject. As an
outgrowth, the American Society for Quality Control was formed in
1946 and currently numbers some 20,000 members.

Applications of statistical quality control have been profit-
ably made in virtually all segments of industry and in procurement.
The reason is not difficult to find, that is, variation in even the
most refined production is universally present. Thus control of
processes and decision making on incoming lots is based on data
subject to variation, and therefore it is natural to use the very
methods developed for handling problems involving variation, namely,
statistical methods.

After an introduction to fundamental statistics and probability, a major portion of the book is spent on control charts for
process control, and on acceptance sampling for decision making on

lots or processes. Following this material is a chapter on statistical tolerancing for assemblies, and brief chapters on studying
relationships by correlation and on concepts of reliability. All
data in examples and problems are from actual cases, except for
data obtained by experiments with known populations. The reader
should not be concerned if he or she is not familiar with a product
or production process mentioned. The author may not be either. If
not familiar, just call the product or piece a "widget" and pay
attention to the numerical data.

Statistical quality control methods are of increasing importance in industry because of (1) ever-refining requirements, making
measurement and control of variation more important, (2) the need
for obtaining the best possible performance from production processes, (3) the necessity of saving material and avoiding spoilage,
(4) the need for economy and cost cutting in production, (5) height-
ened competition both at home and abroad, (6) the demands of warran-
tees and consumerism, (7) increasing use of national and inter-
national standards, (8) the increasing needs for reliability, and
(9) the need to make sure that production methods for quality are
sound, for maximum customer satisfaction, and at the extreme, for
defense in court in liability cases.

The present book is more elementary in character than the
author's *Statistical Quality Control Methods* (Dekker, New York,
1976). The latter book assumes a bit of statistical background
and some elementary calculus, and includes derivations and proofs,
as well as some more specialized techniques. The present book aims
toward a wider readership.

Answers to the odd-numbered problems are provided for the
reader's benefit, especially for self-study.

As in the other book, this book is the outgrowth of a long
association with industrial people, both as students and as those
with whom I consulted or whose plants I visited; and also with the
many professors and industrialists with whom I have taught.

Irving W. Burr

CONTENTS

Elementary Statistical Quality Control

Chapter One

WHY STATISTICS?

Throughout industry, measuring, inspection, and testing is being
done every minute, by people and by automatic devices. The results
are in the form of numbers, for example, measurements or a count of
the number of nonconforming pieces in a sample. These are numbers
or data. They are obtained in order to take action on a manufacturing process for improving and rectifying it, or for a decision
on incoming products. An ever present character of such data or
results is that they vary: from time to time, piece to piece, sample to sample. And this is true even if the production process is
held as constant as is humanly possible. Wherever we have variation we have a statistical problem, whether we like it or not,
whether we know it or not. Thus in such problems we ought to be
using the very methods designed for analysis of data, that is, statistical methods. They can be used to minimize the chances of wrong
decisions. Fortunately there are many statistical techniques which
can be readily learned by anyone with even quite limited background.
They will greatly aid in process improvement and control, and in
decision making on lots of products.

1.1 STATISTICAL QUALITY CONTROL

Control of quality of output is very old, certainly going back to
the building of the pyramids in Egypt, and even to the making of
arrowheads in stone age times. But *statistical* quality control,
that is, the use of statistical methods in quality control, is a
relatively recent development, which began at the Bell Telephone
Laboratories in the 1920s.

The objectives of statistical quality control are to collect
data in a sound, unbiased manner and to analyze the results appropriately so as to obtain satisfactory, dependable, and economic
quality. A mirrorlike finish on all pieces, within a total tolerance of .0002 in., may not be necessary for the use at hand, so we
say "satisfactory." "Dependable" quality is obtained when processes are adequately controlled, and we can rely upon them to produce the desired quality consistently. Moreover the production of
quality must be done "economically" so as to hold costs down, to be
competitive and able to sell at a profit. Statistical quality control methods enable us to obtain maximum benefit out of production
and inspection data and at lowest cost.

1.2 DATA: STATISTICS FOR ACTION

The more closely we try to work, the more trouble we have with
variation. This pencil, with which the author is writing, is one-
quarter of an inch in diameter, *to the nearest quarter of an inch*.
Everyone measuring its diameter in any place or angle would obtain
the same result: 1/4 in. The same is true for measurements to the
nearest .1 in.: all measurements would be .3 in. But probably if
measured to the nearest .01 in., and certainly if to the nearest
.001 in., measurements at different places or angular directions,
or even at the same place and direction, would differ. What this
illustrates is that the closer we try to work, the more we are
bothered by variation. Fortunately the diameter of this pencil is
not a very critical dimension. Other characteristics are of greater
importance, such as the gluing.

The very fact that industry has continually tried to do a better job and to work to ever closer tolerances has increased the need
to handle varying results statistically. Statistical quality control methods can be used to avoid or decrease the likelihood of
error. For example, on a production process, we can adjust its
average level (reset it) when we should leave it alone, or we can
fail to adjust when in fact we should adjust. On an incoming lot
we can err by rejecting a lot which in fact is satisfactory, or we

can accept a lot which is unsatisfactory. Statistical quality control methods enable us to determine the risks of such wrong decisions and to control these risks within economic limits.

1.3 PATTERNS OF VARIATION

Numerical results or data tend to form characteristic patterns or forms of variation. For example, if we have a fairly large collection of measurements and make up numerical classes, and then find how many measurements there are in each class, we will find that some classes in the middle have many measurements, while those classes relatively far out in either direction have few measurements. Such a "tabulation" gives us a picture of how the data run, that is, a pattern of the variation. Such patterns are determined by the manufacturing process and measuring technique from which the data were obtained. The pattern or "distribution" for the process may or may not represent satisfactory quality performance. For example, virtually all of the measurements may lie between the specification limits, or a sizable proportion of the measurements may lie beyond one specification limit or outside both limits. Then we have the question of what action to take.

Patterns or distributions of data take various shapes. Through the years, standard distributions or "models" of variation have been developed to enable us to analyze data. In this book we shall be especially interested in three such models or statistical laws. We shall find them of great use in applications and decision making. Fortunately they can be readily understood and applied without extensive use of mathematics.

1.4 WIDE APPLICABILITY

Statistical methods can be used *wherever we have variation*. Variation of production exists everywhere in industry, and some statistical technique can help us in process improvement, control, experimentation, and decision making on product. The relatively few techniques to be presented in this book can be applied advantageously to most industrial production jobs. The author does not know

of any industry where the methods cannot be applied. Don't let
anyone say to you "I can see how they can be applied in Henry's
operation. But our processes are different." In this book we are
not interested in the *differences* between processes, which anyone
can see! We are interested in what is common to processes, and
most clearly one of these common elements to all industrial processes is *variation*.

1.5 SUMMARY

We can obtain satisfactory, dependable, economic quality from our
processes and those of our suppliers by use of statistical quality
control methods. These methods will enable us to gain economical
control of processes and find and eliminate causes of poor quality.
And they will provide the tools for sound decisions on incoming
lots of material. So, on with the show!

Chapter Two

CHARACTERISTICS OF DATA AND HOW TO DESCRIBE THEM

This chapter presents methods of describing and summarizing data so
as to picture their distribution, to analyze them, and to facilitate
the making of decisions. We are here thinking of the description of
a *sample* of data, that is, observed data from some process or lot.

2.1 TWO BASIC CHARACTERISTICS OF DATA

There are two main characteristics of numerical data which we always
consider in describing our data. These are (1) the average or level
of the numbers and (2) the variability between the numbers. Let us
consider an example on the outside diameter of a brass "low-speed
plug." Three of some 20 samples, each of five, were the following
in .0001 in. units:

Sample 1   1921   1919   1924   1924   1925
Sample 2   1922   1923   1921   1923   1924
Sample 3   1923   1926   1926   1929   1929

By inspection of the numbers we can see that the first two samples
seem to be averaging or centered at about the same level, while the
third sample averages much higher. This is the first characteristic. But looking again at the first two samples (which averaged
the same), we see that the variability between the numbers in sample 1 is much greater than for sample 2. In fact the total extent
of variation, which we call the *range*, is 6 = 1925 - 1919 for the
first sample, but is only 3 = 1924 - 1921 for the second sample.
Now consider samples 1 and 3. For each, the range is 6, but the
average for sample 3 is far above that for sample 1. These two

distinct characteristics of data appear perhaps even more clearly
in the accompanying graphical picture using dots on scales.

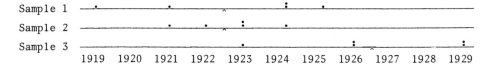

Sample 1
Sample 2
Sample 3
1919 1920 1921 1922 1923 1924 1925 1926 1927 1928 1929

The small arrowpoints, below the lines, indicate the "arithmetic mean," average, or balance point for the sample. The first
two are at 1922.6, while the third is at 1926.6.

In order to describe or summarize a sample of numbers, we need
to provide *two* statistical measures, an average and a measure of
the variability. It is not enough to give only the average, nor
obviously only a variability measure.

2.2 MEASURING AVERAGE LEVEL AND VARIABILITY

Averages. The objective of an *average* is to measure the central or balance point around which a set of numbers seems to cluster. An average is also said to locate the *central tendency*. The
one average with which almost everyone is familiar is called the
arithmetic mean when we might be considering several different
averages and their properties. But we shall in this book call the
arithmetic mean, the average.

We have then (recognizing that there exist other averages)
the following definition.

Definition 2.1. The *average* of a set of n numbers, say, x_1,
x_2, ..., x_n, is the sum of the numbers divided by n. (The three
dots in the definition stand for all x's between x_2 and x_n.) It
is often clearer to use symbols instead of words, as long as the
meaning of the symbols is clear. Let us let $\bar{x}$ stand for the average of the x's. Then we have

$$\bar{x} = \frac{\text{Sum of x's}}{\text{Number of x's}} = \frac{x_1 + x_2 + \cdots + x_n}{n} \qquad (2.1)$$

A most convenient symbol to use in formulas is the capital Greek

letter sigma, that is, Σ. It is used to mean the sum of whatever
is written after it. Thus Σx means "sum of the x's," or ΣR means
"sum of the R's." Then, if we use this useful symbol, (2.1) can be
written

$$\bar{x} = \frac{\Sigma x}{n} \tag{2.2}$$

which is briefer and just as clear.

Let us now use the definition for the three samples of diameters. For sample 1 we have

$$\bar{x} = \frac{1921 + 1919 + 1924 + 1924 + 1925}{5} = \frac{9613}{5}$$

$$= 1922.6 \text{ or } .19226 \text{ in.}$$

For the other samples we have

$$\bar{x} = \frac{9613}{5} = 1922.6$$

$$\bar{x} = \frac{9633}{5} = 1926.6$$

These three averages, $\bar{x}$, locate the arrowpoints on the graphical
picture and describe the level at which the respective samples of
diameters tend to center.

Variability measures. As with averages, there are many measures of variability which could be discussed. But in this book we
shall only need to include two such measures. We have already introduced one, the "range."

Definition 2.2. The *range* R of a set of numbers is the difference between the highest number and the lowest. The formula is
thus

$$\text{Range} = R = \text{maximum } x - \text{minimum } x \tag{2.3}$$

The range therefore measures the total amount of variation within
the sample of n numbers. For the three samples, we have

$$R_1 = 1925 - 1919 = 6$$
$$R_2 = 1924 - 1921 = 3$$

$$R_3 = 1929 - 1923 = 6$$

and so, as we have seen, the first and third samples have twice as
much total variation as does the second sample. The range is a
simple measure of variability, is readily understood, is easily obtained, and is widely used for small samples in statistical quality
control.

There is another measure of variability, however, which is of
much importance; in fact its concept is basic in all statistics,
not only statistical quality control. This is the so-called *standard deviation*. Let us develop it from basic needs.

When we have a sample of data, it seems to make much sense to
try to measure how far away from $\bar{x}$ the numbers x lie *on the average*.
That is, how big a deviation from $\bar{x}$ is only to be expected? Some
deviations of x's from $\bar{x}$ will of course be greater than this typical deviation, while others will be smaller. Consider sample 1:

Data (x)	Deviation $(x - \bar{x})$	Deviations squared $[(x - \bar{x})^2]$
$x_1 = 1921$	$x_1 - \bar{x} = -1.6$	$(x_1 - \bar{x})^2 = (-1.6)^2 = 2.56$
$x_2 = 1919$	$x_2 - \bar{x} = -3.6$	$(x_2 - \bar{x})^2 = (-3.6)^2 = 12.96$
$x_3 = 1924$	$x_3 - \bar{x} = +1.4$	$(x_3 - \bar{x})^2 = (+1.4)^2 = 1.96$
$x_4 = 1924$	$x_4 - \bar{x} = +1.4$	$(x_4 - \bar{x})^2 = (+1.4)^2 = 1.96$
$x_5 = 1925$	$x_5 - \bar{x} = +2.4$	$(x_5 - \bar{x})^2 = (+2.4)^2 = 5.76$
$\bar{x} = 1922.6$	$\Sigma(x - \bar{x}) = 0$	$\Sigma(x - \bar{x})^2 = 25.20$

The first number, $x_1 = 1921$, "deviates" 1.6 units from the average
$\bar{x}$, being *below* it, hence the minus sign. This is obtained by
$x_1 - \bar{x} = 1921 - 1922.6 = -1.6$. Similarly the fifth deviation is
$x_5 - \bar{x} = 1925 - 1922.6 = +2.4$, x_5 being *above* $\bar{x}$. Now what we wish
is some sort of average of these five deviations, $x - \bar{x}$. If we
try the same game of averaging them as we did in finding $\bar{x}$, we run
"aground," because the sum of the deviations here is zero. In fact
the sum of the deviations is always zero, because of +'s and -'s
adding out to zero. Thus some other average of the deviations is
in order. To get around the trouble with the signs of the deviations, we square the deviations. Now it would seem logical to

divide this sum of the squared deviations, $\Sigma(x - \bar{x})^2$, by 5, and then to take the square root, so as to get back the original unit (.0001 in.). In fact division by n has been often practiced in the past. But modern usage calls for division of $\Sigma(x - \bar{x})^2 = 25.20$ by $n - 1 = 5 - 1 = 4$, for reasons which will be given later on (Sec. 7.4). Thus we now complete the job of finding the "standard deviation s" by dividing by $5 - 1 = 4$ and taking the square root of the result. Hence the standard deviation is

$$s = \sqrt{25.20/4} = \sqrt{6.30} = 2.51 \text{ or } .000251 \text{ in.}$$

Note that of our five deviations, $x - \bar{x}$, only one -3.6 is larger in size than s = 2.51, while one other +2.4 is nearly the same size. We choose now to define the standard deviation s by an appropriate formula, which is seen to be inherent in the preceding development.

Definition 2.3. The *standard deviation* s of a set of n numbers x_1, x_2, ..., x_n is given by

$$s = \sqrt{\frac{\Sigma(x - \bar{x})^2}{n - 1}} \tag{2.4}$$

In review, for comparison let us find the standard deviation for sample 2:

x	$x - \bar{x}$	$(x - \bar{x})^2$
1922	-.6	.36
1923	+.4	.16
1921	-1.6	2.56
1923	+.4	.16
1924	+1.4	1.96
9613	0	5.20

$\bar{x} = 1922.6$

$$s = \sqrt{\frac{5.20}{4}}$$

$$= \sqrt{1.30}$$

$$= 1.14 \text{ or } .000114 \text{ in.}$$

So here s = 1.14, or a bit less than half of the s for sample 1.
This time two of the five deviations are larger in size than s, and
three smaller.

We now need to make a few comments. The first is that the
technique for finding the standard deviation s may well seem a bit
or even quite complicated to the reader. And certain it is, that
it is more time consuming than finding the range. Indeed the move-
ment toward the use and application of control charting of measurements did not really begin to get off the ground until it was found
that for *small samples* the range was just about as good to use as
the standard deviation, and thus much time could be saved. A second point is that there are several short cuts we can learn to use
in finding s. Moreover, if an electronic computer is used it is
just as easy to find s as to find R, being perhaps more easily programmed. Then too the *concept* of the standard deviation is enor-
mously important. Finally, early workers in statistical quality
control were so proud at having mastered the standard deviation
that its Greek letter symbol, the small sigma σ was incorporated
in the official seal of the American Society for Quality Control!

2.3 CONDENSING DATA INTO A FREQUENCY TABLE

When you have a relatively large sample of data, such as we see in
Table 2.1, it is desirable to condense them into a so-called *frequency table*. About all you get from looking over such a long list
of numbers, as in Table 2.1, is eye strain. A frequency table will
help us picture the data.

Definition 2.4. A *frequency table* gives numerical classes
which cover the range of data, and lists the frequency of cases
within each class.

Let us therefore consider the needed steps in making a frequency table. The first step is to choose limits for the classes.
There would usually be eight to twelve such classes, which must
cover the entire range of the data. "Class limits" are given to
the precision of the measurements or data. In Table 2.1 the charge

TABLE 2.1. Charge Weights for 99 Insecticide
Dispensers in Grams (Specifications 454 ± 27 g)

476	478	473	459	485	454	456	454	451	452
458	473	465	492	482	467	469	461	452	465
459	485	447	460	450	463	488	455	478	464
441	456	458	439	448	459	462	495	500	443
453	457	458	470	450	478	471	457	456	460
457	434	424	428	438	460	444	450	463	467
476	485	474	471	469	487	476	473	452	449
449	477	511	495	508	458	437	452	447	427
443	457	485	491	463	466	459	471	472	472
481	443	460	462	479	461	476	478	454	

weights were to the nearest gram, so the class limits are to the
gram. Each possible measurement must have *one* and *but one* class into which it can go. The extreme charge weights are 424 and 511 g,
so the range R is 511 - 424 = 87 g. If we include 10 possible
charge weights per class, we shall have 9 or 10 classes. Convenient classes are 420-429, 430-439, and so on. (The beginner might
think that such classes have only nine possible measurements each,
but look closely and you will see 10.) Note that every possible
charge weight will have one, and but one, class to which it belongs,
since a weight such as 429.6 g will never occur.

The next step is to list all the classes in a column and take
each charge weight in turn and tally it, just like the tallying of
votes. Thus 476 is tallied in the class 470-479. Then we take the
next, 478, and make a second tally in this class, and so on. This
gives us Table 2.2, after we count the tallies in each class.

Looking at the class frequencies, we see that they rise steadily from 3 to a maximum of 28 and then fall fairly steadily to a
minimum of 1 at the 510-519 class. The midvalue column gives the
midmost charge weight in each class. For this we note that a weight
measured as a 420 g might come from any true weight 419.5 to 420.5
g, and likewise one measured as 429 g might be from a true weight
428.5 to 429.5 g. So the extreme possible true weights in the

TABLE 2.2. Frequency Table for Charge Weights of
Insecticide Dispenser, from Data of Table 2.1
(Specifications 454 ± 27 g)

Class in grams	Frequency, f	Midvalue, x
420-429	3	424.5
430-439	4	434.5
440-449	10	444.5
450-459	28	454.5
460-469	19	464.5
470-479	20	474.5
480-489	8	484.5
490-499	4	494.5
500-509	2	504.5
510-519	1	514.5
Total frequency	99	

420-429 class are 419.5 and 429.5. These are called "class boundaries." The midvalue of this interval is 424.5 as recorded in the
third column. Similarly the other midvalues are as shown in Table
2.2.

By looking at the table, we see that the point of greatest concentration is about 454.5 g, that is, about the middle of the specification limits. However, we note that considerably more than half
of the *remaining* charge weights (namely, 54) lie *above* the 450-459
class. This means that the process average was above the specified
"nominal" weight of 454 g. This overfill above the required average proved to be costing the company about $14,000 per month. Also
note that since the specification limits were 427 and 481 g, there
were, within these 99 charge weights, a few outside of the specifications. By actual count in Table 2.1, one weight was below 427 g
and 14 above 481 g. In the total production, of which the 99
charge weights were but a sample, there would of course be far more
outside the limits. By using statistical control methods, the overfill was soon cut to $2000 per month, and yet with safer meeting of
specifications.

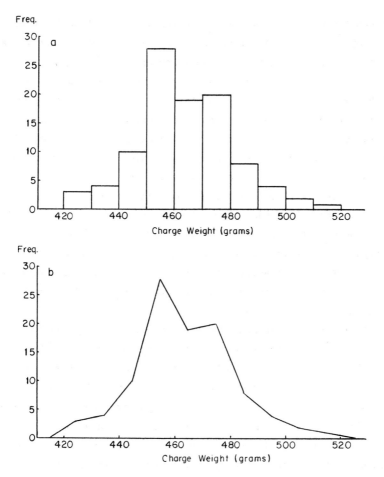

Fig. 2.1. (a) Histogram and (b) frequency polygon for charge weights of an insecticide dispenser. Specifications 427 ± 27 g. Data in Table 2.1.

Two kinds of frequency graphs, which we show in Fig. 2.1, are of further help in visualizing the *frequency distribution* or manner in which the frequencies in the classes are distributed. The *frequency polygon* plots the frequencies f against the midvalues, emphasizing the latter. Note that the broken line graph is connected to the midvalues having zero frequency (414.5, 524.5) which lie next to those with some frequency. The *histogram* has blocks with bases between the class boundaries, and heights representing frequencies. They emphasize the class intervals. Either type of frequency graph may be used according to individual preference.

A frequency table, or even better, one of the two frequency graphs helps us visualize the pattern of variation for the numbers, their average and point of greatest concentration, and their typical variability and total spread. Also we may find out something about the curve shape, especially if we have a substantial sample, say, at least 100 observations. That is, the frequencies may decrease to zero much more slowly on one side of the highest frequencies than they do on the other. For example, eccentricities (distances between two center points), or percent impurity, can never be negative but a few relatively large values may occur. Thus, starting at a zero measurement, the highest frequency may be quickly reached, but then the frequencies decrease more slowly and may "tail out" quite a distance.

A frequency table and/or graph may reveal irregularities. For example, the frequencies may suddenly stop and become zero in one or both "tails," giving a "bob-tailed" appearance. This could be the result of 100% sorting to eliminate extreme values in one or both directions. Such an abrupt stop might thus be at a specification limit. Or there may be alternating higher and lower frequencies. Thus, if the classes are in .00005 in. from some nominal dimension, there might be many more cases measured as whole ten-thousandths of an inch than at half ten-thousandths.

In any case a frequency graph can be of much aid in picturing the pattern of variation to an associate or an executive, or in a report. For example, in pressing out railroad car wheels from heated steel cylindrical blocks, a high rate of rejection (about 35%) was experienced. Upon weighing a substantial sample of blocks, an average of 650 lb and a range of 70 lb showed up. A frequency graph was shown to the man who cut off the blocks. He was amazed and said he thought he could do better. He was able to cut the range to 20 lb, and the rejection rate of wheels dropped to 5-10%.

2.4 SAMPLE DATA VERSUS POPULATION

Up to now, we have been talking about *samples* of industrial product
and the associated data taken by some method of measurement. The
product might be auto carburetors, for example, and we take a sample of them, making a careful inspection of each to seek out any
"defects." These defects may be anything from a critical nature
which could make the carburetor dangerous, down to the mildest kind
of imperfection. Each carburetor upon inspection yields 0, 1, 2,
... defects, that is, a whole number or integer. Such "counted"
data are called *discrete data*.

Definition 2.5. *Discrete data* are the result of counting defects or defective pieces in a sample, so that only whole numbers
or integers may occur.

Now suppose that our sample of carburetors came from a large
lot of carburetors. We could, given enough time, inspect the entire lot of carburetors for defects. This would yield a "population distribution" of counts of defects on carburetors for the lot.
Now we would hope that the "sample distribution" of counts would be
similar to that of the population or lot. In order to make it likely, we should have drawn the sample of carburetors in an *unbiased*
manner. One unbiased method is to draw the sample from the lot in
such a way that all carburetors in the lot have the same chance to
be chosen for the sample. This is called *random sampling*.

Definition 2.6. We say that we are doing *random sampling* when
each item in the lot has the same chance to be chosen for the sample as has every other item in the lot.

Unless we sample unbiasedly, our sample distribution may be
widely different from the population distribution. For example, a
sample of 100 piston ring castings taken from a lot of 3000 yielded
25 defective castings out of the 100. It was decided to sort the
remaining 2900 in the lot, so as to salvage the good ones. Only
four more defective castings were found in the 2900! Question:
Was the sample of 100 chosen at random from the 3000?

In the example just given the "population" was the lot of 3000 castings of which 29 were defective. Thus the so-called *fraction defective* in the lot was 29/3000 or about .01 (1%). Meanwhile the *sample fraction defective* was 25/100 or .25 (25%). Hence the sample proportion was wildly different from that of the population. (The industrial explanation was that the castings were poured in stacks of molds and one stack must have been poured with cool iron, and the sample happened to be chosen from the part of the box containing these bad ones.)

We have just seen two types of discrete data, namely counts of the number of defects or the number of defective items in a sample, or a lot. The other type of data is called *measurement data*.

Definition 2.7. *Measurement data* are obtained whenever we take an item of product and compare against a standard unit of that measurement. This in general involves reading a scale of some kind, and the resulting measurement is so many whole units and some decimal fraction; for example, 1.0626 in. or 31.28 cm^3. Indefinitely many decimal places are theoretically possible, but in practice the number of places is limited by the measuring technique. Such data are also called *continuous data*.

For measurement data the populations are perhaps more complicated than for discrete data. The reason is that each item of product does not have just one measurement, but instead repeated measurements of the item could yield varying results. One example of a measurement population is the distribution of measurements of the diameter of a shaft obtained by very precise repeated measurements in just one place. The variation in this distribution reflects the "repeatability" error in the measuring technique. The population size is not finite, because the number of repeated measurements we could take is unlimited.

On the other hand, if we take a lot of shafts and measure the diameter of each just once, then we have a distribution of measurements. But the variation in these observed measurements reflects not only shaft-to-shaft variation but also variation from place to place within a single shaft, and also measurement variation or errors for repeated measurements on one place on one shaft. And therefore in a real sense we can never actually "get our fingers on" the population distribution for the lot.

The appropriate statistical models for discrete and for measurement data are quite different, and we must therefore be careful as to which type of data we have so as to use the appropriate model, as we shall see.

2.5   INTERPRETATION OF $\bar{x}$ AND s

These notations are for the average and the standard deviation for a *sample* of data. We think of $\bar{x}$ as a *typical* value of the x's in a sample, or as a point around which the numbers x tend to cluster. It is a one-number description for the *level* of the numbers. We may also say that it is a balance point for the x's, since, as we saw in Sec. 2.2, the (algebraic) sum of the deviations x - $\bar{x}$ is zero. See the examples in that section.

As we have seen in Sec. 2.2, the standard deviation s is a typical deviation of an observation x from the average $\bar{x}$. It is an "only-to-be-expected" deviation, descriptive of the variation within the sample. Having by now discussed frequency distributions, we are in a position to add another interpretation. This is for substantial samples of, say, at least 50 observations, and fc_ well-behaved data with frequencies increasing to a maximum and then the frequencies decreasing. Most data from homogeneous conditions will show such a tendency. We can then say that

1. Between $\bar{x}$ - s and $\bar{x}$ + s will lie 65 to 70% of the cases.
2. Between $\bar{x}$ - 2s and $\bar{x}$ + 2s will lie about 95 or 96% of the cases.
3. Between $\bar{x}$ - 3s and $\bar{x}$ + 3s will lie at least 99% of the cases.

These proportions come from the so-called *normal* distribution. But they also apply to data which are anything like reasonably "well behaved." Looking at the little table just given we can say that a

deviation of 2s or greater in size will occur only about one time
in 20, and a deviation of 3s or more in size will be quite rare.

2.6 EFFICIENT CALCULATION OF $\bar{x}$ AND s: CODING

Let us consider sample number 1 of Sec. 2.1. The x's were 1921,
1919, 1924, 1924, and 1925. Suppose that we subtract the lowest
value (1919) from each and call the difference y. Then the five
y values are 2, 0, 5, 5, 6. Is it not quite apparent that the y's
vary just the same amount as do the x's? But if not clear, let us
find the mean and standard deviation:

$y = x - 1919$	$y - \bar{y}$	$(y - \bar{y})^2$
2	-1.6	2.56
0	-3.6	12.96
5	+1.4	1.96
5	+1.4	1.96
6	+2.4	5.76
18	0	25.20

$$\bar{y} = \frac{18}{5} = 3.6$$

$$s = \sqrt{\frac{25.20}{5 - 1}} = \quad 6.30 = 2.51$$

and hence $s_y = s_x$, the subscript showing of what variable we have
found the standard deviation. Moreover we can find $\bar{x}$ and $\bar{y}$ by
merely adding back on the 1919 which we subtracted from each x to
find y. Thus $\bar{x} = \bar{y} + 1919 = 3.6 + 1919 = 1922.6$.

There was no magic in 1919. We could as well have subtracted
say, 1923 which is a "guess" at $\bar{x}$. Then the y's would have been

$y = x - 1923$	$y - \bar{y}$	$(y - \bar{y})^2$
-2	-1.6	2.56
-4	-3.6	12.96
+1	+1.4	1.96.
+1	+1.4	1.96
+2	+2.4	5.76
-2	0	25.20

$$\bar{y} = \frac{-2}{5} = -.4$$

$$s = \sqrt{\frac{25.20}{4}} = 2.51$$

$$\bar{x} = -.4 + 1923 = 1922.6$$

We mention that in finding the deviations for these y's one must be quite careful with signs. Thus for the first deviation $y - \bar{y} = -2 - (-.4) = -2 + .4 = -1.6$ and for the last $y - \bar{y} = +2 - (-.4) = 2 + .4 = 2.4$.

So far we have not seen much gain, except that for $\bar{x}$ we have avoided the addition of all of the 191 figures in the x's. There is a shortcut we can use for finding s_y from the y's, however. Since the y's are such simple numbers, they are easily squared. We may then use the following formula:

$$s_y = \sqrt{\frac{n\Sigma\,y^2 - (\Sigma\,y)^2}{n(n-1)}} \tag{2.5}$$

which may be proved from (2.4). Our work then looks like the following:

y	y^2
2	4
0	0
5	25
5	25
6	36
18	90

$$s_y = \sqrt{\frac{5 \cdot 90 - 18^2}{5 \cdot 4}} = \sqrt{\frac{450 - 324}{20}} = \sqrt{\frac{126}{20}}$$

$$= \sqrt{6.3} = 2.51$$

This approach is a considerable help, especially when $\bar{y}$ is a messy decimal instead of a number as simple as 3.6 or -.4. It means we do not have to round off any numbers until we do the final division.

Finding s, however, for small samples is not often done (unless an electronic computer is being used), the range being used instead. But when a large sample is being analyzed, as in a frequency table, then we have a large gain in calculational time available as follows.

TABLE 2.3. Frequency Table for Charge Weights of Insecticide
Dispenser, from Table 2.2, and Calculations by Coding for $\bar{x}$ and s

		Coded variable		
Midvalue, x	Frequency, f	v	vf	v^2f
424.5	3	-4	-12	48
434.5	4	-3	-12	36
444.5	10	-2	-20	40
454.5	28	-1	-28	28
464.5	19	0	0	0
474.5	20	+1	+20	20
484.5	8	+2	+16	32
494.5	4	+3	+12	36
504.5	2	+4	+8	32
514.5	1	+5	+5	25
	$\Sigma f = 99$		$\Sigma vf = -11$	$\Sigma v^2f = 297$

$$\bar{v} = \frac{\Sigma\ vf}{\Sigma\ f} = \frac{-11}{99} = -.11$$

$$s_v = \sqrt{\frac{n\ \Sigma\ v^2f - (\Sigma\ vf)^2}{n(n-1)}} = \sqrt{\frac{99(297) - (-11)^2}{99(98)}} = \sqrt{\frac{29282}{9702}}$$

$$= \sqrt{3.018} = 1.737$$

$x_0 = 464.5 \qquad d = 10$

$\bar{x} = 464.5 + 10(-.11) = 463.4$ g

$s_x = 10(1.737) = 17.37$ g

Let us consider again the data on charge weights, as given in
Table 2.2. See Table 2.3, which gives the midvalues x in the first
column and the corresponding frequencies f in the second column.
For $\bar{x}$, we want the average of 99 charge weights and could treat the
three cases in the 420-429 class as though all were at the midvalue
424.5, and so on. Then we would proceed as follows:

$$\bar{x} = \frac{\Sigma\ fx}{\Sigma\ f} = \frac{3(424.5) + 4(434.5) + 10(444.5) + \cdots + 1(514.5)}{3 + 4 + 10 + \cdots + 1}$$

$$\bar{x} = \frac{45,875.5}{99} = 463.39$$

This is not too troublesome for $\bar{x}$, but finding s for the x's along this line would give much trouble. Using this approach for s in (2.4) we would now need to find

$$\Sigma\ f(x - \bar{x})^2 = 3(424.5 - 463.39)^2 + 4(434.5 - 463.39)^2$$
$$+ \cdots + 1(514.5 - 463.39)^2$$

which is basically $\Sigma(x - \bar{x})^2$, taking account of the frequencies.

So instead we "code" the x's into, say, v values. To do 'this we choose a midvalue somewhere in the middle. See Table 2.3. It is fun to try to guess the one nearest $\bar{x}$. Here we chose 464.5. Then for the "coded" variable v we subtract 464.5 from each mid-value x, in turn, and also divide by the difference, 10, between consecutive midvalues. Thus for x = 474.5, v = (474.5 - 464.5)/10 = +1; x = 494.5, v = (494.5 - 464.5)/10 = +3; x = 424.5, v = (424.5 - 464.5)/10 = -4, and so on. This gives the v column in the table. Then we say there are three v's of -4, four v's of -3, and so on. To find the sum of all 99 v's, we multiply each v by its frequency; thus

$$\Sigma\ vf = (-4)3 + (-3)4 + (-2)10 + \cdots + (+5)1 = -11$$

Likewise for the sum of 99 v^2's we have

$$\Sigma\ v^2 f = (-4)^2 3 + (-3)^2 4 + (-2)^2 10 + \cdots + (+5)^2 1 = 297$$

(The $v^2 f$ column is perhaps most easily found by multiplying the two previous columns, because v times vf = $v^2 f$, for example, $(-4) \cdot [(-4)3] = (-4)(-12) = 48$.)

Then we have for the coded variable v,

$$\bar{v} = \frac{\Sigma\ vf}{\Sigma\ f} = \frac{-11}{99} = -.11$$

Using Σ vf as analogous to Σ y in (2.5), $\Sigma\ v^2 f$ like $\Sigma\ y^2$ in (2.5), and Σ f like n in (2.5) we have

$$s_v = \sqrt{\frac{99(297) - (-11)^2}{99(98)}} = 1.737$$

as shown at the bottom of Table 2.3.

Then to find $\bar{x}$ and s_x, we merely decode as shown there also. We first *subtracted* 464.5 from each, and next *divided* by the class interval 10 g. To decode we first *multiply* $\bar{v}$ by 10 g, then add 464.5 g back on, as shown. To find s_x from s_v, we merely multiply by 10 g (464.5 g does not enter in) as shown. The 10 g gives back the unit grams in each case.

Let us now summarize this calculation by coding as follows, using d for the class interval and x_0 for the guessed average which we subtract from each midvalue:

$$v = \frac{x - x_0}{d} \quad \text{or} \quad x = x_0 + dv \tag{2.6}$$

$$\bar{v} = \frac{\Sigma vf}{\Sigma f} = \frac{\Sigma vf}{n} \tag{2.7}$$

$$s_v = \sqrt{\frac{n \Sigma v^2 f - (\Sigma vf)^2}{n(n - 1)}} \tag{2.8}$$

$$\bar{x} = x_0 + d\bar{v} \tag{2.9}$$

$$s_x = ds_v \tag{2.10}$$

Note that if we have small samples and treat each frequency f as 1, and also do not divide by any d to find y's, then we have v = y, and (2.7) is $\bar{y}$, and (2.8) becomes (2.5).

2.7   CURVE SHAPE

Sometimes a frequency distribution shows frequencies decreasing from the maximum frequency much more rapidly in one direction than in the other. That is, the distribution is not symmetrical around the point of greatest concentration. There are available objective measures of the extent of this lack of symmetry. But we shall not consider them here. For information on measures of curve shape the reader is referred to Ref. 1.

2.8 POPULATION VERSUS SAMPLE CHARACTERISTICS

The arithmetic mean or average of a *population distribution* and
also the standard deviation are analogous to those of a sample. In
order to distinguish between sample and population characteristics,
we use different symbols, however. In particular we shall use the
following:

	Sample	Population
Average	$\bar{x}$	μ = mu (the Greek letter m)
Standard deviation	s	σ = sigma (the Greek letter s)

Instead of mu = μ, the American Society for Quality Control Standard, A1-1971, uses $\bar{x}'$. The author has chosen to use μ instead, in
line with the widespread use of Roman letters for sample characteristics and Greek letters for population characteristics. It is unfortunate that there is no one universal notational system in statistics. The reader must try to keep things straight, with all the
help the author can give him, including the glossary.

There is one difference in calculation of s for a *sample* and
of σ for a *finite population*. Let us suppose that we have a lot of
400 plummets whose weights are the quantities in question. Let
each plummet be very accurately weighed giving 400 weights in, say,
grams: $x_1, x_2, \ldots, x_{400}$. Then we have for the mean

$$\mu = \frac{\Sigma\ x}{N} = \frac{\Sigma\ x}{400}$$

where N is the lot or population size, analogous to (2.2). But now
for the standard deviation, we use

$$\sigma = \sqrt{\frac{\Sigma(x - \mu)^2}{N}} = \sqrt{\frac{\Sigma(x - \mu)^2}{400}}$$

where the sum of the squares of the deviations from the population
mean μ is divided by N, not N - 1, N being the size of the
population.

When the population is of unlimited or infinite size, then
other methods of calculation are used, which we shall not describe

here. But in any case, for a population arising under some set of
homogeneous conditions of production and measurement, there will at
least theoretically be an average μ and a standard deviation σ.
And we can regard our sample $\bar{x}$ and s as estimates of them, if the
sample is chosen in an unbiased manner.

The relationship between samples and populations is the basic
problem of statistics, both theoretical and applied.

2.9 SUMMARY

In this chapter we have discussed one average, the arithmetic mean
for finding a typical or representative level in a sample of numbers. Although this is but one of several types of averages it is
by far the most important one for our purposes. For measures of
variability we covered the range and the standard deviation. The
former is commonly used for small samples up to about 10 and the
latter is used fairly exclusively for larger samples. These three
measures $\bar{x}$, R, and s are objective descriptions of the characteristics of a sample of data. For a population, the characteristics,
mean μ and standard deviation σ, correspond to $\bar{x}$ and s for a sample.
In addition to using $\bar{x}$ and s to describe a sample, it is often desirable to construct a frequency table and draw a frequency graph,
if the sample is of at least, say, 25 numbers. In addition we have
also discussed shortcuts to the calculation of $\bar{x}$ and s, both for
ungrouped data and for data in a frequency table. Finally a dis-
tinction was made between two general types of data: (1) discrete
or counted data and (2) continuous or measured data. Somewhat different approaches are used in the analysis of the two general types
of data.

PROBLEMS

Find for each of the following small samples $\bar{x}$, R, and s:

2.1 Hardness Rockwell 15T: 72, 72, 66, 67, 68, 71, specifications
being 66-72, on carburetor tubes.

2.2 Tube diameters: .2508, .2510, .2506, .2509, .2506 in.

2.3   Dimension on a rheostat knob: 142, 142, 143, 140, 135 .001-in. units.

2.4   Dimension on an igniter housing: .534, .532, .531, .531, .533 in.

2.5   Eccentricity or distance between center of cone and center of triangular base on needle valves: 50, 35, 36 .0001-in. units.

2.6   Number of defects on subassemblies: 2, 8, 3, 3, 7, 1.

2.7   Density of glass: 2.5037, 2.5032, 2.5042, 2.5042 g/cm^3.

2.8   Percent of silicon in steel castings: .94, .89, .98, .87.

For the following two samples of raw data, choose appropriate numerical classes and tabulate the data into a frequency table. Draw an appropriate frequency graph. Comment on the data.

2.9   Data on eccentricity of needle valves, conical point to base, in .0001 in. units. Maximum specification limit .0100 in. Suggest 0-9, 10-19, and so on for classes.

32	30	30	37	18	37	50	35	36	57	24	75	49	6	24
67	25	25	52	56	53	18	39	47	40	51	51	31	61	28
15	10	35	27	49	19	51	34	40	19	32	10	39	16	50
15	30	50	32	46	29	39	19	34	42	40	30	70	16	57
12	19	23	34	14	40	58	36	41	7	11	8	40	12	12
66	58	19	29	37	74	20	9	15	30	38	88	83	57	90

2.10 Chemical analyses for manganese in 80 heats of 1045 steel. Data in .01% units. Suggest 66-68, 69-71, and so on for classes. Specifications: 70-90.

74	79	77	81	72	66	75	80	76	86	84	70	80	62	74	71
68	79	81	76	79	79	84	78	74	88	71	80	79	74	76	75
81	80	80	78	76	81	70	76	79	80	79	84	75	75	76	83
88	83	79	91	73	78	82	74	81	75	76	72	83	97	76	90
79	75	74	73	93	92	70	75	86	87	79	69	79	77	76	82

Find $\bar{x}$ and s for the following:

2.11 The frequency distribution obtained in Prob. 2.9.

2.12 The frequency distribution obtained in Prob. 2.10.

2.13 Density of glass in g/cm^3

Density	Frequency	Density	Frequency
2.5012	2	2.5052	19
2.5022	6	2.5062	10
2.5032	25	2.5072	4
2.5042	33	Total:	99

2.14 Over-all height of bomb base

Height (in.)	Frequency	Height (in.)	Frequency
.830	1	.834	13
.831	3	.835	11
.832	11	.836	6
.833	14	.837	6
		Total:	65

2.15 Spring tension in pounds

Tension (lb)	Frequency	Tension (lb)	Frequency
50.0	2	53.0	32
50.5	2	53.5	12
51.0	4	54.0	18
51.5	6	54.5	2
52.0	9	55.0	1
52.5	12	Total:	100

REFERENCE

1. I. W. Burr, *Statistical Quality Control Methods*, Marcel Dekker, New York, 1976.

Chapter Three

SIMPLE PROBABILITY

Some basic knowledge of probability is needed in statistical quality control. But the amount needed is not great. Moreover, the
reader is sure to find it of interest for its own sake, not only
for use in quality control, but also in every day life. The concepts are of much importance in your thinking processes!

3.1 LIKELIHOOD OF AN EVENT OCCURRING

Probability is concerned with the likelihood of an event occurring.
The event is perhaps quite likely to occur on a given trial or
opportunity, or it may be most unlikely. Here at the author's home
it is quite likely that some rain will fall on any given day, because this was true on about four-fifths of the days last year.

The scale for the probability of an event is 0 to 1. If an
event *cannot occur* on a trial, then the probability of its occurrence is 0. Or, if an event is *certain* to occur, the probability
is 1. If in the long run an event will occur half of the time then
the probability is .5, and so on. We emphasize that a "trial" or
experiment must be clearly defined, and also the "event" clearly
defined.

As an example suppose that a trial is drawing a piece at random from some production line, and that the event in question is
that the piece drawn is a defective or nonconforming one. Let us
suppose that the probability of a defective is .05. This means
that 5% of the time when we draw a random piece from the line, it
is defective. The complementary event is that the piece drawn is
a good one. What would you say is the probability of a piece drawn

being a good one? The answer is .95, because when we draw a piece,
it is certain to be either a defective or a good one.

P(defective) = .05
P(good) = 1 - .05 = .95
P(good or defective) = 1

95% of the time, a piece drawn will prove to be a good one. Similarly a golf ball might be either a good one, a "second," or a reject (always in one of the three classes). Suppose the respective
probabilities are .97, .02, .01 (the sum being 1) for each ball
produced must be in *one and but one* of the three classes. Then

P(good) = .97   P(second) = .02   P(reject) = .01

3.2 OCCURRENCE RATIO

Suppose that we have a production process for which the probability
of a piece containing at least one "minor defect" is constantly .08.
We then say that the probability is .08 for a "minor defective,"
that is, a piece containing one or more of the class of defects
called *minor*. Now what happens to the *observed proportion* of minor
defectives as we continue to sample, that is, to the occurrence
ratio? Let us give a bit of notation to help the discussion. Let

p' = constant probability of a (minor) defective   (3.1)

where the letter p stands for probability and the prime means a
population probability.

d = number of defectives observed   (3.2)

n = number of pieces inspected or tested   (3.3)

$p = \dfrac{d}{n}$ = sample proportion of defectives   (3.4)

With these notations available to us, let us ask the question
again. How does p = d/n behave as we sample more and more, that
is, increase n? Would we not expect that the observed occurrence
ratio p = d/n would tend to approach p' = .08? Let us learn by an
experiment how the approach proceeds. We start with five samples

of 10, then samples of 50.

| Sample | | Totals | | Occurrence ratio |
n	d	Σn	Σd	p = Σd/Σn
10	0	10	0	.0000
10	1	20	1	.0500
10	1	30	2	.0667
10	1	40	3	.0750
10	0	50	3	.0600
50	5	100	8	.0800
50	4	150	12	.0800
50	6	200	18	.0900
50	4	250	22	.0880
50	8	300	30	.1000
50	1	350	31	.0886
50	3	400	34	.0850
50	5	450	39	.0867
50	3	500	42	.0840
50	5	550	47	.0855
50	5	600	52	.0867
50	5	650	57	.0877
50	4	700	61	.0871
50	3	750	64	.0853
50	6	800	70	.0875
50	4	850	74	.0871
50	7	900	81	.0900
50	4	950	85	.0895
50	4	1000	89	.0890

The first two columns are for the current sample of 10 or of 50.
The third and fourth columns are for the total sample size and the
cumulative total number of defectives. The fifth column is based
on the third and fourth columns and gives the current overall proportion defective or occurrence ratio, total defective over total
inspected.

Now how does the proportion defective behave? It only "tends" to approach p' = .08. Sometimes it gets closer; sometimes it backs away from p'. Before the total sample size was 100, the occurrence ratio or fraction defective was below .08. At 100 and 150 it was exactly .08 and thereafter above.

What will happen in the next 1000? Gamblers like to say the "law of averages" says that because p = .0890 was above .08, the next 1000 will yield a fraction below .08 to make up the discrepancy. The law of averages says no such thing. It does say that as the number of trials or sample size increases, the *probability* of the occurrence ratio lying in any range around p', for example, .75 to .85 increases, and approaches certainty. Misinformation about the law of averages has cost gamblers untold amounts of money.

Principle. Take any limits around the true constant probability p'. Then by taking n sufficiently large, we can make the probability of p = d/n lying between the limits to be as close to certainty as we wish.

In the above sense then we can say that p = d/n is an estimate of the constant population probability p'. How close the estimate is depends upon the sample size n, the value of p', and also upon luck or chance. But we can learn to "play the odds" well, and will do so.

3.3 EXAMPLE 1 AND PROBABILITY LAWS

In this example and the two following ones, we shall be illustrating some laws of probability, which the reader will find make very good sense.

Consider again the production line producing pieces with a constant probability .08 of the piece being a (minor) defective, and such that each piece is independent of the others produced. What is meant by the latter is that the probability of the next piece being a defective is .08 and of being a good one is .92, regardless of what the preceding pieces have been like. (This property is called *independence*, and is typical of a production process which is "in control," as we shall be seeing.)

For a single draw of one piece, there are just two possible outcomes or events. The piece can either be a good or a defective one. The respective probabilities are obviously .92 and .08.

Now let us suppose that we draw a sample of two pieces and inspect them. What now are the possible outcomes? There are three: namely, the sample may contain 0, 1, or 2 defectives. Let us find the probabilities of these outcomes or events.

For the probability of the sample containing no defectives, we must have good pieces on both draws. Now 92% of the times or trials, the first piece will be a good one, in the long run. Of all of these 92% of the trials, we will follow by drawing *another* good one 92% of the time. Thus the probability of the sample of two containing both pieces good is

$$P(2 \text{ good}) = .92(.92) = .92^2 = .8464$$

Next find the probability of both being defectives. On the first draw 8% of the time we have a defective. Now 8% of *these* times we will draw a second defective. So the proportion of times *both* pieces are defectives is (8%)(8%) = .08(.08) = .0064. Finally for the probability of the sample containing exactly one defective and one good we can proceed as follows. Since there are only three *possible* outcomes: 2 good, 1 good, and 0 good, we must have

$$P(2 \text{ good}) + P(1 \text{ good}) + P(0 \text{ good}) = 1$$

because some one of these three events is certain to occur. Thus

$$P(1 \text{ good}) = 1 - P(2 \text{ good}) - P(0 \text{ good})$$
$$= 1 - .8464 - .0064 = .1472$$

But there is also a *direct* way to find the probability of just one being good. The event of one being good may be achieved in *either* of two ways or orders (good, defective) or (defective, good). For the first, we start off with a good one 92% of the time, and of these 8% of the time follow with a defective. So the probability is .92(.08) = .0736. Similarly P(defective, good in this order)

= .08(.92) = .0736. Adding these probabilities for the two nonover-
lapping or *mutually exclusive* ways, we have as above

P(1 good) = .1472

Let us summarize the situation for the sample of two by finding
the sample fraction defective, p = d/n

P(p = 0)   =   .8464
P(p = .5)   =   .1472
P(p = 1.0) =   .0064

1.0000

Since p' = .08 for the process, 84.64% of the time the sample has
p = 0 and *favorably* misrepresents the process, while the remainder
of the time p exceeds p' = .08 and the sample *unfavorably* misrepresents the process. A sample of n = 2 cannot give a p of .08, to
perfectly mirror the process.

Let us now state more formally some laws of probability, which
we have just been illustrating.

Definition 3.1. If two events A and B might occur on a trial
or experiment, but the occurrence of either one prevents the occurrence of the other, then events A and B are called *mutually
exclusive*.

Law 3.1. If two events A and B are mutually exclusive, then
we have

P(A or B, mutually exclusive events) = P(A) + P(B)   (3.5)

We have seen this law in finding the probability of just one defective in the sample of 2, that is, .0736 + .0736 = .1472. Also
P(2 good) + P(1 good) + P(0 good) = P(0 to 2 good). But the last
event is certain to occur. So its probability is 1.

Definition 3.2. If one of two events A and Ā is certain to
occur on a trial, but both cannot simultaneously occur, then A
and Ā are called *complementary events*. (A and Ā are mutually
exclusive.)

Law 3.2. For any event A and its complementary event Ā we have

$$P(A) + P(\bar{A}) = 1 \qquad\qquad (3.6)$$

We have seen this in the example of a single draw where p' was given as .08

P(1 defective in 1 draw) = .08 = p'
P(1 good in 1 draw) = 1 - P(1 defective in 1 draw)
= 1 - .08 = .92

The two events were complementary.

We also saw another case in n = 2, where the event of exactly one defective in the sample was complementary to the event of there being either 0 or 2 defective in the sample:

P(1 defective) = 1 - P(0 or 2 defectives)
= 1 - .8528 = .1472

Finally we have a law for independent events.

Definition 3.3. Two events A and B are *independent* if the occurrence or nonoccurrence of A does not affect the probability of B occurring.

Whenever a process produces defectives independently or at random, so that the probability of a defective on the next piece does not depend upon what the preceding pieces were like, then we have the case of independent events. Such a process is said to be in control, that is, stable, even though some defectives or nonconforming pieces are being produced.

Not all processes do behave this way. For example, we may consider the case of the piston ring castings. The sample of 100 contained 25 defective castings, while the remaining 2900 had only 4 defectives. This was because the defectives occur "in bunches" from a defect-producing condition. The sample of 100 was not drawn at random but instead got into a bunch of bad castings. (The castings were made from stacks of molds, and if the iron is not hot enough when poured into a stack, many castings may be defective.) Under such conditions, whether a piece is good or defective *does* have an influence on the probability of the next one being defective.

Law 3.3. If two events A and B are independent, then we have

P(both A and B occurring) = P(A)·P(B)   (3.7)

This law was used in finding the probabilities of the possible outcomes for a sample of n = 2 from a process with p' = .08 for a defective. Thus, P(good, good) = P(good)·P(good) = .92(.92), P(good, defective) = P(good)·P(defective) = .92(.08), P(defective, good) = P(defective)·P(good) = .08(.92). But these are only true for a process yielding defectives independently or randomly.

3.4 EXAMPLE 2 AND EQUAL LIKELIHOOD, DEPENDENCE

As a second example of probability, let us consider drawing *without replacement* from a lot of N = 6 meters, of which D = 1 is defective. Here we have introduced the notation:

N = number of pieces in a lot   (3.8)
D = number of defective pieces in a lot   (3.9)

These are capital letters directly corresponding to the small letters n and d for a sample.

Now consider the very simple case in which we just draw a *random* sample of 1 from the lot of 6. "Random" means that each of the six meters is equally likely to be chosen for the sample, that is, the probability for each is 1/6. There are only two kinds of samples possible: (1) that the sample yields a good meter or (2) that the sample yields a defective meter. Five of the samples are in class (1) and only one sample in class (2). So we have

$$P(good) = \frac{5}{6} \qquad P(defective) = \frac{1}{6}$$

We might also reason that there are five chances in six of a good meter and only one chance in six of a defective meter.

Now next consider drawing a sample of n = 2 from the lot having N = 6, of which D = 1 is defective. How many defectives can we have in the sample of two? We might have d = 0 or d = 1 in the sample, but not more. Why?

This is a case of two consecutive drawings which are not independent. Let us see. Take first the case of the sample yielding
no defectives, that is, two good meters. We need

P(2 good) = P(good, good)

   = P(good on first draw)·P(good on 2nd draw,

      given good on first)

   = $\frac{5}{6} \cdot \frac{4}{5} = \frac{2}{3}$

The probability of a good one on the second draw is 4/5, because
after the first draw yielded a good meter, there remain in the lot
five meters, of which only four are good, and each of the five is
equally likely to be chosen.

Next consider the probability of the sample containing one
defective. This can be obtained by drawing a good one followed by
the defective or by drawing the defective followed by drawing a
good one.

P(good, defective)

   = P(good first)·P(defective, if good on first)

   = 5/6 · 1/5 = 1/6

P(defective, good)

   = P(defective)·P(good, if defective on first)

   = 1/6 · 5/5 = 1/6

Note that on the second draw in the second case we are sure to find
a good meter because that is all that is left in the lot. Since
these two cases are mutually exclusive, we add their probabilities
to find the probability that the sample contains one defective:

P(one defective and one good) = $\frac{1}{3}$

So two-thirds of the time our sample contains all good meters and a
third of the time there will be a defective.

Again we emphasize that the probabilities on the second draw
depend upon what happened on the first draw. This is always the

case when we draw *without replacement* from a lot.

There is another approach to finding the foregoing probabilities. This lies in *counting* possible *equally likely samples*. Let us call the five good meters g_1, g_2, g_3, g_4, g_5, and the defective one d. Then we can in this simple case write down all possible samples. There are 15 as follows:

$$g_1g_2 \quad g_1g_3 \quad g_1g_4 \quad g_1g_5 \quad g_2g_3 \quad g_2g_4 \quad g_2g_5 \quad g_3g_4 \quad g_3g_5 \quad g_4g_5$$
$$dg_1 \quad dg_2 \quad dg_3 \quad dg_4 \quad dg_5$$

where we are considering *unordered* samples, that is, g_2g_4 is the same sample as g_4g_2. (If we were to call them different we would have to write out 30 samples instead of 15.)

In the first row are ten samples with both meters good and in the second row five samples with one defective meter and one good. Therefore,

$$P(2 \text{ good}) = \frac{10}{15} = \frac{2}{3}$$

$$P(\text{one defective and one good}) = \frac{5}{15} = \frac{1}{3}$$

just as before.

Here we have used the following definition of probability:

Definition 3.4. If an event E can occur in any one of e equally likely ways, but there are n equally likely ways for the experiment to occur (E occurring or failing to occur), then the *probability* of E occurring is

$$P(E) = \frac{e}{n} \tag{3.10}$$

3.5   EXAMPLE 3, ANOTHER LOT PROBABILITY PROBLEM

In order to illustrate further the last type of calculation of probabilities, let us consider a lot of ten radios, of which two contain minor defects and are therefore defectives. So we have g_1, g_2, $\ldots$, g_8 and d_1, d_2 for the 10 radios.

Take a random sample of one radio from the lot. There are just 10 possible samples, each of one radio. Of these eight will contain a good one, two a defective. Following (3.10):

$$P(1 \text{ good}) = \frac{8}{10} \qquad P(0 \text{ good}) = \frac{2}{10}$$

Now consider samples of n = 2 from the lot. The *unordered* samples with both radios good are

$$g_1g_2 \quad g_1g_3 \quad g_1g_4 \quad g_1g_5 \quad g_1g_6 \quad g_1g_7 \quad g_1g_8 \quad g_2g_3 \quad g_2g_4$$
$$g_2g_5 \quad g_2g_6 \quad g_2g_7 \quad g_2g_8 \quad g_3g_4 \quad g_3g_5 \quad g_3g_6 \quad g_3g_7 \quad g_3g_8$$
$$g_4g_5 \quad g_4g_6 \quad g_4g_7 \quad g_4g_8 \quad g_5g_6 \quad g_5g_7 \quad g_5g_8 \quad g_6g_7 \quad g_6g_8$$
$$g_7g_8$$

that is, 28 such samples. For the samples with 1 good and 1 defective we have 16 samples:

$$d_1g_1 \quad d_1g_2 \quad d_1g_3 \quad d_1g_4 \quad d_1g_5 \quad d_1g_6 \quad d_1g_7 \quad d_1g_8$$
$$d_2g_1 \quad d_2g_2 \quad d_2g_3 \quad d_2g_4 \quad d_2g_5 \quad d_2g_6 \quad d_2g_7 \quad d_2g_8$$

And finally the *one* sample without a good radio and both defectives is

$$d_1d_2$$

Since we will assume sampling to be at random, all samples are equally likely and we use (3.10):

$$P(2 \text{ good}) = \frac{28}{28 + 16 + 1} = \frac{28}{45}$$

$$P(1 \text{ good}) = \frac{16}{45} \qquad P(0 \text{ good}) = \frac{1}{45}$$

Now such an approach of enumerating all possible outcomes is *theoretically* possible in this type of probability calculation. But it rapidly becomes far too long. For example, if we had taken samples of 3 from the lot of 10, we would have had to list 120 different samples. Thus we need some counting machinery such as is given in the next section.

3.6 COUNTING SAMPLES: COMBINATIONS AND PERMUTATIONS

It is quite likely that the reader has already come across combinations, and perhaps permutations, in high school or college mathematics. We here give a little introduction to combinations and permutations. In *combinations* we consider, for example, that we have n objects which we can distinguish between, that is, tell them apart. In industry each meter or radio might carry a serial number. But if not we still assume that they are distinguishable one from another.

Now how many distinct samples, each of one, can we draw from a lot of N = 10? Quite obviously this is N = 10. We call this the *number of possible combinations* of N objects taken 1 at a time, or in symbols:

$$C(N, 1) = N$$

We had two examples of this in examples 2 and 3: $C(6, 1) = 6$, $C(10, 1) = 10$, for the number of possible samples of 1 from 6 or 10.

Next consider samples of two, from say four good pieces. Then the number of distinct *unordered* samples may be obtained from the number of distinct *ordered* samples. For ordered samples we have

$$g_1 g_2 \quad g_2 g_1 \quad g_1 g_3 \quad g_3 g_1 \quad g_1 g_4 \quad g_4 g_1$$
$$g_2 g_3 \quad g_3 g_2 \quad g_2 g_4 \quad g_4 g_2 \quad g_3 g_4 \quad g_4 g_3$$

There are 12 such samples. But the number of possible unordered samples is only half as much, that is, six, because, for example, the one unordered pair $g_2 g_4$ corresponds to two ordered pairs $g_2 g_4$ and $g_4 g_2$. Thus there are twice as many ordered pairs as there are unordered pairs.

Next let us consider lots of 10 distinct pieces. The number of possible ordered samples, each of two, is $10 \cdot 9$, because there are 10 possible choices for the first piece, and having made a choice there remain nine choices for the second piece. Thus we have 90 such *ordered* samples. But just as before, for each *unordered* sample of two there are two *ordered* samples, hence the number of distinct unordered samples is 90/2 = 45, as we saw in example 3.

Now let us go to samples of 3 from 10. The number of distinct *ordered* samples is 10·9·8; 10 choices for the first, 9 for the second, leaving 8 possibilities for the third choice. But now six of these *ordered* samples correspond to just one *unordered* sample. For example, $g_1g_4g_6$, $g_1g_6g_4$, $g_4g_1g_6$, $g_4g_6g_1$, $g_6g_1g_4$, $g_6g_4g_1$ all contain the same three pieces and thus correspond to the same one *unordered* sample. Hence the number of distinct *unordered* samples or combinations is 10·9·8/6 = 120. (It would already be quite a chore to write out all of them.)

Now let us give names to the two kinds of samples. We call the number of distinct ordered samples a *permutation*, using the letter P and two numbers in parentheses following. Thus

$$P(10, 3) = 10 \cdot 9 \cdot 8 = 720$$

is the number of *distinct* permutations or ordered samples each of three objects, chosen from 10. Likewise P(3, 3) = 3·2·1 = 6 is the number of distinct orderings of three objects out of three.

We call the number of distinct unordered samples a *combination*, using the letter C and two numbers in parentheses following. Thus

$$C(10, 3)$$

is the number of distinct unordered samples of 10 objects taken three at a time. But as we have seen this is 120, that is,

$$C(10, 3) = 120 = \frac{10 \cdot 9 \cdot 8}{6} = \frac{P(10, 3)}{P(3, 3)}$$

This should be enough background to enable the reader to understand the general case: For the number of permutations or ordered samples of n distinct objects taken r at a time we have

$$P(n, r) = n(n - 1)(n - 2) \cdots (n - r + 1) \qquad (3.11)$$

where there are r numbers multiplied together. Likewise

$$P(r, r) = r(r - 1)(r - 2) \cdots 2 \cdot 1 = r! \qquad (3.12)$$

in which *all* of the objects are used. Here we used r! ("r factorial") to stand for the product of all whole numbers from r down to 1 inclusive. Then for combinations or number of unordered samples of objects, we have in general

$$C(n, r) = \frac{P(n, r)}{P(r, r)} = \frac{n(n - 1)\cdots(n - r - 1)}{r(r - 1)\cdots 1}$$

$$= \frac{n(n - 1)\cdots(n - r + 1)}{r!} \qquad (3.13)$$

there being r factors in both numerator and denominator.

As an example, consider the number of combinations of 12 distinct objects taken four at a time:

$$C(12, 4) = \frac{12 \cdot 11 \cdot 10 \cdot 9}{4 \cdot 3 \cdot 2 \cdot 1} = 495$$

The machinery we have just built up for counting can enable us to calculate many probabilities efficiently. Since there are tables of the logarithms of factorials (for example, our Table E), it is sometimes convenient to use, for example,

$$P(8, 5) = 8 \cdot 7 \cdot 6 \cdot 5 \cdot 4 = \frac{8 \cdot 7 \cdot 6 \cdot 5 \cdot 4 \cdot 3 \cdot 2 \cdot 1}{3 \cdot 2 \cdot 1} = \frac{8!}{3!}$$

where we have merely completed the factorial 8! in the numerator by multiplying by $3 \cdot 2 \cdot 1$ and then "paying for it" by dividing by $3 \cdot 2 \cdot 1$ or 3!. Therefore,

$$\log P(8, 5) = \log 8! - \log 3!$$

In general, we have

$$P(n, r) = \frac{n!}{(n - r)!} \qquad (3.14)$$

Using the expression in (3.13) gives

$$C(n, r) = \frac{n!}{r!(n - r)!} \qquad (3.15)$$

giving as an example

$$C(8, 5) = \frac{8!}{5! \; 3!}$$

As an example of the use of combinations in counting possible samples consider again example 3 where N = 10, D = 2, and n = 2. The total number of possible ordinary or unordered samples of 2 is

$$C(10, 2) = \frac{10 \cdot 9}{2 \cdot 1} = 45$$

Now of these, how many will contain nothing but good ones? Since there are eight good radios in the lot, we have

$$C(8, 2) = \frac{8 \cdot 7}{2 \cdot 1} = 28$$

Next, how many will contain one good and one defective radio? We must pair up one good with one defective. We can choose one good from the eight good ones in the lot in $C(8, 1) = 8/1 = 8$, that is, "Take one." Likewise, we can take one of the two defectives in $C(2, 1) = 2$ ways. Multiplying, we have $8 \cdot 2 = 16$ possible samples, because for each one of the eight ways to choose a good one there are always two ways to choose the defective to go with it.

Finally, for a sample containing two defectives we have $C(2, 2) = 2 \cdot 1/2 \cdot 1 = 1$, that is, we must take both of them.

As noted before the sum of the numbers of each kind of sample is $45 = 28 + 16 + 1$, and the probabilities are 28/45, 16/45, 1/45 for 2, 1, 0 good ones in the sample.

3.7 APPROACH OF OCCURRENCE RATIO d/n = p TO p'

Let us now carry Sec. 3.2 a step further, making use of Sec. 2.5. Suppose again we draw pieces from a process with p' = .08 for a defective, .92 for a good one, constantly. Suppose that we draw randomly 400 pieces from the process. Would we not "expect" that 8% of the 400, or 32 would be defectives? That is, if we repeatedly take samples of 400, the average number of defectives in the 400 would be 32. But of course we would only observe a distribution of outcomes around the average of 32, sometimes 34, 35, or maybe 29. How much variation can we expect around the expected or average of 32? The expected variation is the standard deviation of the number of occurrences d, which may be shown to be

$$\sigma_d = \sqrt{np'(1 - p')} \qquad\qquad\qquad (3.16)$$

where p' is the true or population fraction defective and 1 - p'
the fraction good (like .92), and n the sample size. Then

$$\sigma_d = \sqrt{400(.08)(.92)} = 5.43$$

Therefore we can reasonably expect to miss the expected number of
defectives, 32, by five or so, perhaps observing 27 or 37 defectives
in 400.

By reference to Sec. 2.5 we can also say that about 65 to 70%
of the time we will observe an occurrence of defectives between
32 ± 5.4, that is, 27 to 37 defectives, since it can be shown that
the distribution is quite well behaved or normal in such a case.
Moreover by taking a ±2 standard deviation range, that is,

$$32 \pm 2(5.43) = 22 \text{ to } 42$$

we can expect to lie within such a range 95 to 96% of the time.
Also we can expect d to lie within

$$32 \pm 3(5.43) = 16 \text{ to } 48$$

at least 99% of the times we draw 400 from the process, that is,
almost always.

These ranges for the observed number of defectives can be
translated into ranges for the proportion d/n, by merely dividing
by n = 400. Thus

    27/400, 37/400 give .0675, .0925
    22/400, 42/400 give .0550, .1050
    16/400, 48/400 give .0400, .1200

Finding ranges for d, then for d/n, help us visualize how close
to the true process defective p', we can expect our observed sample
fraction defective p = d/n to come.

3.8 FURTHER EXAMPLES OF PROBABILITY

Example 4. A balanced cubical die has six faces numbered 1 to
6, with each face as likely to show on top as any other. Thus the
probability of each face showing is 1/6 for a perfectly balanced
die. So the distribution of points is as follows:

Point	Probability
1	1/6
2	1/6
3	1/6
4	1/6
5	1/6
6	1/6

Such a distribution is often called a *uniform distribution.*

Now suppose we roll two such balanced dice. What is the distribution of the total of the faces showing on the two dice? To
solve this, let us assume the dice are distinguishable, for example,
a white die and a red one. Then we diagram them as follows:

White	Red
1	1
2	2
3	3
4	4
5	5
6	6

The dice being assumed balanced and rolling independently, then for
each one way the white die comes up, the red may come up in any one
of six ways. Thus the dice may fall in any one of 6·6 = 36 distinct
(and mutually exclusive) ways. We have now but to count the number
of ways any given possible total may occur. For example, looking at
the preceding table, we see

Total	White	Red	No. of ways
2	1	1	1
3	1	2	
3	2	1	2
4	1	3	
4	2	2	3
4	3	1	

Then taking the number of ways each possible total 2 to 12 can occur, we have but to divide by 36 to obtain the probabilities for each total. This gives the distribution:

Total	Probability
2	1/36
3	2/36
4	3/36
5	4/36
6	5/36
7	6/36
8	5/36
9	4/36
10	3/36
11	2/36
12	1/36
	1

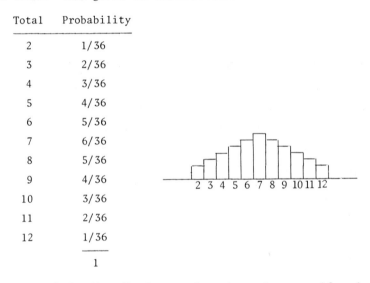

The shape of the distribution as shown is no longer uniform but like an A-shaped roof.

The foregoing probabilities are useful to know in various games.

Example 5. Three coins are tossed independently, each having a "head" and a "tail" face, and they are perfectly balanced. Then what is the distribution of the number of coins showing a head, out of the three coins? There can be 0, 1, 2, or 3 heads showing. Are these four outcomes equally likely? No.

Let us consider the coins distinguishable, say, numbered 1, 2, 3. Then there are eight possible ways for the coins to fall and these *are* equally likely, by 2·2·2. They are

1	2	3
H	H	H
H	H	T
H	T	H
T	H	H
H	T	T
T	H	T
T	T	H
T	T	T

Of these eight distinct ways, only one results in three heads and only one in no heads. But there are three ways to have two heads and three ways to have one head. Thus dividing by 8 gives the distribution of probabilities:

Heads	Probability
3	1/8
2	3/8
1	3/8
0	1/8

Example 6. From a 52-card deck of playing cards, consisting of four suits and each suit containing 13 kinds of cards A, K, Q, J, 10, 9, ..., 2, a hand of five cards is to be randomly drawn. What is the probability of the hand containing (a) four cards of any one rank and one other nonmatching card, (b) three cards of one rank and two of another rank?

First we find how many distinct five-card hands there are. This is a combination because the order of drawing the cards is immaterial. It is

$$C(52, 5) = \frac{52 \cdot 51 \cdot 50 \cdot 49 \cdot 48}{5 \cdot 4 \cdot 3 \cdot 2 \cdot 1} = 2,598,960 = N$$

(How would the reader like to write down all of these?)

Now for the type of hand in (a), we can choose the *rank* of the four cards in any one of 13 ways. Having chosen the rank, there remain in the deck 48 other cards, any one of which might be chosen to go with the four cards. So the number of possible hands of type (a) is $13 \cdot 48 = 624$. Hence we have

$$P(a) = \frac{624}{2,598,960}$$

Next we consider a hand of type (b). How many are possible? We can choose the rank of the three cards in 13 different ways. Now having chosen the rank, how many ways are there to choose the three suits (from the four in the deck)? This is

$$C(4, 3) = \frac{4 \cdot 3 \cdot 2}{3 \cdot 2 \cdot 1} = 4$$

Then for the rank of the two cards, we are left with 12 possible ranks. (Why not 13?) Having chosen the rank for the pair or two cards, we next must choose two suits from the four possible. This is

$$C(4, 2) = \frac{4 \cdot 3}{2 \cdot 1} = 6$$

Multiplying the four numbers together we arrive at the number of possible hands of type (b), namely

$$13 \cdot 4 \cdot 12 \cdot 6 = 3744$$

Thus we have

$$P(b) = \frac{3744}{2,598,960}$$

(Probabilities for other more common types of hands are a bit more difficult to calculate.)

It is often helpful for a quality control worker to be able to solve probability problems of interest to his friends and fellow workers. Such an ability can well be an entrée helpful in "selling" himself and quality control methods to others.

3.9   SUMMARY

Some knowledge of probability is basic in interpreting results of
quality control analysis on industrial processes and on incoming
product, and also on research results. We have defined a probability in two ways: (1) as a fixed likelihood for an event to occur,
lying somewhere between 0 and 1 inclusive, and determining the long-
run ratio of occurrences of the event in question to the number of
trials, (2) in terms of equally likely outcomes being the number of
ways the event can happen, divided by the number of ways it can
occur or fail to occur. We have observed that the occurrence ratio
d/n tends to approach the true probability p', as the number of
trials increases. But this is not a fixed mathematical approach.
We can in fact expect that after n trials we may well be off from
p' by an amount

$$\frac{\sigma_d}{n} = \sqrt{\frac{np'(1 - p')}{n}}$$

and may be even two or three times as far away as this.
    We also gave two basic probability laws:

1. If events A and B are mutually exclusive (cannot both
    occur),

    $$P(A \text{ or } B) = P(A) + P(B)$$

2. If two events A and B are independent (occur without
    influence from each other),

    $$P(A \text{ and } B \text{ both}) = P(A) \cdot P(B)$$

These laws were used in examples for finding probabilities and distributions. Also counting machinery, combinations and permutations
were discussed and used for counting cases, in order to use the
second definition of probabilities.

PROBLEMS

3.1 Toss a coin 100 times, keeping track of the number of heads
thrown. After each 10 tosses, calculate the occurrence ratio of

heads up to that point. Describe the approach to the assumed p'
= .50 for a balanced coin. About how far from the expected 50 heads
(in 100 tosses) might we expect our observed count of heads to lie
(that is, $1\sigma_d$)?

3.2   Proceed as in Prob. 3.1, but roll a die and count the number
of times 6 shows, using p' = 1/6.

3.3   Given a process with a fraction defective p' = .02, and taking
a random sample of n = 2 pieces, find P(2 good), P(1 good), P(0 good).

3.4   Same as Prob. 3.3, but with p' = .01.

3.5   Given, as in example 1, that p' = .08 for the fraction defective and we take n = 3. Find P(3 good), P(2 good), P(1 good),
P(0 good).

3.6   From a lot of N = 8 gauges, of which D = 1 is slightly off,
find the distribution of d for results from a random sample of n = 2.
Show graphically.

3.7   Same as Prob. 3.6, but with n = 3. Show graphically.

3.8   Same as Prob. 3.6, but with D = 2, n = 2. Show graphically.

3.9   Same as Prob. 3.8, but with n = 3. Show graphically.

3.10   If the fraction defective for a process is p' = .10 on minor
defectives, and a random sample of 400 is taken, how many can be
expected to be defectives. How far from this expectation would be
a reasonable deviation? (σ_d) What is about the very worst discrepancy we could experience? $(3\sigma_d)$

3.11   Find the value of C(10, 4), C(10, 6), P(10, 4).

3.12   From a standard deck of 52 cards, four are drawn without
replacement.
    a.   How many different possible hands are there?
    b.   In how many of these will all four be of the same one rank?
    c.   In how many will there be three cards of the same rank and
         one of some other rank?
    d.   What then is the probability of the kind of hand in b and
         in c?

Chapter Four

THREE BASIC LAWS FOR ATTRIBUTE DATA

4.1 COUNTED DATA: DEFECTS OR DEFECTIVE PIECES IN A SAMPLE OF n PIECES

We shall discuss in this chapter three important distribution models for attribute or counted data. These are basic in control charts for process control and for acceptance sampling for decision making on lots received.

A piece, article, sample, or subassembly has the "attribute" of being free from defects or else possessing one or more defects. Hence counts are called "attribute data." We must be careful to distinguish between two kinds. On the one hand, we can carefully inspect a sample of n units or pieces, say, and count the total number of defects found on all n units. This might be five defects, for example. Now it is possible that all five are on just one of the n units and that there are no defects at all on the remaining n - 1 units. But this still counts as five defects. On the other hand, we may again inspect the n units or pieces for defects, but now count how many units possess one or more defects. Each such piece having at least one defect is called a *defective*. The others are good or nondefective pieces. Now if a sample of n pieces contains five defects, these could possibly be all on one piece, yielding one defective and n - 1 good pieces in the sample. But the five defects are much more likely to be scattered over several pieces, perhaps even making five pieces count as defectives. Let us summarize by giving two definitions.

Definition 4.1. Inspecting or testing n pieces, we may search for defects or nonconformances in the n pieces and r⌐ ⌐d the total of such defects. This is measuring quality by a *count of defects*.

Definition 4.2. In inspecting or testing n pieces, we may consider whether each of n pieces does contain any defects. Each piece having one or more defects is called a *defective*. We record the total number of such defectives in the n. This is measuring quality by a *count of defectives*.

We shall consistently use the following symbols here and in later chapters:

n = number of units in a sample   (4.1)

d = number of defective units in the sample of n units   (4.2)

$p = \dfrac{d}{n}$

    = sample fraction defective

    = proportion of defective units in the sample   (4.3)

$q = 1 - p$

    = sample fraction good   (4.4)

$d = np$

    = number of defective units in the sample   (4.5)

p' = population or process fraction defective   (4.6)

q' = population or process fraction effective or good

    = $1 - p'$ (4.7)

c = number of defects in a sample

    (the sample might be of n units or just one)   (4.8)

c' = population average number of defects in a sample   (4.9)

In this chapter we shall consider three distribution models, two for counts of defectives and one for counts of defects. All three are widely applicable. Moreover, they are intimately related.

4.2 THE BINOMIAL DISTRIBUTION FOR DEFECTIVES

We have already discussed a special case of the binomial distribution in Sec. 3.3. There we had a process or population fraction defective of $p' = .08$, and we considered n = 1 and n = 2. In the latter case the number of defectives d in the sample could be 0, 1, or 2. The probability for each kind of sample was worked out.

Let us "sneak up" on the general case by taking a little bit more complicated case, namely

$$p' = .10 \qquad q' = .90 \qquad n = 4$$

We now have five possible kinds of samples, that is, with d = 0, 1, 2, 3, 4. What is the probability for each kind of sample, that is, about how often could we expect each different kind of sample outcome? Obviously d = 0 will prove commonest, and d = 4, the rarest. We shall again be assuming that p' is constant (at .10) and that the results on each draw for the sample are independent.

First find the probability of drawing a sample with all pieces good. This is found by using Law 3.3 for probability.

$$
\begin{aligned}
P(d = 0) &= P(4 \text{ good}) \\
&= P(\text{good})P(\text{good})P(\text{good})P(\text{good}) \\
&= .9^4 = .6561
\end{aligned}
$$

Thus about two-thirds of the time, we draw a sample of n = 4 pieces from the process, all four will be good ones. Now next we seek P(3 good, 1 defective). This event can take place when the defective occurs on the first, second, third, or fourth draw, and all other drawings yield good pieces. Thus using Laws 3.1 and 3.3 for mutually exclusive events

$$
\begin{aligned}
P(d = 1) &= P(3 \text{ good}, 1 \text{ defective}) \\
&= P(g,g,g,d) + P(g,g,d,g) + P(g,d,g,g) + P(d,g,g,g) \\
&= (.9)(.9)(.9)(.1) + (.9)(.9)(.1)(.9) \\
&\quad + (.9)(.1)(.9)(.9) + (.1)(.9)(.9)(.9) \\
&= 4(.9)^3(.1) \\
&= .2916
\end{aligned}
$$

Note particularly that the numerical probabilities for the four mutually exclusive ways in which the sample can show d = 1 are all equal.

Next consider samples with d = 2; they have two defectives and two good ones. How many distinct orders are there for such samples? Six: ggdd, gdgd, gddg, dggd, dgdg, ddgg. But the probability for each one of these sample outcomes is $(.9)^2(.1)^2$. Thus we have for the probability for d = 2 the following

$$P(d = 2) = 6(.9)^2(.1)^2 = .0486$$

Continuing, there are only four orders of sampling results which yield d = 3, namely gddd, dgdd, ddgd, dddg. The probability for each order being $(.9)(.1)^3$, we have

$$P(d = 3) = 4(.9)(.1)^3 = .0036$$

Finally for d = 4, all four must be defective, and so

$$P(d = 4) = (.1)^4 = .0001$$

These being the only five possible outcomes, the sum of the five probabilities should be 1, which is easily verified here. (This provides a useful check on our calculations.)

Now what about the coefficients 4, 6, 4 of the products of powers of .9 and .1? The 4 occurring for P(d = 1) was the number of ways in which we could choose just one of the four consecutive drawings of pieces to yield the one defective. This is thus

$$C(4, 1) = \frac{4}{1} = 4$$

The coefficient 6 was the number of distinct orders of drawings which yield two good and two defective, namely

$$C(4, 2) = \frac{4 \cdot 3}{2 \cdot 1} = 6$$

Likewise the latter coefficient 4 was the number of distinct orders of drawings which yield one good and three defective, namely

$$C(4, 3) = \frac{4 \cdot 3 \cdot 2}{3 \cdot 2 \cdot 1} = 4$$

This reasoning enables us to write all five probabilities in one formula:

$$P(d) = C(4, d)(.9)^{4-d}(.1)^d$$

By substituting in turn d = 0, 1, 2, 3, 4, we have the five desired probabilities, if we take by definition that C(4, 0) = 1 and C(4, 4) = 1, since there is but one way for each to occur, namely, take none from four, or take all four.

Before continuing we tabulate our results and then picture them.

Binomial distribution

(p' = .1, n = 4)

d	P(d)	p = d/n
0	.6561	.00
1	.2916	.25
2	.0486	.50
3	.0036	.75
4	.0001	1.00
	1.0000	

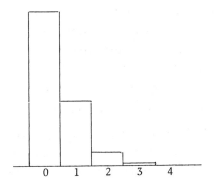

Such a distribution is sometimes described as J-shaped, since as d increases, P(d) continually decreases.

General case. Probably rather fortunately, we shall not need to make much use of the general formula in this book. But we do give it here. It is justified by generalizing the discussion, which has just been given for the special case p' = .1, n = 4. Thus we have for given p' and n

$$P(d) = C(n,\ d)(p')^d(q')^{n-d} \qquad (4.10)$$

which gives the probability for the sample outcome of d defectives among the n pieces, when on each draw the probability of a defective is constantly p'. The coefficient C(n, d) gives the number of distinct orders possible in which d defectives can occur among the n drawings. The probability for each one order is $(p')^d(q')^{n-d}$.

Tables. Instead of working out any desired cases of (4.10), it is best to try to find any needed probabilities in published tables. See Refs. 1 through 5 at the end of this chapter. The table entries may be in the form of particular probabilities, such as, p' = .06, n = 50, P(d = 2) = .22625. Or we may have to subtract *cumulative* entries. For example, again for p' = .06, n = 50, P(d = 2) = P(d = 2 or more) - P(d = 3 or more), since what the event "2 or more" has which the event "3 or more" lacks is precisely the event d = 2. Thus we have P(d = 2) = P(2 or more) - P(3 or more) = .81000 - .58375 = .22625. The approach to use merely requires careful thought, depending upon the entries available in whatever tables you can find to use. It is also quite easy to program an electronic computer to calculate entries such as (4.10), to any de-sired precision. Fortunately, we can often approximate probabili-ties for the binomial distribution by using the so-called Poisson distribution to be discussed in Sec. 4.3.

The name "binomial distribution." If you have studied the bi-nomial theorem in algebra and recall it or at least recognize it, you will see that (4.10) is simply the term of the binomial $(q' + p')^n$ which has p' raised to the d power. Thus $(q' + p')^n$, when expanded, generates all the probabilities P(d). Hence the name *binomial distribution.*

Population characteristics of a binomial distribution. It may be shown that whenever we have a binomial distribution for a sample of n, with fraction defective p', then for the number of defectives d we have

$$E(d) = np'$$
$$= \text{theoretical average number of defectives}$$
$$\text{per sample} \tag{4.11}$$

where the symbol E stands for the theoretical average of whatever is written within parentheses following the E.

Thus when n = 4 and p' = .10, the average number of defectives per sample of 4 is np' = 4(.1) = .4. This result can also be found from the five probabilities as follows in much the same way we would find a *sample* average from a frequency tabulation:

d	P(d)	dP(d)
0	.6561	.0000
1	.2916	.2916
2	.0486	.0972
3	.0036	.0108
4	.0001	.0004
	1.0000	.4000 = E(d)

E(d) is simply a "weighted average" of numbers 0, 1, 2, 3, 4 with weights respectively P(0), P(1), ..., P(4).

We have already seen the formula for σ_d for the binomial, that is, for a typical departure from the expected E(d). This was given as formula (3.16) but it is here repeated:

$$\sigma_d = \sqrt{np'q'}$$

= standard deviation of number of defectives (4.12)

In our example, where n = 4 and p' = .10, $\sigma_d = \sqrt{4(.1)(.9)}$ = .6. Thus this is sort of an average of deviations from the expected E(d) = .4.

For the case in which n = 50 and p' = .06, we expect to have np' = 50(.06) = 3 defectives per sample. This is the average number. But, of course, we will not find 3 every time. How much variation from the average of 3 is only to be expected?

$$\sigma_d = \sqrt{50(.06)(.94)} = 1.68$$

Thus 2 or 4 will also be common occurrences, and 1 or 5 not at all rare. But $3\sigma_d$ = 5.04, and so a discrepancy of over 5, that is, 6 or more from the expected 3 will be rare; that is, rarely will we find 3 + 6 = 9 or more defectives in a sample for such a binomial distribution. Of course, we *cannot find* 3 - 6 = -3 defectives in a sample!

Examples. We now give some examples of binomial distributions including the true fraction defective, sample size, expected or average number of defectives per sample E(d), and only-to-be-expected deviation from E(d), namely σ_d. The probabilities are given to three decimal places in Table 4.1. They are shown graphically in Fig. 4.1.

We have already discussed Case (b) a bit. As noted before, occurrences of three or more standard deviations above the average of np' = 3 will be quite rare. Since np' + $3\sigma_d$ = 3 + 3(1.68) = 8.04, it actually takes nine or more defectives to be above a 3σ limit. The probability of 9 or more is seen to be .003, three times in 1000, in Table 4.1. Cases (a) and (c) have the same

TABLE 4.1. Examples of Binomial Distributions for Various Process
Fractions Defective and Sample Sizes

	(a)	(b)	(c)	(d)	(e)	(f)
p'	.30	.06	.01	.08	.05	1/6
n	10	50	300	20	100	6
E(d) = np'	3	3	3	1.6	5	1
σ_d	1.45	1.68	1.72	1.21	2.18	.91

d	P(d)	P(d)	P(d)	P(d)	P(d)	P(d)
0	.028	.045	.049	.189	.006	.335
1	.121	.145	.149	.328	.031	.402
2	.233	.226	.224	.271	.081	.201
3	.267	.231	.225	.141	.140	.054
4	.200	.173	.169	.052	.178	.008
5	.103	.102	.101	.015	.180	.001
6	.037	.049	.050	.003	.150	
7	.009	.020	.021	.001	.106	
8	.002	.007	.008		.065	
9		.002	.003		.035	
10		.001	.001		.017	
11					.007	
12					.003	
13					.001	

expected number of defectives as Case (b), but differing standard
deviations. The comparison shows up best in Fig. 4.1. Case (f) can
be experimentally tested by tossing six honest dice and counting the
number which yield some one face chosen in advance, for example, 6.

Conditions for a binomial distribution.

1. *Constant probability* p' of a defective on each draw of a
    "piece" is established.

2. Results on drawings are *independent* of preceding results.

3. n draws of a piece are taken.

4. Number of defectives in the n draws are counted.

These conditions may well be fulfilled when drawing a sample of
pieces from a process. When they are met, then the process is stable or said to be "in control," even though it may not be as good as
we want or need it to be. The conditions may also be very nearly
fulfilled if we take random samples from a large lot, for then the

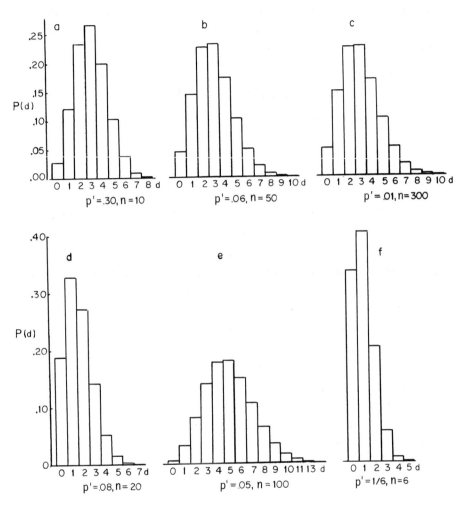

Fig. 4.1. Six examples of binomial distributions from Table 4.1. Examples (a), (b), and (c) all have the same mean np' = 3, but respective σ_d values are 1.45, 1.68, 1.72. For (d) np' = 1.6, σ_d = 1.21; for (e) np' = 5, σ_d = 2.18; for (f) np' = 1, σ_d = .91.

probability for a defective on each draw will be nearly constant (even though it will vary slightly depending on previous draws).

Characteristics for the variable p (sample fraction defective). If, instead of the number of defective pieces d in a sample, we

consider the fraction defective p = d/n as the variable in question,
then we can find its population characteristics by dividing (4.11)
and (4.12) by n. This gives

$$E(p) = p'$$
= theoretical average fraction defective   (4.13)

$$\sigma_p = \sqrt{\frac{p'(1 - p')}{n}}$$
= standard deviation of fractions defective   (4.14)

4.3 THE POISSON DISTRIBUTION FOR DEFECTS

We now take up the standard model for the distribution of defects
on a sample. Examples are counts of the number of (1) typographical
errors on a printed page, (2) breakdowns of insulation in 1000 m of
insulated wire at a test voltage, (3) leaks in 20 radiators, (d) gas
holes in castings, (5) defects found in inspection of a subassembly,
and (6) pinholes in a test area of painted surface.

If the defects occur independently and not in "bunches," then
the Poisson distribution may well apply. However, the number of
possible defects should greatly exceed the average number. Also
the area of opportunity for defects in each sample should be the
same. Then, if we know c', the average number of defects per sample, we can find by the Poisson distribution the probability of a
sample containing no defects, of it having exactly one defect, exactly two, and so on. These are, of course, called P(0), P(1),
P(2), ...

Use of tables. Later on we shall give a formula for Poisson
distribution probabilities. However, usually we do not use the
formula, but instead use a table, such as Table B in the back of
this book. In Table B we first find the row determined by c', the
average number of defects per sample. Then for any given desired
number of defects c as given in the column heading, we find the
probability of c or fewer defects. For example, if the theoretical
process average is c' = 3 defects per sample, then in the c = 5 column we find 916. Since decimal points were omitted in printing the

table, this is .916. Thus

P(5 or less defects, given c' = 3) = .916

Likewise

P(4 or less defects, given c' = 3) = .815

These mean that we expect to find five or fewer defects in 916 samples out of 1000 samples, and also we expect to find four or fewer defects in 815 samples out of 1000 samples. Now by subtracting .815 from .916 we get .101, which is the probability of *exactly* five defects in a sample. This is because the event "5 or less defects" or 0, 1, ..., 5 defects, contains just the event c = 5, which is not contained in the event "4 or fewer defects," or 0, 1, ..., 4. Thus about 101 samples out of 1000 samples will contain exactly c = 5 defects. In this way we can use the cumulative probability Table B to find the individual probabilities. Thus for c' = 3, we have

			P(0)	=	.050
P(0)	=	.050	P(1)	=	.149
P(1 or fewer defects)	=	.199	P(2)	=	.224
P(2 or fewer defects)	=	.423	P(3)	=	.224
P(3 or fewer defects)	=	.647	P(4)	=	.168
P(4 or fewer defects)	=	.815	P(5)	=	.101
P(5 or fewer defects)	=	.916	P(6)	=	.050
P(6 or fewer defects)	=	.966	P(7)	=	.022
P(7 or fewer defects)	=	.988	P(8)	=	.008
P(8 or fewer defects)	=	.996	P(9)	=	.003
P(9 or fewer defects)	=	.999	P(10)	=	.001
P(10 or fewer defects)	=	1.000			

Some published tables, for example in Ref. 6, give individual terms like those in the second column, as well as cumulative probabilities. Reference 7 gives a very complete set of individual

TABLE 4.2. Examples of Poisson Distributions
for Various Process Averages of Defects c'

	(a)	(b)	(c)	(c)	(e)
c'	.25	1.0	1.6	3	5
$\sigma_c = \sqrt{c'}$	.50	1.00	1.26	1.73	2.24
c	P(c)	P(c)	P(c)	P(c)	P(c)
0	.779	.368	.202	.050	.007
1	.195	.368	.323	.149	.033
2	.024	.184	.258	.224	.085
3	.002	.061	.138	.224	.140
4		.015	.055	.168	.175
5		.003	.018	.101	.176
6		.001	.005	.050	.146
7			.001	.022	.105
8				.008	.065
9				.003	.036
10				.001	.018
11					.009
12					.003
13					.001
14					.001

probabilities up to c' = 10.

One other possibility we sometimes encounter is the need for
a probability such as P(3, 4, or 5 defects). For this we could add
P(3) + P(4) + P(5), obtaining .493, or, more directly, we could use

P(3, 4, or 5 defects)
= P(5 or fewer defects) - P(2 or fewer defects)
= .916 - .423 = .493

Note that we did *not* subtract P(3 or fewer defects). If we had,
this event would have left only 4 or 5 and would have given
P(4 or 5 defects). It merely requires a little care to decide which
cumulative probabilities to subtract for any desired probability.

Examples. We now give some examples of the Poisson distribution, in order to show some typical shapes. These are all taken from Table B, by subtraction as was shown for the case, c' = 3. See Table 4.2. Histograms are given in Fig. 4.2, picturing these distributions.

The distributions for an average of c' = .25 is strongly J-shaped with over three-fourths of the probability at c = 0. Thus

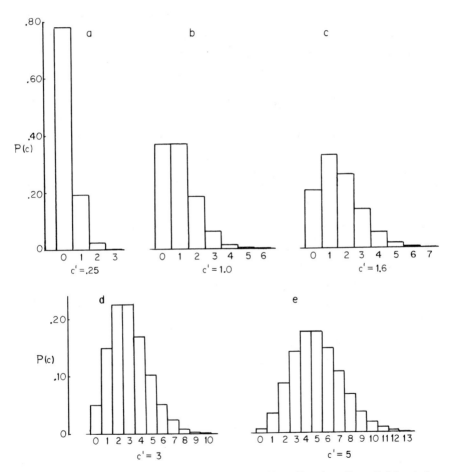

Fig. 4.2. Five examples of the Poisson distribution from Table 4.2. Distribution means are listed by the histograms, namely, the c' values. The standard deviations ($\sqrt{c'}$) are, respectively, .50, 1.00, 1.26, 1.73, 2.24.

77.9% of the samples can be expected to yield no defects at all. For c' = 1, the average being one defect per piece or sample, no defects and one defect are equally common, and together make up nearly three-fourths of the cases. However, there is scattering and even six defects can occur in about one sample out of 1000. For c' = 1.6, the most probable value is now one defect, and cases up to seven defects, as a practical limit, can occur. For c' = 3 and 5, as averages, we find a tie for the most probable numbers of defects at 2 and 3, and at 4 and 5, respectively, with considerable tailing out to high numbers of defects. These distributions are a bit more symmetrical than the others shown.

Population characteristics of the Poisson distribution. The theoretical process average number of defects per sample is c' as we have been emphasizing. We now give the standard deviation for the count c of defects. It can be shown to take the very simple form

$$\sigma_c = \sqrt{c'}$$

= population standard deviation of counts c   (4.15)

For the examples given in Table 4.2, these σ_c values are given just below the c' values. As usual they are an amount of departure from c' which is only to be expected. As with the binomial distribution, there will be very few occasions when the observed c count of defects will be above $c' + 3\sigma_c$. Thus, for example, when c' = 5, σ_c = 2.24 and $c' + 3\sigma_c$ = 5 + 3(2.24) = 11.72, and the probability of 12 or more is .005.

All Poisson distributions are unsymmetrical around the two sides from c', i.e., below and above. But as c' increases, the Poisson distribution becomes more symmetrical. The maximum probability always occurs at c', or within one of it.

The Poisson distribution as an approximation to the binomial. The Poisson distribution is very useful for approximating a binomial distribution. We shall especially make use of this fact in Chapter 9 when we discuss decision making for acceptance or rejection of incoming lots.

To see the approach of probabilities by the binomial to those of the Poisson distribution, look again at examples (a), (b), and (c) of Table 4.1. In each of the three the average number of defectives, namely np', is 3. In (a) we have n = 10, p' = .30, so that we expect 3 defectives among the 10, and can easily have as many as 5 or 6 in the 10. In (b) with n = 50, p' = .06 there is the same expected number of 3, but this is in a sample of 50, not 10. Again 5 or 6 defectives may readily occur, but among the 50. The probabilities for each d count of defectives are rather similar in examples (a) and (b), but there is rather more variability in distribution (b) than in (a) (σ_d = 1.68 to 1.45). Now in example (c) with n = 300, p' = .01, once more the expected count of defectives d is 3. But now σ_d is a bit larger yet. But compare the probabilities for cases (b) and (c). Are they not quite noticeably similar? We could go further and work out the probabilities for n = 3000, p' = .001 with np' again 3. But even in case (c), we are already near the Poisson limit. From Table 4.1(c) and Table 4.2(d) we have the following probabilities.

Count d or c	0	1	2	3	4	5
(c) Binomial P(d)	.049	.149	.224	.225	.169	.101
(d) Poisson P(d)	.050	.149	.224	.224	.168	.101

Count d or c	6	7	8	9	10
(c) Binomial P(d)	.050	.021	.008	.003	.001
(d) Poisson P(d)	.050	.022	.008	.003	.001

These two sets of probabilities are strikingly similar. Thus we see that if we consider a series of binomial distributions each of which has the same expectation or average count np' but in which n is increased continually while p' is correspondingly decreased, so as to preserve np' at some constant, say, c', then the Poisson distribution with expectation c' is approached as a limit. (Read this again!)

```
n       50   300   3000   10,000
p'     .06   .01   .001    .0003   Poisson
np'     3     3     3        3     c' = 3
```

Now how good is the approximation? It begins to be of "practical use" for $n \geq 10$ and $p' \leq .10$, and the approximation becomes "quite satisfactory" for $n \geq 20$ and $p' \leq .05$.

But why bother with an approximation; why not use the formally correct binomial distribution when it is correct? The reason is that the Poisson distribution requires only *one* constant or *parameter* to determine it completely, namely c'. On the other hand, the binomial distribution requires that we know *two* constants n and p'. Thus for Poisson probabilities we enter a table, such as Table B, with c' and a desired value of count c (a double-entry table). But a binomial table, such as those in the references, must be entered with n and p' as well as a desired count of d (a triple-entry table). Thus for substantial coverage of cases, the binomial tables require vastly more entries than does a Poisson table. And so even a small table, like Table B, is quite adequate, but a binomial table to be as adequate might well need to be 50 times as large.

As an example, suppose for a binomial distribution n = 100, p' = .04 and we desire the probability of $d \leq 3$. With such a sizable n and small p' we can have much confidence in the approximation. Using the Poisson distribution in Table B for np' = 100(.04) = 4, we find

P(3 or less, given c' = 4) = .433

From an appropriate binomial table, we find the probability

P(3 or less, given n = 100, p' = .04) = .429

Or if we desire the probability of exactly four defectives, we can use the Poisson in Table B with c' = 4 finding

P(4) = P(4 or less) - P(3 or less)

      = .433 - .238

      = .195

whereas a binomial table yields

P(4) = .197

Clearly these approximations by the Poisson are entirely satisfactory.

Conditions for a Poisson distribution.

1. Samples provide *equal areas* of *opportunity* for defects (same size of unit, subassembly, length, area, or quantity).

2. Defects occur *randomly* and *independently* of each other.

3. The average number c' remains *constant*.

4. The possible number of defects c is far greater than the average number c'.

These conditions may well occur in a production process. The quality control person can usually arrange for 1 to be true. Condition 4 also is commonly true. Condition 2 is, however, very possibly not true; that is, defects may occur "in bunches" (for example, dust particles on areas of paper or inspection errors on gas holes in castings). In process control using control charts for defects, we actually make a test as to whether, for a process, c' is remaining constant, as we shall be seeing in Chapter 6.

Formula for P(c) for the Poisson distribution. Now for the sake of completeness, we give the formula for P(c), even though we shall commonly use tables, such as Table B.

$$P(c \text{ defects, given average } c') = \frac{e^{-c'}(c')^c}{(c!)} \qquad (4.16)$$

where e = the natural base for logarithms, namely e = 2.71828···. It can be used in programming an electronic calculator, especially well by beginning with $P(0) = e^{-c'}$, and then successively finding each P(c + 1) from the preceding P(c). But tables are usually entirely sufficient for our purposes.

4.4* THE HYPERGEOMETRIC DISTRIBUTION FOR DEFECTIVES

This distribution carries quite a fearsome name! But the situation
to which the distribution applies is really quite simple. It is
present when we inspect a sample of articles or pieces for defectives, choosing them randomly from a lot without replacement. Thus
if we draw our sample of pieces one at a time from the lot, at each
draw giving each of the pieces remaining in the lot an equal chance
to be drawn next, we have the conditions for the hypergeometric
distribution.

We have in fact already seen two examples of this distribution
in the second and third examples of Chapter 3. Thus we have already
made a "dent" in the problem.

Let us work out the general formula after going over another
example, using combinations as we did for example 3 in Sec. 3.6.
Suppose we are given a lot of N = 10 pieces of which D = 3 are defective and we choose a random sample of n = 4 from the lot. The
random drawing makes all possible samples equally likely. How many
of them are there?

$$C(10, \ 4) \ = \ \frac{10 \cdot 9 \cdot 8 \cdot 7}{4 \cdot 3 \cdot 2 \cdot 1} \ = \ 210$$

Now, of these, how many will contain none, one, two, or three defectives? (We could not have all four of the sample pieces defective,
since there are only three defectives in the lot.)

For no defectives in the sample, we must have chosen four good
ones. See the diagram. Since there are seven good ones in the lot,

we need to find how many ways we can draw four good ones from the
seven present in the lot. This is

---

*Sections designated by an asterisk can be omitted in a first read-
ing of the book or chapter.

$$C(7, 4) = \frac{7 \cdot 6 \cdot 5 \cdot 4}{4 \cdot 3 \cdot 2 \cdot 1} = 35$$

Hence by (3.10)

$$P(0) = \frac{35}{210} = \frac{1}{6}$$

so about one time in six the sample will yield no defective. Next we find the number of possible samples having three good and one defective. We first choose three of the seven good ones present in the lot, which can be done in

$$C(7, 3) = \frac{7 \cdot 6 \cdot 5}{3 \cdot 2 \cdot 1} = 35$$

distinct ways. Now with each one of these ways we might pair up one of the three defectives. Therefore, the number of such distinct samples (3g, 1d) is

$$3 \cdot 35 = 105$$

and so

$$P(1) = \frac{105}{210} = \frac{1}{2}$$

Continuing for P(2), we must have a sample of (2g, 2d). We may choose the two good in

$$C(7, 2) = \frac{7 \cdot 6}{2 \cdot 1} = 21$$

ways, whereas the two defectives can be chosen in

$$C(3, 2) = \frac{3 \cdot 2}{2 \cdot 1} = 3$$

ways. Hence we have 21·3 = 63 ways in which the sample will yield (2g, 2d) giving

$$P(2) = \frac{63}{210} = \frac{3}{10}$$

Finally for (1g, 3d), we have for the one good

$$C(7, 1) = \frac{7}{1} = 7$$

and for the three defectives, taking all that there are,

C(3, 3) = 1

Therefore, the number of distinct such samples is only 7·1 = 7 and

$$P(3) = \frac{7}{210} = \frac{1}{30}$$

The reader may readily check that the sum of the four probabilities is 1.

Probably the reader can now see that the general law for P(d) is the following for a lot of N pieces containing D defectives from which n are randomly drawn:

$$P(d) = \frac{C(D, d)C(N - D, n - d)}{C(N, n)} \qquad (4.17)$$

Thus we count the number of *possible* samples of n pieces taken from a lot of N pieces, all being equally likely. This is C(N, n). Then since there are D defective pieces in the lot we count the number of possible choices of d defectives for the sample, C(D, d). This leaves N - D "good" pieces in the lot of which we must choose n - d to complete the sample. The number of possible choices is thus C(N - D, n - d). Then we use the multiplication principle that for each possible choice of the d defectives, we have all possibilities for drawing n - d good ones, thus giving the numerator in (4.17).

It is quite a chore to calculate even one probability P(d) if N is at all large. There is a substantial table of P(d) values and cumulative probabilities, designated p(x) and P(x), respectively, in Ref. 8. It is quite complete up to N = 100. For larger N's one must resort to tables of logarithms of factorials such as Table E in the back of the book, using (3.15) for the combinations. Or, if available, Ref. 9 provides a table of C(N, D) up to N = 100.

Approximations. The foregoing "exact" methods of finding P(d) for a hypergeometric distribution are fortunately not likely to be necessary. Because, if N is at least 8 or 10 times n, then the successive drawings for n, out of N, will not cause the original fraction defective D/N to vary appreciably. To be more specific, let us

suppose that on the first selection of a piece from the lot, it
turns out to be a good one. Then there are still D defectives in
the remainder of the lot, but only N - 1 pieces left. So the fraction defective on the second draw is

$$\frac{D}{N - 1}$$

But if we happen to draw a defective on the first piece, then there
remain D - 1 in the N - 1, and the fraction defective on the second
draw is

$$\frac{D - 1}{N - 1}$$

When N is large these are both close to D/N = p'.

Thus, when N is at least 8 to 10 times n, we may obtain a good
approximation to hypergeometric distribution probabilities by assuming a constant fraction defective p' = D/N and using the *binomial
distribution* (4.10). Moreover, if *in addition* p' ≤ .05 and n ≥ 20,
we can go further and use the Poisson to approximate with c' = np'
= n(D/N). Then Table B becomes feasible to approximate the hypergeometric distribution. This will give satisfactory approximations
in most industrial situations.

Example of approximation. Suppose that we desire to find the
probability of a sample of n = 100 to contain d = 3 defectives, when
chosen randomly from a lot of N = 800, of which D = 16 are defectives. Direct use of (4.17) along with (3.15) and tables of logarithms of factorials gives after *considerable* calculation

P(3) = .194   (hypergeometric distribution)

However, since N = 8n, we can usefully approximate this probability
by using the binomial (4.10), with n = 100 and p' = D/N = 16/800
= .02. Putting these in, we find after some calculation

P(3) = .182   (binomial approximation)

But, we see *also* that p' is much below .05 and n above 20, so we may
go even further and use a Poisson distribution to approximate the

hypergeometric probability. The one to use is that with

   c' = np' = 100(.02) = 2

For P(3) we may now merely use Table B as follows with c' = 2:

   P(3) = P(3 or less) - P(2 or less)
        = .857 - .677
        = .180   (Poisson approximation)

A reference. A more complete treatment of each of the three
distributions and their interrelationships and approximations are
given in Ref. 10.

4.5 SUMMARY

We have been studying in this chapter two distribution models for
defectives. The first was the binomial distribution in which we
draw randomly from a process with constant fraction defective p'.
It may also be thought of as drawings from an infinitely large lot,
or as drawings *with replacement* from any size of lot. (The latter
is, of course, a stupid thing to do in industrial inspection.) Then
we also studied the hypergeometric distribution for defectives for
random drawings *without replacement* from any size of lot. We say
that the binomial distribution could commonly be used for approximating the hypergeometric distribution, fortunately.

Further, we studied the Poisson distribution model for counts
of defects on samples of product. Table B includes good coverage
of the Poisson distribution. Moreover, it was shown that under
appropriate conditions the Poisson distribution could be used to
approximate both of the binomial and hypergeometric distributions.
Thus the Poisson distribution is useful for counts of *defectives* as
well as counts of *defects*. (Most fortunately!)

The binomial and Poisson distributions are the background
models for process control charts for defectives and for defects,
respectively.

The reader would do well to review periodically the conditions
of applicability of the three distributions. Also he should be

careful to distinguish between counts of defectives and of defects.

PROBLEMS

4.1   Past data have shown an average of $c' = 1.2$ typographical errors per magazine page of print. (a) Find the probability of one or fewer typographical errors on a page. (b) Find the probability of three or more errors on a page.

4.2   Past data have shown an average of $c' = 3.6$ leak points per radiator on the initial check test. (a) Find the probability of three or fewer leaks on a radiator. (b) Of five or more.

4.3   At final inspection of trucks an average of $c' = 2.0$ defects are found. (a) Find the probability of two or fewer defects. (b) Of exactly two defects. (c) Of two or more.

4.4   Over a test area of painted surface an average of $c' = 5.4$ pinholes is standard for a process. (a) Find the probability of two or fewer pinholes. (b) Of over seven pinholes.

4.5   An executive picks up a sample of five brass bushings from a tote pan containing 1000 of them, of which 100 have at least slight burrs. (a) What is the probability that none in the sample have any burrs? (b) Which distribution is formally correct to use? (c) Which is the simplest distribution you can justifiably use for the calculations?

4.6   An automatic screw machine is producing spacers, with $p' = .01$ for off-diameter pieces. A sample of $n = 10$ is taken randomly. (a) What is the probability of no off-diameter pieces in the sample? (b) Which distribution is formally correct? (c) Can you approximate with the Poisson distribution?

4.7   A process is producing temperature controls for automatic hot water tanks. The fraction defective has been running at $p' = .0020$. In a sample of 100, approximate the probabilities of no, one, and two defective controls by the Poisson distribution.

4.8 The fraction defective considered acceptable for a certain minor defective is p' = .05. Sampling acceptance of lots calls for random samples of 100, from lots of 4000 pieces, with acceptance, if eight or fewer in the sample have this defect. Suppose p' actually is .05, use the Poisson approximation to find the probability of such a lot of being accepted.

For the following problems find σ_c and find the probability of $c \geq c' + 3\sigma_c$.

4.9   Problem 4.1.

4.10 Problem 4.2.

4.11 Problem 4.3.

4.12 Problem 4.4.

4.13 If p' = .20, n = 4, find the probability of none, one, two, three, and four defectives in the sample.

4.14 If p' = .10, n = 5, find the probability for each of 0, 1, ..., 5 defectives in the sample.

4.15 If from a lot of seven clocks of which two are defective, a random sample of two is drawn, find P(0), P(1), P(2).

4.16 If from a lot of nine gauges, three of which are inaccurate, a random sample of three is drawn, find P(0), P(1), P(2), P(3).

REFERENCES

1. U. S. Department of the Army, *Tables of the Binomial Probability Distribution*, National Bureau of Standards, Appl. Math. Ser. 6, U. S. Government Printing Office, Washington, D. C., 1950.

2. Harvard Univ., Computing Lab., *Tables of the Cumulative Binomial Probability Distribution*. Harvard Univ. Press, Cambridge, Mass., 1955.

3. W. H. Robertson, *Tables of the Binomial Distribution Function for Small Values of p*, Tech. Services, Department of Commerce, Washington, D. C., 1960.

4. H. G. Romig, *Fifty to 100 Binomial Tables*, Wiley, New York, 1947.

5. S. Weintraub, *Cumulative Binomial Probability Distribution for Small Values of p*, Free Press of Glencoe, Collier-Macmillan, London, 1963.

6. E. C. Molina, *Poisson's Exponential Binomial Limit*, Van Nostrand-Reinhold, Princeton, N. J., 1947.

7. T. Kitagawa, *Tables of Poisson Distribution*, Baifukan, Tokyo, 1952.

8. G. J. Lieberman and D. B. Owen, *Tables of the Hypergeometric Probability Distribution*, Stanford, Calif., 1961.

9. T. C. Fry, *Probability and Its Engineering Uses*, Van Nostrand-Reinhold, Princeton, N. J., 1928

10. I. W. Burr, *Statistical Quality Control Methods*, Marcel Dekker, New York, 1976.

Chapter Five

CONTROL CHARTS IN GENERAL

We now begin the study of a most powerful and versatile set of tools for the analysis and improvement of a production process. These tools are called "control charts." They were first invented by Dr. Walter A. Shewhart of the Bell Telephone Laboratories in 1924, and developed by him and his associates in the 1920's. He published a complete exposition of the theory, practical application, and economics of control charts in 1931 [1]. Seldom has a whole field of knowledge been so well explored and its applications so well pointed out in the first publication. So basic and applicable are control charts that new uses are continually being found in all sorts of products and industries. And yet control charting methods are simple enough to be learned and applied by one with even very modest mathematical background.

5.1 RUNNING RECORD CHARTS OF PERFORMANCE

It is a common practice of top executives, superintendents, foremen, directors of quality, chief inspectors, engineers, quality supervisors, and research workers to keep track of performance by means of plotted figures measuring such performances. This is done by periodically plotting an appropriate descriptive number of the current results. Such a number could be a measurement or some sort of count (attribute data). Although we shall be emphasizing various data indicating quality of performance of production processes, running record graphs can be made on any numbers of interest. Nonquality examples are gross production, absenteeism, unit costs, late shipments, hospital calls, and number of customer returns.

TABLE 5.1. Running Record for Four Quality Characteristics on Three Kinds of Product

Sample number	Diameter of uncoated wire		Defect: "Plug holes," final inspection on planes	Packing nuts for tractor, defectives in samples of 50, January	
	Average x̄	Range R	Defects c	Defectives d	Fraction defective p
1	.0985	.0030	2	12	.24
2	.0980	.0025	3	10	.20
3	.0970	.0010	5	5	.10
4	.0985	.0030	6	4	.08
5	.0973	.0025	5	0	.00
6	.0972	.0005	6	3	.06
7	.0980	.0040	5	1	.02
8	.0965	.0010	4	0	.00
9	.0970	.0010	6	3	.06
10	.0975	.0010	9	8	.16
11	.0980	.0025	9	7	.14
12	.0980	.0015	6	0	.00
13	.0975	.0010	5	0	.00
14	.0980	.0040	9	14	.28
15	.0984	.0030	9	13	.26
16	.0985	.0030	8	14	.28
17	.0980	.0040	7	4	.08
18	.0970	.0010	6	5	.10
19	.0968	.0010	11	0	.00
20	.0975	.0010	12	10	.20
21	.0972	.0010	12		
22	.0970	.0010	8		
23	.0980	.0010	8		
24	.0978	.0010	7		
25	.0985	.0010	5		
26	.0970	.0020			
27	.0985	.0010			
28	.0974	.0020			
29	.0972	.0015			

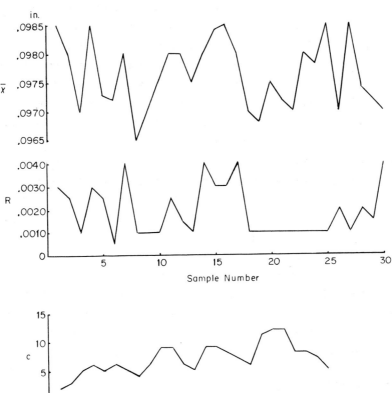

Fig. 5.1. Four running record graphs of data in Table 5.1: x̄'s and R's for diameters of uncoated wire, counts c of "plug hole" defects on planes, and fractions defective p for packing nuts.

In Table 5.1, we give quality data on three products. On wire diameter, we list the average $\bar{x}$ and the range R for samples of n = 7. All such diameters are supposed to lie between specification limits for individual x's of .096 to .102 in. The second product is a series of large aircraft, inspected at the end of production for all types of defects, one of which was called "plug holes." Naturally if production were perfect there would be no plug hole defects (nor any other). But in practice they do occur, forming a distribution. The third product was packing nuts for a farm implement, where considerable trouble was encountered. An intensive program was begun in October, and the present samples were taken the following January. Some substantial improvement had already occurred.

Running record graphs of the four quality characteristics are shown in Fig. 5.1.

Which of these four charts shows the best control? That is, which chart shows the most nearly random pattern such as might be obtained by drawing chips randomly from a bowl? Which chart shows the poorest control, that is, clearest evidence of some changes of production conditions, effecting changes in quality? Try to guess.

5.2 PERFORMANCE VARIES

One thing perfectly obvious in all four graphs in Fig. 5.1 is *variation*. Performance does vary from sample to sample, time to time. Some variation in production processes is only natural and must be lived with; such variation is unavoidable no matter how much effort is expended on the process. The more closely we try to work, the more variation shows up. But on the other hand, there are some extreme variations which are not unavoidable, and their causes can be identified and appropriate action taken.

5.3 UNUSUAL PERFORMANCE CALLS FOR ACTION

A significant rise in the sample fractions defective p can signal the presence of some change in the process. We want to find out what the cause is so that we may eliminate it and obtain a lower process fraction defective p'.

Or in the case of a process being studied for defects c, we may note a significant drop in defects. We want to know whether it is caused by lax inspection or by some change in production conditions, such as raw material or a better method, which, if identified and incorporated, will continue to give better product.

When a quality characteristic is measurable, we have two charts to watch, $\bar{x}$ and R, reflecting level and variability. Either or both may give evidence of process improvement or deterioration. Ranges may show worse or better within-sample variation. If better, we want to find out why so we can "do it again." If worse, we want to find the cause so we can eliminate it and obtain a more consistent, less-variable product. Records of averages $\bar{x}$ are very important, for example, to dimensional control, strength, or chemical impurities. They may point the way to action for improvement.

The big word here is ACTION.

5.4 WHAT IS UNUSUAL?

This is the basic question in following a performance record. Just when does a point give a reliable indication that a change has come into the process? In Fig. 5.1, is the low $\bar{x}$ point of sample 8 a clear indication of some cause in the process? What about the five $\bar{x}$'s of .0985 in.? Is the pattern of ranges R such as can be attributed to chance variation?

Are the high or low points on the running records for plug hole defects c meaningful? Is there reliable evidence of a trend upward? Are the fractions defective of the packing nuts homogeneous, or do they provide reliable evidence of an erratic production process?

These are the kinds of questions to which we need answers. Can you tell by examining the running record charts? And would you get the same decision if we were to use a different scale, say smaller? The author has often shown in lectures a slide with two running records plotted one above the other. The listeners are asked to vote on which shows the better control, the most freedom from extraneous process changes. They all vote for the one drawn to a smaller scale (so it looks more homogeneous). The author then tells them that it is the same data merely drawn on two different scales. Even though both scales are shown, nearly everyone votes for the one drawn to the smaller scale. If you, in using graphs, make decisions that depend *to any degree* on the scale used in plotting the data, then your decisions cannot be very objective!

Now the truth of the matter is that one cannot be sure when looking at a graph whether a given extreme-looking point is really a reliable indication of a process cause or whether it should just be considered part of the natural unavoidable variability of the process. A good part of the trouble is that when you look at a running record graph you cannot readily take account of the *sample size*. Thus in the fraction defective record for the packing nuts, the sample size was n = 50. Had it been 20 or 25, probably none of the points would clearly indicate a cause. Or if n were 200, then a great many of the 20 points would have provided clear evidence of some cause in the process. We therefore need a way to "play the odds" well by taking account of the sample size and using appropriate distribution models.

5.5 TWO KINDS OF CAUSES

We shall soon be discussing how we can use the control chart approach to greatly sharpen our decision-making ability, and thereby to get the most out of our production processes and manufacturing.

In the preceding section we have begun to consider two rather distinct types of causes of process variation. There are, on the one hand, the few factors that have a relatively large influence upon quality performance. They are very well worth seeking and doing something about. Moreover they are findable, that is, can be identified. On the other hand, there is a large number of causes, each of which had an almost negligible effect, but between them they give rise to a *distribution* of outcomes. This is the random pattern due to "chance."

Let us summarize these distinctions by a little table:

	Number present	Effect of each	Worth seeking?
Assignable causes	Very few, perhaps only one or none	Marked	Well worth seeking out, and possible to find
Chance causes	Very large number	Slight	Not worth looking for

It is worthwhile for you, the reader, to think about a production
process with which you are familiar, then to make a list of chance
causes and potential assignable causes. By using your knowledge
and imagination you should be able to list 20 or 30 chance causes
which may slightly affect the result, say, a dimension. For example, variation in stock, vibration, tool speed, coolant, lubricant,
positioning in a gauge, dust present, and reading of the gauge.
Assignable causes might be a change in stock or a change of process
setting or a broken cutting tool. Try listing assignable causes
for your process.

   One word more: factors affecting production quality are not
really classifiable into two perfectly distinct categories. To
some extent, they may be usefully thought of as forming a distribution with their relative strength or influence on the outcome going
from a *few* very influential ones, we call assignable causes, on
down to the host of *small chance* causes. (Such a distribution is
sometimes called a Pareto distribution.) Our job is best done by
trying to find when an assignable cause is operating and what change
in conditions brought it into action.

5.6 CONTROL CHARTS

Walter Shewhart, in wrestling with a problem made complicated by
variation, came to realize that it should be possible to determine
when a variation in product quality really is the result of some
process change, and when the apparent variation should be merely
attributed to chance. That is, to decide whether some assignable
cause is operating. This required that one should be able to measure how much variation is only to be expected in the process.

Then variations outside this distribution or pattern become clear indicators that an assignable cause has become operative, and we should seek to find it, whether it produces an undesirable or desirable effect.

Shewhart and his associates proceeded to experiment. They decided upon the use of limits set at three standard deviations above and below the average. The standard deviation is estimated for the distribution of product quality when chance causes only are operating. Then if chance causes only are at work, very rarely will we find a sample result outside the limits:

Average ± three standard deviations

Hence if we do find such a point or result, it is a much better bet that it is due to the presence of some assignable cause, rather than being a result of chance causes only. Then we look into the process conditions for the assignable cause.

Let us now reconsider the data of Table 5.1 as plotted into running record graphs in Fig. 5.1. We show in Fig. 5.2 these same data now made up into control charts by the inclusion of a central or average line (solid) and control limits (dotted). Points outside the limits are to be regarded as clear indications of the presence of an assignable cause. On the other hand, points inside the control limits are attributed to chance causes and no action is taken. We are likely to be wasting our time looking for the cause of any fluctuations between the control limits. It is like looking for a needle in a haystack without using a magnet! In this way we save our time from fruitless search, to use in more thorough study, when warranted, for assignable causes.

Looking at the control chart for $\bar{x}$'s we see that there are five $\bar{x}$ points on or above the upper control limit and two on or below the lower control limit. Each is to be regarded as a clear indication of some factor affecting the wire diameter. Process level is quite out of control. In the chart for ranges R, there are three points above the upper control limit, indicating causes of excessive within-sample variability, that is, of too much

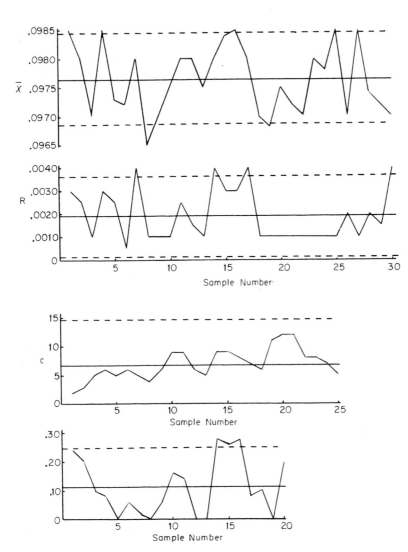

Fig. 5.2. Four control charts of data in Table 5.1: x̄'s and R's
for diameters of uncoated wire, counts c of "plug hole" defects on
planes, and fractions defective p for packing nuts.

jumping around. There is also quite a long run of R's of .0010 in.
below the central line, which is a hopeful sign.

Looking at the chart for c's, plug hole defects, we see no
points above the upper control limit. So under that criterion, the
process is in control, and no action is indicated. However, there
does seem some evidence of a trend toward higher counts of plug
holes. The first nine counts are all below the central line. Then
near the end there are three *consecutive* points all fairly close to
the upper control limit. Further study is indicated.

Finally we examine the p chart. The three high fractions defective lying above the upper control limit are a clear indication
of an assignable cause. The other points are in control. The
relatively large number of p = .00 points is an encouraging sign,
however.

The details of figuring central lines and control limits will,
of course, be given in subsequent chapters, specifically Chapters
6 to 8.

5.7 INTERPRETATIONS OF POINTS AND LIMITS

Since these interpretations are so basic and important, let us set
them down explicitly.

Interpretation of a point beyond a control limit. Any point
beyond either control limit on a control chart is to be regarded as
a clear indication of the presence of an assignable cause among the
conditions under which the point arose. This includes production
and measurement or test conditions.

Interpretation of a point between control limits. Any point
between the control limits of a control chart is attributed to
chance causes, since it provides no reliable indication of the presence of an assignable cause. Although an assignable cause may have
lain behind the point, we are in general better off to attribute it
to chance and wait for clear evidence.

There is one other indication of the presence of an assignable
cause used by many quality control workers. If a long run of

successive points lie on one side of the central line, this may be taken as evidence of an assignable cause. How long is "long?" Opinion varies. The author recommends the following:

Assignable cause by a run. When a run of at least *eight consecutive* points, all lying on the same side of the central line, occurs this gives clear evidence of the presence of an assignable cause (a change of average in the characteristic in question).

5.8 TWO PURPOSES OF CONTROL CHARTS

There are two purposes of control charts, which we consider briefly here.

Purpose 1 (Analysis of past data). The objective of analyzing past data is to test for control or uniformity. Could such results as have been observed *readily* have come by chance variations alone from a homogeneous process? Or is there reliable evidence of some non-chance-factor assignable cause entering into the process? If the latter, then action is suggested in order to improve control of the process.

Purpose 2 (Control against standards). For this control chart purpose, some standard or standards are available. They may be from analysis of past data leading to a long run of good control. Or they may be from some goals or requirements. Then each sample result is compared with the standard via control limits to see whether it is compatible. If inside the limits, no action is indicated. But if outside the control limits, then appropriate action is taken.

Thus in Purpose 1, at least in the beginning, a preliminary run of, say, 25 samples is analyzed for homogeneity. Then each additional sample is plotted and interpreted at once, in relation to the previous data. In Purpose 2, we can set up control limits immediately, plot each sample as it arises and take appropriate action. After a while it may prove necessary or desirable to change the standards to reflect current process capabilities. Both purposes have their place in applications to process control.

5.9   PROCESS IN STATISTICAL CONTROL

A definition basic to our work is the following:

Definition 5.1.   Process in statistical control.   A process is
said to be in statistical control if the sample results from the
process behave like random drawings of chips from a bowl.

In this we usually think of the chips in a bowl as having numbers on each and forming a distribution for the bowl as a whole.
Drawings are best done with replacement so that at each drawing the
distribution of numbers in the bowl is always the same.   Or the bowl
could contain two different colors of chips or balls, one for good
ones another for defectives, so as to simulate a binomial population.

By use of control charts, it has commonly been found possible
(if sufficiently desirable) to get production processes into statistical control.

5.10   ADVANTAGES OF A PROCESS IN CONTROL

There are quite a few practical and lucrative advantages of a process in control (that is, in statistical control).

1.   Ordinarily, when beginning to analyze a production process
by a control chart and a preliminary run of data, we will find a
lack of control.   This is an indication that further improvement is
possible through finding assignable causes and removing undesirable
ones, or incorporating desirable ones such as, say, improved strength.
Thus the operation of getting a process into good control commonly
brings a marked improvement in quality.

2.   A process in control is predictable.   For example, we do
not need to measure or test every piece or unit.   We can accurately
predict that the current production is like the past, as long as
current sample results continue to show control relative to the limits.   This is not the case if there is lack of good control; we are
in a much weaker position relative to the prediction of quality.

3.   With regard to advantage 2, when the test is destructive,
for example, blowing times of fuses or tensile strength of rivets,
the only practical way to know what the untested product can do is

to obtain a good control so that we can predict reliably from test results from the destroyed pieces, as to how the remaining product would perform, if tested.

4. In line with advantage 2, we can often greatly reduce the amount of inspecting or testing needed, when we obtain good control of a process at a satisfactory quality level. If the process is not in statistical control, then we may well need to retain 100% inspection or at least heavy inspection, instead of small samples to compare with control limits.

5. If product from the supplier shows good statistical control at satisfactory quality performance, then receiving inspection may well be reduced to a minimum. The supplier's process is predictable.

6. In a similar manner, if our own final inspection or test results show good control at a satisfactory quality level, we can much more safely guarantee our output to the consumer.

7. A record of good statistical control of our processes is very helpful in dealing with customer complaints or, at the extreme, in liability cases.

8. A control chart on measuring instruments on *homogeneous* material, showing good control, provides confidence in the instrument and the technician. Also it enables an accurate determination of the measurement error. But if there is lack of control in measuring homogeneous material, there can be little confidence in the measurements.

9. Good control of laboratory or research results give confidence in the adequate control of the experimental conditions. Poor statistical control of results indicates poor control of the experimental conditions.

5.11 SUMMARY

In this chapter we have been describing a powerful decision-making tool for process improvement and control. Various types of control charts are available for analyzing process data. In fact, virtually all varying data may be analyzed by some type of control chart.

The objective is to determine how much variation can be attributed
to chance causes and thus to determine limits of normal variation.
Then results outside such limits can be taken as reliable indicators of the presence of some assignable cause, which is to be sought
among the process conditions operating while that sample was being
produced and inspected or tested. Appropriate action can then be
taken. In this way processes can be brought into statistical control, bringing advantages such as those in Sec. 5.10.

A short booklet very well worth reading to supplement this
chapter is Ref. 2.

PROBLEMS

5.1 For some production process with which you are familiar, make
a list of chance causes and of potential assignable causes.

For Problems 5.2 to 5.4, plot the quality variables given; make any
comments or speculations you think might justify further study.
Preserve your running record charts for problems in Chapters 6 and 7.

5.2 Inspection of fiber containers for contamination, resulting
from gluing, from samples of n = 25.

Sample number	1	2	3	4	5	6	7	8	9
Fraction defective, p	.04	.20	.04	.04	.00	.08	.08	.08	.08

Sample number	10	11	12	13	14	15	16	17	18
Fraction defective, p	.12	.00	.04	.04	.00	.08	.00	.00	.08

Sample number	19	20	21	22	23	24	25
Fraction defective, p	.00	.00	.00	.12	.04	.00	.00

5.3 For 21 production days the total defects c found on the days
production of 3000 switches ran as follows for December:

    30 56 47 86 44 23 16 64 80 54 73
    65 76 69 53 58 30 91 90 36 57

5.4 A transmission main shaft bearing retainer carried specifications of 2.8341 to 2.8351 in. In a production run, measurements
were made in .0001 in. from the "nominal" 2.8346 in. Thus the

specifications were -5 to +5 in the coded units. The following data
on averages and ranges were recorded. Plot the two running record
graphs, $\bar{x}$'s above the corresponding R's.

$\bar{x}$	+3.4	+1.4	+5.6	-2.6	+1.2	-2.0	-8.2	-5.8	-7.8	-4.0
R	6	3	24	7	6	3	7	6	2	6

$\bar{x}$	-5.0	-10.2	+2.2	+.8	+1.0	+1.6	-7.8	-3.8	-2.0	0
R	5	9	7	8	2	4	3	7	2	0

REFERENCES

1. W. A. Shewhart, *Economic Control of Quality of Manufactured Product*, Van Nostrand-Reinhold, Princeton, N. J., 1931.

2. American Society for Quality Control Standard (B3) or American National Standards Institute Standard (Z1.3 - 1969), *Control Chart Method of Controlling Quality During Production*, ASQC, 161 W. Wisconsin Ave., Milwaukee, Wis. 53203.

Chapter Six

CONTROL CHARTS FOR ATTRIBUTES: PROCESS CONTROL

This chapter covers the use of control charts for process control, where quality is measured by either a count of defective pieces or a count of defects in a sample of product. There are two distinct cases, as pointed out in Chapter 5: (1) analysis of past data and (2) control with standards given. We first consider the analysis for counts of defective pieces or units.

6.1 CHARTS FOR DEFECTIVES OR NONCONFORMING PIECES

A sample of n pieces or units is to be inspected for any units which contain defects or nonconformances. There may be just one kind of defect being looked for. Or we may be looking for any defect within a class of defects. For example, we may have a class of six different "major" defects, a piece being a major defective if it has one or more major defects. Or the class might be of "minor" defects or of "incidental" defects. If the defect in question is a "critical" one, then 100%, 200%, or more inspection is used to prevent any critical defects getting out, and thus no sampling would be used.

As you can see then, "defects" span a wide range of significance and importance. But in any case we may well wish to analyze the process for decision making and take appropriate action.

6.1.1 *The np or d Chart*

As given before we use the notations:

$$n = \text{number of pieces in the sample} \tag{6.1}$$

$$d = np = \text{number of defective pieces in the sample} \tag{6.2}$$

$$p = \frac{d}{n} = \text{fraction defective for the sample} \qquad (6.3)$$

$\bar{p}$ = average fraction defective for a run of samples $\qquad$ (6.4)

p' = true process fraction defective, or a standard $\qquad$ (6.5)

Let us now take up the case of the standard given and then of analysis of past data, although in practice the latter analysis is likely to be used initially in the analysis of a process.

Standard given, p'. We assume that somehow we have a *standard* process fraction defective, p'. This may be from a substantial run of homogeneous past production or a goal or a requirement of some kind. (Much as we might like, p' cannot in general be 0.) We assume also that we have a clear definition of each defect in the class being studied and that the defects act independently. In general, we will also take a random sample from production. (But sometimes we take all pieces arising from production, forming them into samples of n consecutive pieces.) Further, accurate inspection is assumed and must be arranged for through adequate training and good tools.

Then in this case we are testing with each sample count d, as to whether (1) the sample result is compatible with the assumed standard p' or (2) the sample result gives reliable evidence that the *process fraction defective* p' was not at the assumed p', while this sample was being produced. This is done by setting appropriate control limits and noting whether d = np lies (1) between the limits or (2) on or outside the limits. The control chart, therefore, provides a decision-making tool.

Now how are the limits set? Obviously, the assumed standard p' and the sample size n should both be taken into account. Let us see how. The conditions given here justify our using the binomial distribution discussed (perhaps even cussed) in Chapter 4. There we saw (4.11) that the expected or theoretical average number of defectives in a sample is np'. This we use for the *central line*,

$$\mathcal{E}_{np} = np'$$
$$= \text{central line for np} \qquad \text{(standard p' given)} \qquad (6.6)$$

Next, we have seen in Chapter 4 that the only-to-be-expected deviation from this average is $\sigma_d = \sqrt{np'q'}$ (4.12), where $q' = 1 - p'$. Now np or d counts will quite seldom lie as much as $3\sigma_d$ away from np'. We thus take as limits

$$\text{Limits}_{np} = np' \pm 3\sqrt{np'q'} \qquad \text{(standard p' given)} \qquad (6.7)$$

In practice, be sure to remember the 3 and the radical. You do not want the assignable cause to be YOU. Since these are 3σ limits for d, as long as p' remains constant observed counts d will seldom lie outside the limits (6.7). In practice, therefore, whenever a point does go outside the limits, we regard this as clear evidence that p' is no longer at the assumed value. Thus we conclude that there is an assignable cause which has changed p'.

Example 1. Let us analyze the following data obtained by experiment. They were taken randomly from a box, containing 1702 beads 1 cm in diameter, of which 1600 were yellow (good pieces) and 102 red (defective pieces). Therefore

$$p' = \frac{102}{1702} = .060$$

A paddle with 50 holes was used for n = 50. (Technically, the "hypergeometric distribution" of Section 4.4 applies rather than the binomial distribution. But since N = 1702, which is 34n, the latter distribution provides a nearly perfect approximation to the former.)

The sample results are given below for 30 samples:

Sample number	1	2	3	4	5	6	7	8	9
Defectives, np = d	1	3	2	3	3	3	2	3	3
Fraction defective, p	.02	.06	.04	.06	.06	.06	.04	.06	.06

Sample number	10	11	12	13	14	15	16	17	18
Defectives, np = d	4	3	5	3	4	4	2	3	6
Fraction defective, p	.08	.06	.10	.06	.08	.08	.04	.06	.12

Sample number	19	20	21	22	23	24	25	26	27
Defectives, np = d	3	7	2	3	3	3	3	3	4
Fraction defective, p	.06	.14	.04	.06	.06	.06	.06	.06	.08

Sample number	28	29	30
Defectives, np = d	2	4	4
Fraction defective, p	.04	.08	.08

We could plot either of the two quality characteristics: number of defectives np = d or the fraction defective p. In Section 6.1.1, we are concerned with the number np of defectives. Thus we have the first 30 points of Fig. 6.1.

Let us now figure the control lines for the points, using the standard p' = .06 and n = 50, from (6.6) and (6.7):

$$\pounds_{np} = np' = 50(.06) = 3$$
$$\text{Limits}_{np} = np' \pm 3\sqrt{np'q'}$$
$$= 3 \pm 3\sqrt{50(.06)(.94)}$$
$$= 3 \pm 5.04 = \text{---}, 8.04$$

The average is thus three defectives per sample. How far off from this average can we be and still be in control? A count of eight defectives would be so very close to the upper control limit, 8.04, that in practical work we would investigate the conditions under which the product was produced and inspected or tested. Certainly np ≥ 9 would be investigated. On the other hand, the arithmetic in carrying out (6.7) for the lower control limit would give -2.04. Since all counts np of defectives are 0 or more, the limit -2.04 is meaningless. All it means is that a count of no defectives would not be sufficiently rare to justify carrying on any investigation. But when a lower control limit is above 0, then points below such a lower control limit are worth investigating and can lead to real process improvement (unless it should prove to be due to lax inspection).

In Fig. 6.1, we see that all points lie below the upper control limit. Therefore we say that the process is in control with respect to the standard p' = .06. We have no reliable evidence

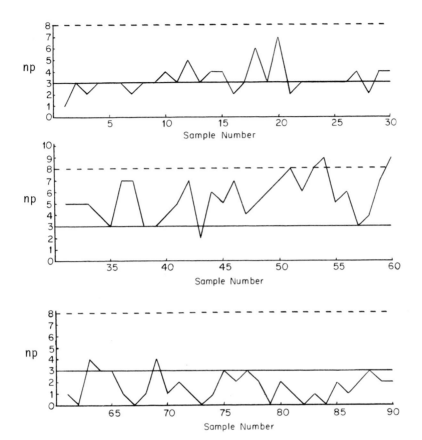

Fig. 6.1. Chart for 90 samples, each of n = 50 pieces, with the number of defectives np = d plotted. Central line and upper control limits given for p' = .06. For samples 1 through 30, p' = .06; 31 through 60, p' = .10; 61 through 90, p' = .03. Note increased average and variability in second part, and decreased average and variability in third part.

that p' is not .06. This is the conclusion we would expect since we know that p' was .06 for these first 30 samples. We might mention that at the start we had quite a run of points on or below the central line, but two-thirds of these, namely six, were right on the central line.

For the next 30 samples, we have the following:

Sample number	31	32	33	34	35	36	37	38	39
Defectives, np = d	5	5	5	4	3	7	7	3	3
Fraction defective, p	.10	.10	.10	.08	.06	.14	.14	.06	.06

Sample number	40	41	42	43	44	45	46	47	48
Defectives, np = d	4	5	7	2	6	5	7	4	5
Fraction defective, p	.08	.10	.14	.04	.12	.10	.14	.08	.10

Sample number	49	50	51	52	53	54	55	56	57
Defectives, np = d	6	7	8	6	8	9	5	6	3
Fraction defective, p	.12	.14	.16	.12	.16	.18	.10	.12	.06

Sample number	58	59	60
Defectives, np = d	4	7	9
Fraction defective, p	.08	.14	.18

Let us suppose that the quality control person still compares these results against p' = .06, thus continuing the central line at 3 and the upper control limit at 8.04. He will find no single point indicative of a process change increasing p' until he reaches samples 51, 53, and 54. These are not compatible with p' = .06 and indicate a shift in p' toward a higher value. Actually all of the first 12 points are on or above the central line of 3, and this "run" is also significant of an increase in p'. For samples 31 through 60, p' was made equal to .10 by adding 76 red beads to the box, giving

$$p' = \frac{178}{1778} = .100$$

Now suppose that p' were to decrease. How will this show np on the chart? The points will tend to be lower. But in this example where there is no lower limit, we cannot get an indication by a point *below* the lower control limit. However, a long run of points below, or possibly on, the central line can be used as a clear indication of a lowering of p'.

The data for samples 61 through 90 were obtained with

$$p' = \frac{49}{1649} = .030$$

and the np counts are plotted as the last 30 points in Fig. 6.1.

Sample number	61	62	63	64	65	66	67	68	69
Defectives, np = d	1	0	4	3	3	1	0	1	4
Fraction defective, p	.02	.00	.08	.06	.06	.02	.00	.02	.08

Sample number	70	71	72	73	74	75	76	77	78
Defectives, np = d	1	2	1	0	1	3	2	3	2
Fraction defective, p	.02	.04	.02	.00	.02	.06	.04	.06	.04

Sample number	79	80	81	82	83	84	85	86	87
Defectives, np = d	0	2	1	0	1	0	2	1	2
Fraction defective, p	.00	.04	.02	.00	.02	.00	.04	.02	.04

Sample number	88	89	90
Defectives, np = d	3	2	2
Fraction defective, p	.06	.04	.04

Starting with sample number 70, all of the remaining 20 points are either on or below the central line. We may regard the first 10 such points for samples 70 through 79 as providing a significant run.

This example illustrates a process in control relative to the actual true process average p'. It also shows how we may obtain an indication of (1) an increase in p' and (2) a decrease in p'. Further, it may be noted by the reader that the clear indication of a process change did not come very quickly. An np chart is not very sensitive to moderate changes in p', unless the sample size n is large. (How large is "large" depends somewhat upon the size of p'.)

Analysis of past data by np charts. The objective is to test whether a run of samples from a production process is in control with respect to itself. Could the observed counts of defects np = d have readily come from a process with a constant p', that is, could we attribute all of the variation to chance causes? Or is there evidence of one or more assignable causes coming in to create shifts in p'?

For this we need a preliminary run of at least 20 samples, preferably all with a constant sample size n, although we can handle the data if n varies. (Sometimes it is desirable to regard all of an hour's production as a sample or a shift's or a day's production. Then n will vary somewhat.) Then we run an np chart on the preliminary series of counts of defectives, np. We look for points out of the control band for evidence of the presence of some assignable cause which should be hunted down from the process conditions at the time the point went out. If there are no such points then we say that the process is in control, that is p' appears to be constant. This does not mean that p' is necessarily satisfactory. It could in fact be poor. But if the process is in control, and p' is too high for our purposes, then it will take a fundamental change in the process. This is because there is no evidence of any assignable cause we can work on.

Now let us see the details of the analysis. We have just been talking about p' for the process. But when we are in the case of "analysis of past data," we do not have any known p'. All we have is a series of np = d counts over samples of n pieces. We are testing whether the process fraction defective p' is constant. So until we disprove this, we will act preliminarily as though p' were constant and proceed to estimate p' from the data. We use as the estimate $\bar{p}$ based on all the data as follows:

$$\bar{p} = \frac{\text{total no. defective}}{\text{total no. inspected}} = \frac{\Sigma \ d}{\Sigma \ n} \tag{6.8}$$

Thus for Example 1, for samples 1 through 30, there were 98 defectives found in 30 samples, each of 50, or in 1500 pieces inspected. Thus

$$\bar{p} = \frac{98}{1500} = .0653$$

This provisional estimate $\bar{p}$ of the assumed constant process fraction defective p' is now substituted in for p' in (6.6) and (6.7). Then

$$\mathcal{E}_{np} = n\bar{p} = \text{central line for np (analysis of past data)} \tag{6.9}$$

$$\text{Limits}_{np} = n\bar{p} \pm 3\sqrt{n\bar{p}(1 - \bar{p})} \quad \text{(analysis of past data)} \tag{6.10}$$

As an illustration, let us calculate the control lines for samples 1 through 30 of Example 1. We have

$$\xi_{np} = 50(.0653) = 3.27$$

$$\text{Limits}_{np} = 50(.0653) \pm 3\sqrt{50(.0653)(.9347)}$$
$$= 3.27 \pm 5.24$$
$$= \text{---} , 8.5$$

So an np of 9 or more would be out of control; of 8, in control.

Example 2. As an illustration, let us consider the following data on the final inspection of auto carburetors for all types of defects, on a single production line for a single day January 15. Each carburetor containing any defects was called a defective. A record sheet was being used upon which to record the various types of defects, with 32 types listed, and further space was available to write in any defect type not among the 32 types. Samples consisted of 100 consecutive carburetors, which, while not chosen at random, can be considered a sampling of production under the conditions in force while they were produced and inspected. There was one incomplete sample at the end of the day which we omitted here.

Sample number	1	2	3	4	5	6	7	8
Sample size, n	100	100	100	100	100	100	100	100
Defectives, d	10	4	6	12	6	8	10	12

Sample number	9	10	11	12	13	14	15	16
Sample size, n	100	100	100	100	100	100	100	100
Defectives, d	8	7	3	4	3	4	4	10

Sample number	17	18	19	20	21	22	23
Sample size, n	100	100	100	100	100	100	100
Defectives, d	7	5	3	6	8	10	4

The numbers of defectives d = np are plotted in Fig. 6.2. For the control lines, we first find $\bar{p}$:

$$\bar{p} = \frac{154}{2300} = .0670$$

from which

$$n\bar{p} = 100(.0670) = 6.7 \quad \text{central line}$$

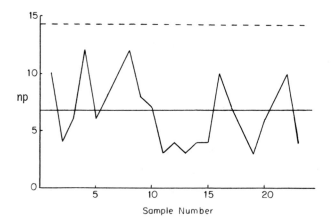

Sample Number

Fig. 6.2. Numbers np of carburetors containing at least one defect of any kind. Consecutive samples, each of n = 100, inspected at end of production line for all types of defects represent a single day's production. Central line $n\bar{p}$ = 6.7, upper limit 14.2.

$$\text{Limits}_{np} = 100(.0670) \pm 3\sqrt{100(.0670)(.9330)}$$
$$= 6.7 \pm 7.5$$
$$= \text{---}, \ 14.2$$

These lines showed the process to be in control for defect types taken as a whole. That is, there was no reliable evidence of any assignable cause affecting a large number of or all defect types. But the $\bar{p}$ value was regarded as quite unsatisfactory.

Careful study of the tally sheet on types of defects brought to light potential causes. Steps were taken to eliminate these causes. Also the defectives from these causes were eliminated from the data as being no longer typical of the process. This gave a revised $\bar{p}$:

$$\bar{p} = \frac{64}{2300} = .0278$$

as a provisional standard p', for a goal for subsequent production. Later on in this book we shall show some of the subsequent data.

This is a very desirable approach to periodically revise $\bar{p}$, eliminating defectives considered taken care of by appropriate action and thus to set a provisional standard p' for the next week or month. One plant making small motors, with which the author is well acquainted, achieved much success in this way.

If sample size varies. For an np chart, whether using $\bar{p}$ or p', if the sample size varies, it is theoretically required that new limits be calculated for each different n. This is because n affects both the central line and the width of the control band, as may be seen in (6.6), (6.7), (6.9), and (6.10). However, if the n's are fairly uniform, we may use $\bar{n}$ in these formulas instead of using each individual n. Now what do we mean by "fairly uniform?" For a rough "rule of the thumb," we first find the average sample size $\bar{n}$, and form $.95\bar{n}$ and $1.05\bar{n}$ (5% variation from $\bar{n}$). Then, if all of our n's lie within these limits, we are safe in using the limits from $\bar{n}$ for all the n's in the set. This can save us quite a bit of calculating and plotting time. If desired, we could find the exact limits for any point close to the $\bar{n}$ limit, whether inside or out.

6.1.2 *The Fraction Defective or p Chart*

We now take up the fraction defective chart, that is, where the quality variable being plotted is p rather than d = np. Basically, this is merely a change of vertical scale from the np chart. Thus consider the data of Example 2 of the preceding subsection. There we plotted the d counts 10, 4, 6, 12, and so on. Now we can find the fractions defective p by dividing d by 100, which gives .10, .04, .06, .12, and so on. Each of these numbers is simply one-hundredth of the d numbers.

Case--Standard given p'. Here we are in the fortunate position of having a standard value of the process fraction defective, p'. (From past data showing good control or some agreed-upon goal.) Then since we assume independence of defectives and constant p' we have the binomial distribution for p values, and can use (4.13) and (4.14) as follows:

$$\mathcal{E}_p = p' = \text{central line for p's} \quad \text{(standard given)} \qquad (6.11)$$

$$\text{Limits}_p = p' \pm 3\sqrt{\frac{p'(1 - p')}{n}} \quad \text{(standard given)} \quad (6.12)$$

Example 3. If we use the data of Example 2 with those defectives deleted which were a result of the causes presumed to have been identified and removed, the following data emerge:

Sample number	1	2	3	4	5	6	7	8
Sample size, n	100	100	100	100	100	100	100	100
Defectives, d	9	3	5	6	3	4	3	6
Fraction defective, p	.09	.03	.05	.06	.03	.04	.03	.06

Sample number	9	10	11	12	13	14	15	16
Sample size, n	100	100	100	100	100	100	100	100
Defectives, d	2	3	0	1	1	1	0	3
Fraction defective, p	.02	.03	.00	.01	.01	.01	.00	.03

Sample number	17	18	19	20	21	22	23
Sample size, n	100	100	100	100	100	100	100
Defectives, d	2	1	0	1	1	6	3
Fraction defective, p	.02	.01	.00	.01	.01	.06	.03

These p values are plotted in Fig. 6.3. Now we use the agreed-upon standard value p' = .0278 from the purified data. Then (6.11) and

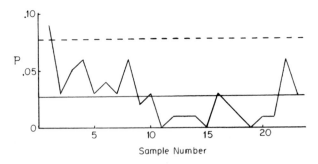

Sample Number

Fig. 6.3. Data on carburetors containing one or more defects from unknown causes. Defects from known and removed assignable causes were eliminated from data of Example 2 to give data of Example 3. Fractions defective p were plotted for the samples of n = 100. A standard value of p' = .0278 was the central line, giving an upper control limit of .077.

(6.12) give

$$\ell_p = p' = .0278$$

$$\text{Limits}_p = .0278 \pm 3\sqrt{\frac{.0278(.9722)}{100}}$$

$$= .0278 \pm .0493$$

$$= ---, \ .077$$

Drawing in these control lines in Fig. 6.3, we find that one point, for the first sample, is out of control. Six different kinds of defects made up the total of nine defectives, which does not exactly "pin-point" the trouble! Perhaps it was general laxness at start-up time in the shift.

As a further illustration of the p chart being essentially a change of scale from the np chart, consider the limits for the first 30 samples of the experimental data of Example 1. The control lines set from p' = .060 were

$$\ell_{np} = 3$$

$$\text{Limits}_{np} = ---, \ 8.04$$

Let us now use (6.11) and (6.12) for control lines for p values:

$$\ell_p = p' = .06$$

$$\text{Limits}_p = p' \pm 3\sqrt{\frac{p'(1 - p')}{n}}$$

$$= .06 \pm 3\sqrt{\frac{.06(.94)}{50}}$$

$$= .06 \pm .1008$$

$$= ---, \ .1608$$

Now since the p values are each d/50, we ought to be able to divide the control lines (central line and limits), for np by 50 and find those for p. Thus

$$\frac{3}{50} = .06$$

$$\frac{8.04}{50} = .1608$$

Case--Analysis of past data. Again we consider the common case of analysis of a preliminary run of sample fractions defective p. Since we do not know p' we use the provisional estimate $\bar{p}$ found by use of (6.8). Then substitute $\bar{p}$ for p' in (6.11) and (6.12), obtaining

$$\mathcal{C}_p = \bar{p} = \text{central line for p's (analysis of past data)} \quad (6.13)$$

$$\text{Limits}_p = \bar{p} \pm 3\sqrt{\frac{\bar{p}(1 - \bar{p})}{n}} \quad \text{(analysis of past data)} \quad (6.14)$$

Example 4. Table 6.1 gives the fractions defective for an electrical equipment given a "B test" and failing. The p values shown are plotted in Fig. 6.4.

Now let us figure the control lines. Overall, the original data showed 485 failures or defectives out of 29,690. By (6.8) this gives

$$\bar{p} = \frac{485}{29,690} = .0163$$

Using the average production of 1650 as n, we find by (6.14)

TABLE 6.1. B-Test Failures of Electrical Equipment for September. Eighteen Fractions Defective for the Month. Daily Production about 1650.

Day	Fraction defective	Day	Fraction defective
1	.0256	11	.0082
2	.0245	13	.0084
3	.0225	14	.0139
4	.0104	15	.0090
6	.0191	16	.0356
7	.0042	17	.0186
8	.0086	18	.0221
9	.0195	20	.0180
10	.0183	21	.0110

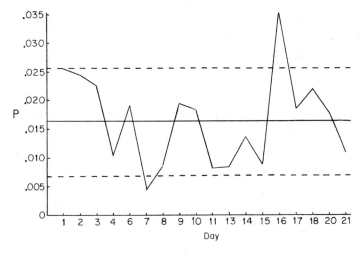

Fig. 6.4. Fraction defective control chart on electrical equipment from B-test failures in September. Daily production is about 1650. Control lines set from data in Table 6.1.

$$\text{Limits}_p = .0163 \pm 3\sqrt{\frac{.0163(.9837)}{1650}}$$

$$= .0163 \pm .0094$$

$$= .0069, .0257$$

Here it can be seen that we have two real control limits, the lower one being above zero. Thus we might have points out of control on either the bad side or the good side; that is, there is a possibility of finding evidence of a "good" assignable cause on the seventh.

If sample size varies. Sometimes sample sizes vary, for example, the p values may represent the entire production over equal blocks of time. Then we may have to take this into account. However, variations in n are less upsetting to the p chart than to the np chart. This is because for the p chart the central line is either

$\bar{p}$ or p' (central line for p chart)

which does not involve n at all, whereas for the np chart the central line varies directly with n, thus

$n\bar{p}$ or np' (central line for np chart)

Therefore, no matter how much n varies, the p values will always
have the same central line about which they vary. On the other
hand, the central line for the np or d counts of defectives will
have differing central lines, if n varies much at all.

Now it is true that the *width* of the control band will vary as
n varies, for both kinds of charts. For the p chart, the width
varies *inversely* with $\sqrt{n}$ as we see in (6.12) or (6.14). For the np
chart, the width of the control band varies *directly* with $\sqrt{n}$ as (6.7)
and (6.10) show. Thus the same percentage variation in n will bring
similar percentage changes in the widths of the control bands of the
two kinds of charts, but in opposite directions. So the big difference is that the central line is unaffected in a p chart by changes
of n but is affected in the np chart.

Let us reconsider Example 4. Actually, the n's ran from a minimum of 1390 to a maximum of 2000. Let us figure the limits from $\bar{n}$
of 1650 and then for n = 1390 and 2000, for both kinds of charts,
and see whether we could reasonably use $\bar{n}$ instead of the actual n's
in each, despite the considerable variation in the n's. We have

$$.0163 \pm 3\sqrt{\frac{.0163(.9837)}{n}} = .0163 \pm \frac{.380}{\sqrt{n}}$$

into which we substitute n = 1390, 1650, 2000

$.0163 \pm .0102 = .0061, .0265$
$.0163 \pm .0094 = .0069, .0257$
$.0163 \pm .0085 = .0078, .0248$

In spite of the sizable variation in the sample sizes, the limits
are not radically different, and they all have the same central
line, $\bar{p} = .0163$. On the other hand, for an np chart we use (6.10)
for limits as follows:

$22.66 \pm 14.16 = 8.5, 36.8$
$26.90 \pm 15.43 = 11.5, 42.3$
$32.60 \pm 16.99 = 15.6, 49.6$

Certainly these differ too much for us to use just one set of limits based on $\bar{n}$ = 1650. The trouble is largely from variation in the central line.

As a rough *rule of thumb* for a *p chart* when the n's vary, we can find $\bar{n}$, and then if all n's lie within $.7\bar{n}$ and $1.3\bar{n}$, use $\bar{n}$ to set one pair of limits for all samples. Sometimes we may need to make up several groupings of n's, within each grouping the n's being fairly uniform. This saves on arithmetic and plotting time.

Percent defective chart. This variation of a p chart is sometimes desired, especially by executives who like to deal with percentages. The author suggests that the quality control person make all the calculations in decimals, as we have been doing for a fraction defective chart, but then merely revise the vertical scale of the chart to read in percents rather than decimals.

Chart for defectives np or for fraction defective p? Which of these two kinds of charts to use in your plant is really a matter of personal preference. Use the type of chart which will be most easily understood by your associates and those who will be taking action on the processes.

Advantages of np chart for defectives

1. There is no need to divide each d by n to find p for the sample.

2. It is easier and quicker to plot whole numbers np, than decimals p.

3. The arithmetic for limits is a trifle easier.

Advantages of p chart for fraction defective

1. It is probably more easily understood, either in decimal form or in percentages.

2. When sample sizes vary, the central line remains the same, and only the width of the band varies, whereas in an np chart even the central line varies.

3. We can permit much more variation of n's around $\bar{n}$ and still use limits based only on $\bar{n}$.

6.1.3. *Instructions for Use of p and np Charts*

1. The point in the process to take *samples* of pieces or possibly to inspect runs of the *entire production* must be decided. This should usually be as near as feasible to where trouble may well be arising. This enables you to be closer physically and timewise to changes in the process which may prove to be assignable causes.

2. Workable definitions of each defect which will cause a piece to be called a "defective," must be delineated. This is a vital and often difficult step. But insofar as possible, the definitions should be so clear-cut that everyone involved would call each given imperfection a defect, or else everyone would call it a good piece. (For example, the author came across a case where the same 60 food-jar caps were examined by many people for "functional defects." The number declared functional defectives by the various people ran from 3 to 60!) Good gauges or other equipment may be needed. Also a set of "limit samples" for visual inspection may be developed, to show for each type of imperfection a limit, so that a larger imperfection is a "defect," whereas a smaller one is permitted. Or limits may separate major defects, minor defects, and no defect.

3. Individuals should be properly trained for inspecting or testing the samples of product, so that results are reliable.

4. Sample size must be determined. It should not be so large that many changes in process conditions may occur while a given sample of product is being produced. But the sample size should be large enough so that $\bar{d}$ is at least one, say.

5. If a sample is to be drawn from, say, an hour's production, steps should be taken to ensure that the sample is drawn randomly and without bias, insofar as possible.

6. It is helpful to make up a list of potential assignable causes which might be a cause of defectives. Enlist the help of everyone connected with the process.

7. It is useful to have a log of the process conditions, so that for each sample there is a record of the time, production line operator, inspector or quality person, test-set or gauge used, set-up man, source of raw materials, and so on.

8. Previous data may be available and useful, but it is advisable to be skeptical unless you know the conditions under which they were collected. Usually it is desirable to get new sample data, perhaps collecting more rapidly than you expect to in the future.

9. Plot the p or np points for the preliminary run of samples, and calculate and draw in the control lines.

10. Show the chart and explain it with great care to the process operator, foreman, and others involved. Look for any indicated assignable causes. Here is where teamwork counts.

11. If an assignable cause for *bad* product has been found *and* eliminated, delete all the sample results occurring while this cause was at work in the process. This may include points within the control band as well as points above the upper control limit.

12. If an assignable cause for *good* product, for example, the source of raw material, is found, consider incorporating it in the process, if feasible.

13. From $\bar{p}$ possibly revised as in 11, set control lines to be extended for new sample data. Examine each sample point as soon as it arises for possible action on the process.

14. Periodically revise $\bar{p}$, not retaining obsolete data.

15. When reasonable control at a satisfactory quality performance is obtained, we can call $\bar{p}$ a standard fraction defective p'. It may then be possible to take samples less frequently, or to take smaller samples. But some, continued follow-through or control is desirable.

16. It is helpful to develop and print forms with spaces for information and data, and perhaps even a place to plot points.

6.2 CHARTS FOR DEFECTS

We shall now take up the second general type of attribute control chart, namely, *charts for defects,* rather than for defectives. We have a single kind of defect under study, or it may be a class of several different kinds of defects, for example, defects we might class as "major A defects." If a single piece, unit, article, or subassembly may well contain more than one defect, we may wish to count the number of defects per sample of n pieces. Here n could even be one, as in final inspection of trucks or lawn mowers. Or n could be 24 for a case of bottles. By counting the number of defects, we may obtain a better idea of how defective the product is. This is especially true when a large or complicated product or assemblies are being inspected and tested, and nearly every one has one or more defects. It does little good to plot a series of fractions defective p, running 1.00, 1.00, .98, 1.00, .96, .98, 1.00, and so on. Instead we learn more by counts of defects in samples, especially using a tally sheet listing the defects in the class under study.

Once more we emphasize that the defects of interest may vary greatly as to importance and seriousness, from one study to another.

6.2.1. The c Chart for Defects

Let us repeat the notations as given before in Chapter 4:

c = number of defects in a sample of product   (6.15)

c' = standard or true process average of defects

    per sample   (6.16)

$\bar{c}$ = average of a series of counts c   (6.17)

The objective of a c chart is to plot and analyze a series of counts of defects c, taken over uniform sample sizes of product.

Standard given, c'. For the following analysis for a c chart we make these assumptions:

1. The samples provide equal "areas of opportunity" for defects to occur, that is, not both small and large subassemblies or varying numbers n of pieces for a sample.

2. Defects act independently, not tending to come in bunches. Thus the occurrence of one defect does not affect the probability for another defect of that type or some other to occur.

3. All types of defects under study are unequivocally defined.

4. Inspection is thorough insofar as possible.

5. Provisionally, c' is constant for the process.

6. The *possible* number of defects is far above c'.

Granted these assumptions, then the Poisson model for the distribution of defects is applicable.

We may now immediately set up the control lines from the assumptions from Section 4.3 and (4.15). Thus

TABLE 6.2. Counts c of "Seeds" in Cases of 24 Glass Bottles

c' = 4		c' = 4		c' = 8		c' = 2	
Case	c	Case	c	Case	c	Case	c
1	6	26	5	51	4	76	2
2	5	27	10	52	8	77	2
3	8	28	8	53	9	78	2
4	4	29	3	54	10	79	0
5	4	30	1	55	5	80	3
6	3	31	4	56	10	81	0
7	5	32	6	57	10	82	1
8	5	33	4	58	7	83	0
9	2	34	3	59	7	84	7
10	4	35	2	60	9	85	2
11	2	36	2	61	9	86	2
12	5	37	4	62	11	87	0
13	2	38	3	63	7	88	0
14	3	39	5	64	8	89	3
15	9	40	4	65	7	90	6
16	1	41	3	66	7	91	1
17	2	42	2	67	5	92	2
18	4	43	2	68	9	93	4
19	3	44	3	69	4	94	4
20	2	45	5	70	5	95	3
21	4	46	3	71	7	96	1
22	5	47	5	72	9	97	4
23	2	48	2	73	8	98	1
24	3	49	3	74	5	99	2
25	3	50	5	75	10	100	3

$$\pounds_c = c' = \text{central line for c values (standard given)} \quad (6.18)$$
$$\text{Limits}_c = c' \pm 3\sqrt{c'} \qquad \text{(standard given)} \quad (6.19)$$

For an example, let us suppose that considerable past data have shown rather good control for the number of "seeds" in cases

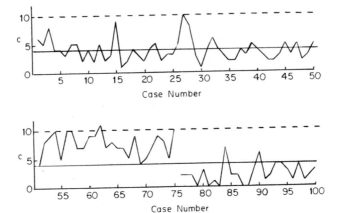

Fig. 6.5. Experimental data from Poisson populations, interpreted as counts of "seeds" in cases of 24 glass bottles. Data are from Table 6.2. The first 50 c values are from c' = 4; cases 51 through 75, c' = 8; and cases 76 through 100, c' = 2. Control lines shown for all 100 cases are from c' = 4.

of 24 glass bottles, with $\bar{c}$ = 4. ("Seeds" are bubbles of at least a certain size.) We might then call c' = 4, against which to check future production. We show 100 c value counts in Table 6.2. The counts are plotted in Fig. 6.5.

Let us use (6.18) and (6.19) to set the control lines with c' = 4:

$$\mathcal{E}_c = 4$$

$$\text{Limits}_c = 4 \pm 3\sqrt{4} = \text{——, } 10$$

Among the first 50 counts of seeds in cases of 24 bottles, there is one count of 10. This is on the control limit and would be considered an indication of an assignable cause. The production conditions under which this case was produced should be studied.

Now the c values of Table 6.2 actually were not counts of seeds; this interpretation was just a possibility for this sampling experiment. Instead the first 50 counts c were from a bowl of 500 chips, numbered according to a Poisson distribution with c' = 4. Thus the c = 10 point being on the control limit was mere chance, an event that will occur by chance only quite rarely. Let us just see how rare it was. For this we consult Table B, entering with c' = 4.

$$P(10 \text{ or more}) = 1 - P(9 \text{ or less})$$
$$= 1 - .992$$
$$= .008$$

Therefore we could expect to draw a c value of 10 or more, eight times in 1000, or once in 125. We had one among the first 50, which is not too surprising. But in practice any such point would be regarded as an indication of an assignable cause.

Now consider cases 51 through 75. They were actually drawn from a Poisson population with c' = 8, that is, twice as high as for the original 50. How soon do we find evidence of an assignable cause? The first definite indication is on case 54, where a count of 10 occurs. Thereafter among these 25 c counts, there are three more 10's and an 11. Each one *by itself* is sufficient to be regarded as evidence of an assignable cause. Note also that all but two counts c lie above the central line. Of course, a jump of c' from 4 to 8 is quite a change. If c' went from 4 to only 5 or 6, it might take quite a few samples before a definite indication shows up.

The last 25 counts were drawn from a Poisson population having c' = 2. Now how can we get an indication of an improvement, that is, of a *lowered* c' value? It cannot come through a point on or below the lower control limit, because with c' = 4 there is no lower limit. (Not until c' = 9 do we have a meaningful lower control limit.) Thus the only way we can obtain an indication of improvement is by a long run of c's on or below the central line, or perhaps by a generally lower set of points. In the last 25 c values, only two counts are above the central line and the first eight are all *below* the central line. Then at the end there are ten in succession all *on or below* the central line; surely a significant run.

We usually use the analysis with standard given, only after a reasonably long run of in-control conditions at satisfactory quality. Then we can set a meaningful c' from the process average $\bar{c}$, and be checking for control against a standard c'.

Analysis of past data. In this case, the purpose is to find
out whether or not the process seems to be in control or whether
there is evidence of assignable causes. If the latter, then we
want to take appropriate action. This will in general yield considerable process improvement. The objective is always to achieve
a process in control at a satisfactory quality level.

In analyzing a run of, say, 25 c values as preliminary data,
we do not have a standard c' from which to work. Thus we cannot
use (6.18) and (6.19). However, we are assuming preliminarily that
the process is in control, that is, has *just one* average c',
even though we do not know what it is. But for our estimate of what
this c' is, we take the average of our c values, namely $\bar{c}$. Thus
instead of (6.18) and (6.19) we must substitute $\bar{c}$ for c' and obtain:

$\ell_c = \bar{c}$ = central line for c values

            (analysis of past data)   (6.20)

Limits$_c = \bar{c} \pm 3\sqrt{\bar{c}}$ (analysis of past data)   (6.21)

Example 5. In inspecting consecutive large aircraft, the following counts of missing rivets were made:

Plane	1	2	3	4	5	6	7	8	9	10	11	12	13
c	11	5	5	5	21	13	17	8	3	2	9	6	8

Plane	14	15	16	17	18	19	20	21	22	23	24	25	26
c	5	10	16	8	6	5	6	7	8	4	1	4	3

These counts of missing rivets are shown in Fig. 6.6. For control
lines we have

$\bar{c} = \dfrac{196}{26} = 7.54$

Limits$_c = \bar{c} \pm 3\sqrt{\bar{c}}$

        $= 7.54 \pm 3\sqrt{7.54}$

        $= 7.54 \pm 8.24$

        $= \text{———}, 15.8$

Three points lie above the upper control limit, indicating the presence of some assignable cause(s). Some process improvement seems to

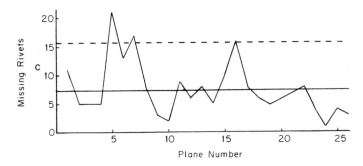

Fig. 6.6. Missing rivets from inspection of 26 large aircraft, analyzed as preliminary data. Central line is 7.54, and upper control limit is 15.8. Three high points are out of control.

have been made toward the end.

The psychological effect of control charts, such as this one, should not be underestimated. It enables everyone to see "How are we doing?" and encourages them to do their best and then improve on that best! Great gains were made in studies of which these data were a part. See also Prob. 6.15.

6.2.2 *Charts for Defects per Unit*

This type of control chart is also concerned with defects. It is especially useful to condense data on defects, in order to give a broad picture, and/or to eliminate too much detail. A second purpose is to take account of varying opportunities for defects, for example, varying areas of cloth, painted surface or paper, or varying lengths of insulated wire for breakdowns of insulation.

Example 6. As an example, in the final inspection of completed trucks just driven off the line, the daily production was 125. It would be possible to make a c chart for the individual trucks, where c would be the number of defects or errors per truck. This would give 125 individual c points to plot per day. Such a chart, while feasible, tends to supply too much detail and fails to give the overall picture. Another possible chart would be to count up the total number of defects or errors on the daily production of 125 trucks, to call this number c, and thus to have just one summary c point per day reflecting the *total* errors. Such a chart gives the overall picture. But it does not show very clearly how much *average* trouble there has been *per truck* during the day. Thus we find it convenient to plot the average errors per truck, found by dividing the total number of errors on all trucks during the day by 125. Now we see Table 6.3. In the third and seventh columns are the total errors c noted for each of the 20 days. These total errors c were then divided by 125 to give the number of errors per truck for the day. For example, for January 13

$$u = \frac{c}{n} = \frac{218}{125} = 1.74$$

We shall use the following notations:

n = number of units of product per sample (6.22)

c = total number of defects in a sample of n units (6.23)

$u = \frac{c}{n}$ = average number of defects *per unit* over n units (6.24)

$\bar{u} = \frac{\Sigma\ u}{k}$ = average values of u for k equal-sized

examples (6.25)

TABLE 6.3. Errors per Truck at A, Assembly Line, and B, Station. Production 125 Trucks per Day. Total Errors and Average Errors per Truck

Day	Total errors, c	Average errors, u	Day	Total errors, c	Average errors, u
Jan 13 M	218	1.74	Jan 27 M	226	1.81
Jan 14 T	200	1.60	Jan 28 T	175	1.40
Jan 15 W	165	1.32	Jan 29 W	181	1.45
Jan 16 Th	195	1.56	Jan 30 Th	191	1.53
Jan 17 F	209	1.67	Jan 31 F	159	1.27
Jan 20 M	250	2.00	Feb 3 M	162	1.30
Jan 21 T	188	1.50	Feb 4 T	171	1.37
Jan 22 W	150	1.20	Feb 5 W	183	1.46
Jan 23 Th	219	1.75	Feb 6 Th	170	1.36
Jan 24 F	252	2.02	Feb 7 F	199	1.59
				3863	30.90

$$\bar{u} = \frac{\Sigma\ c}{\Sigma\ n} = \text{average of u values if n values vary} \qquad (6.26)$$

u' = population or true process average defects
  per unit   (6.27)

The 20 defects per unit from Table 6.3 are plotted in Fig. 6.7. We now need control lines for the errors per truck data. These are provided by the following:

$$\mathcal{E}_u = u' = \text{central line} \quad \text{(standard given)} \qquad (6.28)$$

$$\text{Limits}_u = u' \pm 3\sqrt{\frac{u'}{n}} \quad \text{(standard given)} \qquad (6.29)$$

$$\mathcal{E}_u = \bar{u} = \text{central line} \quad \text{(analysis of past data)} \qquad (6.30)$$

$$\text{Limits}_u = \bar{u} \pm 3\sqrt{\frac{\bar{u}}{n}} \quad \text{(analysis of past data)} \qquad (6.31)$$

Since in our example we do not have u' as a given standard, we must use (6.30) and (6.31). For (6.30) since the sample sizes are equal we may use (6.25), finding

$$\mathcal{E}_u = \bar{u} = \frac{30.90}{20} = 1.55$$

We could also have used (6.26):

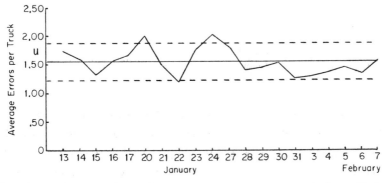

Fig. 6.7. Control chart for average errors per truck from 20 days' production, each of 125 trucks. Data is shown in Table 6.3 and was analyzed as preliminary data for control.

$$\mathscr{E}_u = \bar{u} = \frac{3863}{[20(125)]} = 1.55$$

Then by (6.31),

$$\text{Limits}_u = 1.55 \pm 3\sqrt{\frac{1.55}{125}}$$

$$= 1.55 \pm .33$$

$$= 1.22, \ 1.88$$

These three control lines are shown in Fig. 6.7. Points above the upper control limit indicate the presence of one or more assignable causes of trouble for the day in question. Meanwhile, points below the lower control limit are indicative of a "good" assignable cause, which may help improve quality if incorporated. There are two of the former and one of the latter in the 20 days.

In this plant, defects per unit charts with the accompanying tally table of kinds of errors led to large improvements in the production lines, very big savings on repairing and fixing up trucks, and fostered quality mindedness and pride of workmanship.

A final point worth considering in this example: Would we arrive at the same conclusions if we had plotted the total defects per day in an ordinary c chart, instead of the defects per unit u? Yes. We would use (6.20) and (6.21):

$$\mathscr{E}_c = \bar{c} = \frac{3863}{20} = 193.2$$

$$\text{Limits}_c = \bar{c} \pm 3\sqrt{\bar{c}}$$

$$= 193.2 \pm 3\sqrt{193.2}$$

$$= 193.2 \pm 41.7$$

$$= 151.5, \ 234.9$$

How do these compare with the control lines for u values, namely 1.55, 1.22, and 1.88? If we divide the line values for c's by the sample size n = 125, we obtain exactly the line values for the u's (apart from round-off errors). Thus since the c's divided by 125 are the u values, the two charts give identical decisions.

Example 7. Let us consider briefly the following hypothetical example for breakdowns of insulation under an excessive test voltage for insulated wire:

Length (m)	Breakdowns c	Units (500 m) n	Breakdowns per unit $u = c/n$	Limits$_u$ $= \bar{u} \pm 3\sqrt{\bar{u}/n}$
500	2	1	2.00	——, 6.37
500	0	1	0.00	——, 6.37
1500	5	3	1.67	——, 4.55
500	3	1	3.00	——, 6.37
1000	3	2	1.50	——, 5.10
500	4	1	4.00	——, 6.37
1500	8	3	2.67	——, 4.55
1000	5	2	2.50	——, 5.10
500	1	1	1.00	——, 6.37
1000	4	2	2.00	——, 5.10
	35	17		

For $\bar{u}$ we use (6.26) since the areas of opportunity, n, for defects vary. Here we use a 500-m length as one unit for n. (We could have used 1000 or 1500 m.) We thus assume that in 1000 m there is twice the opportunity for breakdowns as in 500 m, which seems reasonable. The central line for all points is the same:

$$\bar{u} = \frac{35}{17} = 2.06$$

But the *width* of the control bands will depend upon n by (6.31).

$$\text{Limits}_c = 2.06 \pm 3\sqrt{\frac{2.06}{n}}$$

n = 1 $2.06 \pm 4.31 = $ ——, 6.37
n = 2 $2.06 \pm 3.04 = $ ——, 5.10
n = 3 $2.06 \pm 2.49 = $ ——, 4.55

All 10 points lie within the respective limits. Actually they should be expected to because for 500-m lengths we drew from a

Poisson population of chips with $c' = 2$; for 100 m, $c' = 4$; and for 1500 m, $c' = 6$.

In summary, we can effectively use a defects-per-unit control chart if (1) we want an overall chart to summarize a number of units of production, to avoid too much detail or (2) the size of the samples or opportunity for defects varies. Such a chart is really just a change of scale from a c chart for the total defects per sample. The u chart is sometimes called a $\bar{c}$ chart because u is in reality an average of n c values.

6.2.3. *Instructions for Use of c and u Charts*

The instructions given in Sec. 6.1.3 for p and np charts apply with the obvious modification of np to c, and of p to u. In the outline there given, suggestion 2 is important, since an inspector may have to be looking for 15 or 20 different visual defects, for example. Unless the definitions of the various defects are clear and steps taken to be sure that accurate inspection is made, results are likely to be meaningless. In suggestion 4, unless $\bar{c}$ is at least one, the chart is not very powerful at indicating assignable causes. In suggestion 13, revised $\bar{c}$ or $\bar{u}$ is used to extend control lines. In suggestion 15, we would be setting standards c' or u'. In suggestion 16, the development of a list of possible defects, including some blank spaces for tallying defects, is highly desirable.

6.3 SUMMARY

We have described and illustrated two distinct types of control charts each of which finds many applications in industry. Both are concerned with counted or attribute data. Likewise both involve defects, which must be clearly defined and soundly inspected for. On the one hand, we have charts for defectives, in which each piece or unit is declared to be a defective if it contains at least one defect in the class in question. The appropriate charts are the np and p charts. Such charts are mostly used where the process fraction defective is small, say .10 or less, although the mathematics places no such restriction. On the other hand, we have the c and u charts for defects. They involve counts of defects (in some class) found on a sample of units or material. Such counts of defects are especially useful when the chance of a single unit having more than one defect is not negligible. Then counts of defects tell *how defective* the sample or unit is.

We recall that np and p charts are basically the same chart, but merely with a different vertical scale. Similarly, the c and u charts are also basically the same, but simply using a different vertical scale. When sample sizes vary, however, we find it more convenient to use, respectively, the p chart and the u chart, for then, at least the central line will be unaffected by the sample size, and with only the width of the control band depending upon $\sqrt{n}$.

The np and p charts are based upon the binomial distribution, whereas the c and u charts are based upon the Poisson distribution. Both of these distributions assume independence of defects, that is, that the occurrence of a defective or a defect on a piece does not affect the chances on subsequent pieces drawn for the sample. Also if the charts are to be based upon samples, rather than 100% of the output, then the samples are to be randomly drawn. Further, at the outset when analyzing past data, it is provisionally assumed that p' or c' for the process is constant. In the case of c charts, it is assumed also that the maximum *possible* value of c is far greater than the *average* value, c'.

Some further modifications and comments on attribute charts are given in Chapter 8.

Again we mention that "defects" can run anywhere from highly critical to quite incidental. It may be desirable in many circumstances to speak in terms of nonconforming pieces, or nonconformances rather than defectives or defects.

PROBLEMS

6.1   For the experimental data in Example 1, Sec. 6.1.1, for samples 31 through 60, analyze by an np control chart under the case
analysis of past data. How is control? What would the control
lines be for the case control against standard p' = .10. Are the
results compatible with p' = .10?

6.2   For the experimental data in Example 1, Sec. 6.1.1, for samples 61 through 90, analyze by a p control chart under the case
analysis of past data. How is control? What would the control
lines be for the case control against standard p' = .03. Are the
results compatible with p' = .03?

6.3   For the packing nut data of Table 5.1, find the central line
and control limits for a p chart, analyzing as past data. Comment
on control. In March samples of 50 packing nuts gave the following
counts of np = d: 0, 0, 4, 0, 1, 4, 5, 0, 0, 7, 0, 0, 0, 0, 0, 0,
2, 3. Make an appropriate chart and comment. Such charts were of
great value to the company involved in this case.

6.4   For the data plotted in Prob. 5.2, complete the p chart by
finding and drawing in the control lines. Comment. (The next 25
samples showed d = 1, 0, 0, 0, 0, 0, 1, 1, 0, 0, 1, 0, 0, 2, 1, 0,
0, 1, 0, 0, 0, 0, 0, 0, 1.)

6.5   The present problem is on auto carburetors inspected at the
end of the assembly line for all types of defects. These data are
for January 27, and follow the data of Example 2, Sec. 6.1.1. For
a long time a goal of p' = .02 had been set. Plot the given data,
either p or np, and set central line and control limits from p' = .02.
Has the goal been reached? Thirty-five d values with n = 100 follow:
4, 5, 1, 0, 3, 2, 1, 6, 0, 6, 2, 0, 2, 3, 4, 1, 3, 2, 4, 2, 1, 2, 0,
2, 3, 4, 1, 0, 0, 0, 0, 1, 2, 3, 3.

6.6   The data here shown are for 100% inspection of lots of malle-
able castings for cracks by magnaflux. The lot size varies considerably, so a p chart is used instead of an np chart. By some care

one can use only four separate $\bar{n}$'s. Make the control chart and comment.

Lot number	Lot size n	Defectives d	Fraction defective p
1	1,138	143	.126
2	600	30	.050
3	700	298	.426
4	500	70	.140
5	750	141	.188
6	600	26	.043
7	750	130	.173
8	775	141	.182
9	750	17	.023
10	850	66	.078
11	775	126	.163
12	750	20	.027
13	675	100	.148
14	1,395	32	.023
15	116	16	.138
16	750	116	.155
17	780	132	.169
18	1,400	40	.029
19	424	50	.118
20	384	16	.042
21	286	22	.077
22	1,000	52	.052
23	360	51	.142
24	1,100	38	.035
	17,608	1,873	.1064

6.7 For the data of Prob. 6.6, do you think that the occurrences of cracks in castings would be independent, that is, if one casting has a crack, this defect is unrelated to whether the next one has a crack? Might this account for the lack of control noted? Would you still investigate points outside of limits?

6.8   In Prob. 6.6, $\bar{p}$ = .1064 was found by use of (6.8), that is,
1873/17,608. Do you think you would obtain the same $\bar{p}$ by adding
the 24 p values and dividing by 24? This would be an unweighted
average of p's. If we substitute np for d in (6.8) we have
Σ np/Σ n. Thus $\bar{p}$ is a weighted average of p values with weights n.

6.9   Samples of 39 articles each, from an optical company, were
examined for breakage. Forty-three such samples over two months
time yielded d = 2, 1, 1, 0, 2, 3, 2, 1, 1, 2, 1, 1, 1, 2, 1, 1,
0, 1, 0, 1, 0, 2, 0, 1, 1, 0, 1, 0, 7, 10, 6, 1, 2, 2, 2, 1, 1, 0,
2, 2, 2, 5, 2. Analyze by an appropriate control chart. Would you
prefer a p or an np chart? Why? If these samples are in fact lots,
can you think of any practical reason for the size 39?

6.10 For the data on total defects found on daily production of
3000 switches (Prob. 5.3), find the central line and control limits,
analyzing as past data. Comment on control. Make chart if not
previously done.

6.11 The following data were observed just 1 year before the data
of Prob. 5.3 at the start of the control chart program. Analyze as
past data, the 25 production counts of total defects in 3000 switches
per day: 450 454 564 369 294 358 343 227 263 248 692 314
247 521 435 1054 727 282 647 400 372 203 160 275 244
(total 10,143). Comment.

6.12 The following data are for 29 samples each of 100-yd lengths
of woolen goods. The woolen goods pass slowly across a table and
defects noted are marked by passing a bit of yarn of contrasting
color through the goods. Defects: 2, 0, 0, 1, 1, 2, 2, 1, 0, 1,
1, 0, 2, 1, 1, 5, 0, 3, 1, 1, 1, 2, 2, 1, 0, 0, 0, 1, 4. Analyze
the data by a c chart. Why could we not treat as d values with
n = 100 and make an np chart?

6.13 For the experimental data on "seeds" in Table 6.2, analyze
samples 51 through 75 as past data by a c chart. Does the process
show control, that is, homogeneity? Also analyze for control
against the standard c' = 8.

6.14 Follow the same procedure as for Prob. 6.13 but use samples 76 through 100 and c' = 2.

6.15 For large aircraft assemblies, there were 18 categories of defects. (The data were taken subsequently to those in Table 5.1.) Analyze by a c chart for whatever defects are assigned and comment.

Plane	Alignment	Tighten	Cello seals	Foreign matter	Replace	Plug holes
1	5	91	3	26	2	7
2	9	151	7	23	5	4
3	3	149	8	29	4	4
4	4	205	12	20	18	2
5	6	185	5	26	4	3
6	7	106	2	13	9	5
7	14	171	1	13	2	4
8	18	113	8	16	2	2
9	11	171	5	21	8	6
10	11	162	16	16	5	7
11	11	148	8	18	13	6
12	8	117	14	13	6	5
13	10	134	12	11	9	3
14	8	116	7	27	3	7
15	7	112	3	23	5	9
16	16	161	3	34	14	10
17	13	125	14	14	10	7
18	12	145	9	7	17	3
19	9	129	6	25	11	8
20	11	151	5	18	6	3
21	11	140	19	15	15	6
22	8	118	6	18	8	1
23	8	126	3	13	9	5
24	9	101	4	1	1	1
25	4	110	2	8	4	1
26	13	142	7	1	4	0

6.16 The following were the number of breakdowns in insulation in 5000-ft lengths of rubber covered wire: 0, 1, 1, 0, 2, 1, 3, 4, 5, 3, 0, 1, 1, 1, 2, 4, 0, 1, 1, 0. What were the central line and control limits for the number of breakdowns? What would be the central line and limits for the average number of breakdowns per length for samples of five lengths?

6.17 The following data show daily averages of errors per truck over the daily production of about 125 trucks at final inspection. These data precede those of Table 6.3.

Day	Total errors, c	Average errors, u	Day	Total errors, c	Average errors, u
Nov 21 Th	156	1.25	Dec 5 Th	175	1.40
Nov 22 F	198	1.58	Dec 6 F	189	1.51
Nov 25 M	281	2.25	Dec 9 M	135	1.08
Nov 26 T	312	2.50	Dec 10 T	159	1.27
Nov 27 W	256	2.05	Dec 11 W	148	1.18
Nov 29 F	182	1.46	Dec 12 Th	174	1.39
Dec 2 M	192	1.54	Dec 13 F	178	1.42
Dec 3 T	178	1.42	Dec 16 M	260	2.08
Dec 4 W	196	1.57	Dec 17 T	231	1.85
				3600	28.80

(a) Plot the average errors per truck for the days.

(b) Draw in control lines from the previous $\bar{u} = 1.97$ and comment.

(c) Analyze these data as a preliminary run of data, checked for control.

(d) Suggest a $\bar{u}$ value for subsequent data.

Chapter Seven

CONTROL CHARTS FOR MEASUREMENTS: PROCESS CONTROL

We now take up the study of a most powerful set of tools for process
control, when the quality characteristic is measurable. Any of the
huge variety of product measurements may be analyzed by such control
charts. Typical industrial measurements are length, thickness,
diameter, width, weight of an item, density, chemical composition,
percent impurity, hardness, tensile strength, package content weight,
resistance, voltage, color characteristics, angle of bend, length of
life, ultimate strength, corrosion resistance, thickness of coating,
starting torque, horsepower, and acidity. Usually quality measurements carry some unit. Also the production process manufacturing
the product has requirements or goals set in the form of a minimum
or a maximum limit, or else two specifications, between which the
measurements are supposed to lie with high probability. The objective is to obtain "satisfactory, adequate, dependable and economic
quality" from the process, in the words of the quality control
pioneers at the Bell Telephone Laboratories.

7.1 TWO CHARACTERISTICS WE DESIRE TO CONTROL

We have seen in Chapter 2 that when we have a sample of data either
measurement or attribute, we can begin an analysis of the data by
tabulating them into a frequency table and then drawing a histogram
or a frequency polygon, such as Fig. 2.1 shows.

   Now as we discussed before, there are two characteristics of
data in general. For example, with the weights of charge given in
Table 2.1, the specifications were 454 ± 27 g. The average content
weight for the process is to be at least 454 g, and the customer is

anxious to have none below 427 g, while the manufacturer is desir-
ous of having none above 481 g, because of excessive overfill. We
must therefore be concerned with both the *average* charge weight of
the process and the *variability*. It would be undesirable to have
the average charge weight right at 454 g if we have so much variability that perhaps 10% of the charge weights are below the 427-g
minimum. Such a situation would be: *satisfactory average, excessive variability*. Also it would be undesirable to have the average
charge weight be 447 g, even though, say, no charge weights are outside the limits 427, 481 g. This situation would be: *low average,
satisfactory variability*. Thus we will need to pay close attention
to *both* the *average* and the *variability* whenever the quality characteristic is measurable.

In Chapter 2, we were mostly concerned with sample characteristics. Thus for a sample of measurements x, the ordinary average
was called $\bar{x}$ (the sum of the numbers, Σ x, divided by n, the number
of them). The typical amount of deviation of the sample x's from
their average, $\bar{x}$, was called the "standard deviation" s, given in
(2.4). We might also use another measure of sample variability,
the range R = max x - min x. The latter is only used for quite
small samples up to n = 10. Thus we may describe our sample of x's
by $\bar{x}$ and R, or by $\bar{x}$ and s.

But we also must consider the *true process characteristics*.
For the present we shall assume that the process is uniform or
steady, that is, in control, even though there is variability in
the product. How can this be? What we mean is that the process
average and process variability are both constant. By now it must
be apparent that we need symbols for these.

We shall call the true process average, *if constant*, by the
small Greek letter mu, μ:

μ = E(x)

= expectation or average of x's for process   (7.1)

The sample averages, $\bar{x}$, are estimates of μ when the process is in
control, that is, stable. We have already been using the small

Greek letter sigma, σ, for process or population standard deviation: (4.12), (4.14), and (4.15). Thus we naturally use for the population standard deviation of the x's

$$\sigma = \sigma_x$$

$$= \text{standard deviation of x for process or population} \qquad (7.2)$$

Now if we know μ and σ for a process, we know quite a bit. For example, if the distribution is reasonably well behaved, that is, with the greatest frequency toward the middle of the range of x values and with symmetrically decreasing frequencies on each side, we can make the following statements:

$\mu - \sigma$ to $\mu + \sigma$ include about 65 to 70% of the x's

$\mu - 3\sigma$ to $\mu + 3\sigma$ include at least 99% of the x's

These statements come from the so-called normal distribution, a widely applicable model. (See Sec. 7.5.)

Our objective of control charts for measurements is to determine whether the process is in control, that is, has constant μ and σ. If it is in control, we want to estimate μ and σ and compare them with requirements. If the process is not in control, we want to take steps to get it into control, at a satisfactory combination of μ and σ. And then we want to maintain control.

7.2 AN EXAMPLE, $\bar{x}$, R CHARTS FOR PAST DATA

The data on charge weights of an insecticide dispenser, given in Table 2.1, are here reproduced again along with 25 additional samples in Table 7.1. Also for each sample of four charge weights x, there is given the average $\bar{x}$ and the range R. These charge weights were obtained by the quality control man upon his return from a short course in statistical quality control. The problem was that inventories showed that the monthly overfill (above the specified average of 454 g) was running at $14,000. That is, the manufacturer was giving away this much over and above the required average. And yet despite this average overfill, some charge weights were running below the minimum of 427 g. Clearly improvement was desirable.

TABLE 7.1. Charge Weights of Insecticide Dispenser in Grams Taken in Samples of Four. Average $\bar{x}$ and Range R Listed for Each Sample. First 25 Samples, Preliminary Data. Specifications for x: 454 ± 27 g. Stand Number 4.

Sample number	Date	Observed charge weights x_1	x_2	x_3	x_4	Average $\bar{x}$	Range R	Remarks
1	12/13	476	478	473	459	472	19	
2		485	454	456	454	462	31	
3		451	452	458	473	458	22	
4		465	492	482	467	476	27	
5		469	461	452	465	462	17	
6		459	485	447	460	463	38	
7		450	463	488	455	464	38	
8		Lost	478	464	441	461	37	Sample of 3
9		456	458	439	448	450	19	
10		459	462	495	500	479	41	
11	12/14	443	453	457	458	453	15	
12		470	450	478	471	467	28	
13		457	456	460	457	458	4	
14		434	424	428	438	431	14	
15		460	444	450	463	454	19	
16		467	476	485	474	476	18	
17		471	469	487	476	476	18	
18		473	452	449	449	456	24	
19		477	511	495	508	498	34	
20		458	437	452	447	448	21	
21		427	443	457	485	453	58	
22		491	463	466	459	470	32	
23		471	472	472	481	474	10	
24		443	460	462	479	461	36	
25		461	476	478	454	467	24	End of preliminary run
26	12/17	450	441	444	443	444	9	
27		454	451	455	460	455	9	
28		456	463	Lost	445	455	18	Sample of 3
29		447	446	431	433	439	16	
30		447	443	438	453	445	15	
31		440	454	459	470	456	30	
32		480	472	475	472	475	8	
33		449	451	463	453	454	14	
34		454	455	452	447	452	8	
35		474	467	477	451	467	26	
36		459	457	465	444	456	21	
37		465	475	456	468	466	19	
38		458	450	451	451	452	8	
39		447	417	449	445	440	32	
40		453	442	456	453	451	14	
41	2/18	471	467	461	455	464	16	
42		462	454	462	468	462	14	
43		474	471	471	463	470	11	
44		461	454	468	452	459	16	
45		473	453	465	475	466	22	
46		474	455	486	490	476	35	
47		466	471	482	474	473	16	
48		447	454	476	486	466	39	
49		473	488	482	475	480	15	
50		460	450	461	445	454	16	

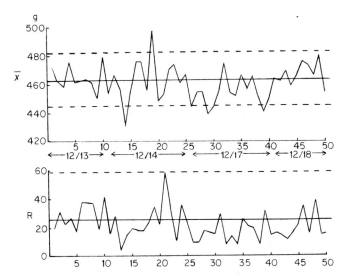

Fig. 7.1. Charge weights in grams of insecticide dispensers. Averages $\bar{x}$ and ranges R plotted for samples of n = 4 from Table 7.1. First 25 samples were the preliminary data. Specifications for x: 454 ± 27 g.

The quality control man used the first 25 samples, that is, December 13 and 14, as the preliminary run. These are plotted as the first 25 $\bar{x}$ points and R points of Fig. 7.1. After plotting the preliminary points the next step is to calculate the control lines (central line and limits) and draw them on the chart.

The case of control here is *analysis of past data*, that is, there were no available standard values, μ and σ, from which to work. Instead the average of averages $\bar{\bar{x}}$ and the average range $\bar{R}$ are used for finding the control lines.

$$\bar{\bar{x}} = \frac{\Sigma \bar{x}}{k} = \text{overall average for k sample } \bar{x}\text{'s} \qquad (7.3)$$

$$\bar{R} = \frac{\Sigma R}{k} = \text{average of k ranges R} \qquad (7.4)$$

The process level is reflected by $\bar{\bar{x}}$, whereas $\bar{R}$ is a measure of process variability. These are taken as central lines, for the respective charts.

$\mathcal{E}_{\bar{x}} = \bar{\bar{x}}$ = central line for $\bar{x}$'s (analysis of past data)   (7.5)

$\mathcal{E}_R = \bar{R}$ = central line for R's (analysis of past data)   (7.6)

For control limits we use these two provisional averages $\bar{\bar{x}}$ and $\bar{R}$ as follows:

$\text{Limits}_{\bar{x}} = \bar{\bar{x}} \pm A_2\bar{R}$ (analysis of past data)   (7.7)

$\text{Limits}_R = D_3\bar{R}, D_4\bar{R}$ (analysis of past data)   (7.8)

The quantities A_2, D_3, and D_4 are "control chart constants," calculated mathematically from the normal distribution model, which is discussed in Sec. 7.5. The quantities depend upon the sample size n for the samples. (See Table C in the back of the book.)

Now for the first 25 samples of Table 7.1, we find

$$\bar{\bar{x}} = \frac{11,589}{25} = 463.6 \text{ g}$$

$$\bar{R} = \frac{644}{25} = 25.8 \text{ g}$$

Using Table C, we also have from (7.7), (7.8), and Table C for the row n = 4:

$\text{Limits}_{\bar{x}} = 463.6 \pm .729(25.8)$
$\quad\quad = 463.6 \pm 18.8$
$\quad\quad = 444.8, 482.4 \text{ g}$
$\text{Upper control limit}_R = D_4\bar{R}$
$\quad\quad\quad\quad = 2.282(25.8)$
$\quad\quad\quad\quad = 58.9 \text{ g}$
$\text{Lower control limit}_R = D_3\bar{R} = 0(25.8) = 0$

(The lower control limit for ranges, while calculated at zero, is really meaningless, just as we had seen for the lower limit for some p, np, and c charts. Only if n exceeds 6, do we have a real lower control limit for R's.)

Upon drawing in these control lines, we find two $\bar{x}$ points were out of the control band--one high, one low. Also there is one very high range, almost on the control limit, indicating excessive variability within sample 21. Investigations were begun at once on the

conditions under which these three samples were produced. Also the
control lines were extended into the next week's data, and watch
kept after each sample $\bar{x}$ and R were plotted. Three more $\bar{x}$ points
below the lower control limit were found. Such points give real
concern as to whether the lower specification 427 g was being met.
In fact, even among the few charge weights of these samples there
were two below 427 g. Some improvement in variability seems to have
occurred because in the second 25 R's only five are above the central line.

After the first 50 samples, the control lines were revised for
subsequent data analysis as follows:

$$\bar{\bar{x}} = \frac{23,066}{50} = 461.3$$

$$\bar{R} = \frac{1,091}{50} = 21.8$$

$\text{Limits}_{\bar{x}} = 461.3 \pm .729(21.8) = 445.4, 477.2$

$\text{Upper control limit}_R = 2.282(21.8) = 49.7$

These control lines would be extended.

Some of the assignable causes which were actually found were
obstructions in the tubes, defective cutoff mechanisms, and poor control in timing and pressure. Within a month the rate of loss from
overfill per month was down to about \$12,000, and inside of three
months it was down to about \$2,000 per month, along with much safer
meeting of the minimum specification.

Some readers may be concerned about the two samples of n = 3,
which we treated as though n = 4 for them. Some variation of treatment is theoretically called for, but there were only two such samples, and the control chart constants are not radically different,
so the easy approximate approach was used.

7.3 AN EXPERIMENTAL EXAMPLE, $\bar{x}$, R CHARTS FOR PAST DATA

Let us now conduct an experiment to illustrate $\bar{x}$ and R charts, by
drawing numbered beads from a bowl. We shall try to estimate the
contents of the bowl from our sampling data of $\bar{x}$ and R. Later on,
the exact contents will be provided.

For the preliminary run of 25 samples, each of n = 5, in Table 7.2, we first plot the x̄'s and R's as shown in Fig. 7.2. We note one seemingly extremely low x̄, for sample 9, at -2.6. In practice, is this a reliable indication of an assignable cause? The R's appear to be rather homogeneous. Let us find the control lines by (7.5) to (7.8)

$$\mathcal{L}_{\bar{x}} = \frac{\Sigma \, \bar{x}}{25} = \frac{+.6}{25} = +.02$$

$$\mathcal{L}_{R} = \frac{\Sigma \, R}{25} = \frac{94}{25} = 3.76$$

$$\text{Limits}_{\bar{x}} = \bar{\bar{x}} \pm A_{2}\bar{R}$$

$$= +.02 \pm .577(3.76)$$

$$= +.02 \pm 2.17$$

$$= -2.15, \, +2.19$$

Upper control limit$_{R}$ = $D_{4}\bar{R}$ = 2.115(3.76) = 7.95
Lower control limit$_{R}$ = $D_{3}\bar{R}$ = 0(3.76) = ——

Drawing in the control lines, we find for sample 9 that x̄ is below the lower control limit. Therefore, in practical application, we would examine the conditions surrounding sample 9 for some assignable cause of off-level performance (although such a point can possibly occur through an increase in the process variability). With the grinding of piston rings by disc grinders, such a low x̄ could be caused by softer stock, thinner piston rings from a previous grind, slowage of feed rate, or by some temporary excessive pressure upon the grinding discs. Since in our experiment all of the numbered beads were being chosen randomly one by one with replacement after each, this excessively low x̄ is simply a rare accident of sampling. (The author makes no apology for such a point occurring. He plunges into such a sampling experiment never knowing what *will* occur and has only probability to guide him, much like the great 19th century preacher, Henry Ward Beecher, who "plunged into every sentence trusting God almighty to get me out of it.")

Not having found the assignable cause (which we cannot find in this random sampling experiment), we extend the limits already found,

TABLE 7.2. Record Sheet for Measurements

Material or product: Piston rings
Characteristic measured: Edge width
Unit of measurement: in .0001 in. from .2050 in.
Specs.: .2050 ± .0005 in.
Plant: No. 1
Data recorded by: I. W. Burr

Series number	Date (and hour?) produced	Measurements of each of five items in series					Average for series	Range for series	Standard deviation
		a	b	c	d	e			
1		+1	0	-2	-1	0	-.4	3	1.14
2		+1	+1	+3	0	-1	+.8	4	1.48
3		+1	+2	-1	+4	+2	+1.6	5	1.82
4		-1	-2	+1	-2	-2	-1.2	3	1.30
5		0	0	+3	0	0	+.6	3	1.34
6		+2	-1	+1	-1	-1	0	3	1.41
7		-1	0	-1	+1	-2	-.6	3	1.14
8		+1	-1	-1	-1	-2	-.8	3	1.10
9		-4	-3	0	-3	-3	-2.6	4	1.52
10		+1	-1	-1	-2	+2	-.2	4	1.64
11		-2	0	+1	-1	+3	+.2	5	1.92
12		0	+2	+2	+1	-2	+.6	4	1.67
13		-2	-2	0	-3	0	-1.4	3	1.34
14		-2	-2	0	+3	+3	+.4	5	2.51
15		+1	+1	0	+2	0	+.8	2	.84
16		0	0	+2	-2	+3	+.6	5	1.95
17		0	0	-3	0	+1	-.4	4	1.52
18		0	+1	+2	+2	+1	+1.2	2	.84
19		-1	+4	+1	0	+1	+1.0	5	1.87
20		0	0	+2	0	0	+.4	2	.89
21		-1	-3	-1	-1	-4	-2.0	3	1.41
22		0	+3	0	-3	+1	+.2	6	2.17
23		+5	+1	-2	0	0	+.8	7	2.59
24		0	+1	0	+1	+3	+1.0	3	1.22
25		0	+2	-1	0	-1	0	3	1.22
26		-1	+1	+3	-1	+2	+.8	4	1.79
27		0	0	0	0	+1	+.2	1	.45
28		+2	0	+3	+1	+1	+1.4	3	1.14
29		+1	-1	+1	-1	-3	-.6	4	1.67
30		+1	+1	-3	+2	+2	+.6	5	2.07
31		+2	+2	-2	-1	+1	+.4	4	1.82
32		-1	+2	-1	+1	+2	+.6	3	1.52
33		+1	+2	-1	+3	0	+1.0	4	1.58
34		-1	-1	0	-1	-2	-1.0	2	.71
35		0	0	-3	+2	+2	+.2	5	2.05
36		+3	+4	-1	+2	0	+1.6	5	2.07
37		+2	+1	0	0	0	+.6	2	.89
38		+2	+3	-2	-1	0	+.4	5	2.07

TABLE 7.2 (continued)

Series number	Date (and hour?) produced	Measurement of each of five items in series					Average for series	Range for series	Standard deviation
		a	b	c	d	e			
39		-1	+1	-1	-2	0	-.6	3	1.14
40		+1	+4	-4	+2	-4	-.2	8	3.63
41		+1	+1	-2	0	-2	-.4	3	1.52
42		0	-5	0	+5	-1	-.2	10	3.56
43		-2	0	+3	+1	0	+.4	5	1.82
44		-1	-1	0	0	0	-.4	1	.55
45		-2	-1	+1	0	+1	-.2	3	1.30
46		+2	+2	-3	-1	+1	+.2	5	2.17
47		-1	+1	0	-1	-1	-.4	2	.89
48		0	0	-2	0	0	-.4	2	.89
49		0	+3	0	0	-2	+.2	5	1.79
50		+1	0	0	-1	-1	-.2	2	.84

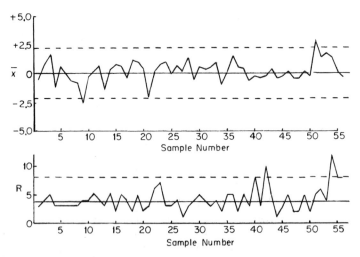

Fig. 7.2. Averages $\bar{x}$ and ranges R from Table 7.2. Experimental data interpreted as though edge-widths of piston rings were subject to specifications of .2050 ± .0005 in. Coded in .0001 in. units from .2050 in. Samples 1 through 50 were from $\mu = 0$, $\sigma = 1.715$; 51 through 53 from $\mu = +2$, $\sigma = 1.715$; and 54 through 56 from $\mu = 0$, $\sigma = 3.47$. Analyzed as past data, preliminary run, 1 through 25. Control lines revised after sample 50 and extended.

and plot the new sample results $\bar{x}$ and R as they come. The extended
limits are shown in Fig. 7.2. On sample 40, we find an R on the
upper control limit, a cause for action to look for something which
might give greater variability. (A change in average *level* would
not increase the process *variability*. Loosening of disc grinders
might increase variability.) Then on sample 42 there is a very high
R point. Again the cause is sought. But we know that it is here
simply a very rare accident in this sampling experiment.

After the first 50 samples, we again calculate control lines.
If, as here, we have not found the cause for out-of-control points,
we use all of the 50 samples.

$$\varepsilon_{\bar{x}} = \frac{+4.6}{50} = +.09$$

$$\varepsilon_R = \frac{190}{50} = 3.80$$

$$\text{Limits}_{\bar{x}} = +.09 \pm .577(3.80)$$
$$= +.09 \pm 2.19$$
$$= -2.10, +2.28$$

$$\text{Upper control limit}_R = 2.115(3.80) = 8.04$$
$$\text{Lower control limit}_R = 0(3.80) = \text{———}$$

Can we at this point estimate what the population in the bowl
was like? Yes, it can be estimated rather tentatively, because control is not perfect. However, it was not so very bad, even though
in practice every $\bar{x}$ and R point outside the limits would be studied.

In Ref. 1 [p. 16], a guide is given as to when in practice we
may regard a high degree of control as having been attained: (1)
none out of 25 successive points, (2) not more than 1 out of 35, or
(3) not more than 2 out of 100, fall outside 3σ limits. Here, despite one $\bar{x}$ point and two R points out of control limits, control is
quite good. Nevertheless, even just one point outside the control
band is worth investigating.

Let us now use $\bar{\bar{x}}$ and $\bar{R}$ to make estimates of the population or
process distribution; that is, we wish to estimate the population
average μ and standard deviation σ.

$\bar{\bar{x}}$ estimates μ if reasonable control shown   (7.9)

$\dfrac{\bar{R}}{d_2}$ estimates σ if reasonable control shown   (7.10)

where d_2 is another control chart constant.   (See Table C.)   Using
these we have the estimates

$$\mu = \bar{\bar{x}} = +.09$$
$$\sigma = \frac{\bar{R}}{d_2} = \frac{3.80}{2.326} = 1.63$$

Thus we can expect at least 99% of the measurements x to lie within
the limits

$$\mu \pm 3\sigma = +.09 \pm 3(1.63) = -4.80, +4.98$$

Therefore we would seem to be meeting specifications of -5 to +5.

In practice we would extend the limits found from the 50 samples and plot each $\bar{x}$ and R as they arise, watching for evidence of
assignable causes.

Now let us change the distribution of numbers in the bowl and
plot $\bar{x}$ and R for each sample and see how soon we obtain a warning
of a change having occurred. Sample 51 is

+1, +1, +6, +2, +5 $\bar{x} = +3.0$, R = 5

Plotting these sample results on Fig. 7.2, we find the $\bar{x}$ point
to lie well above the upper control limit, whereas the R point is in
control. Thus the sample shows evidence of an assignable cause,
that is, something has changed the distribution of individual x's
for the process. Note also the +6, which is out of specifications.
Since $\bar{x}$ went out, we expect it to be a result of a shift upward in
the process mean μ (although it could be from a sizable increase in
the process standard deviation σ, even with no change in μ). It
would seem that the process level should be reset. But by how much?
A sample of 5 is rather too few to determine how far off μ is. If
one makes a practice of resetting or adjusting a process on too few
measurements he will only tend to increase the variability of which
the process is capable. Thus we take two more samples

52 +1 +2 +2 +4 -2   x = +1.4, R = 6

53 +1 +4   0   0 +4 $\bar{x}$ = +1.8, R = 4

These $\bar{x}$'s are also high though not beyond the control limit. We average the three $\bar{x}$'s:

$$\frac{(+3.0 + 1.4 + 1.8)}{3} = +2.07$$

Thus we adjust the process level down by 2. Note that all three ranges are well inside the limit, and so we are confident that the assignable cause affected μ and not σ.

Next the bowl distribution was again changed. The first sample follows:

54 -1 +2 -4 +2 +8 $\bar{x}$ = +1.4, R = 12

The $\bar{x}$ point is in control, but the range is way above the upper control limit, signaling that something has caused an increase in σ. So steps are taken to find the assignable cause. Causes which increase σ are of major importance, in general, and their correction deserves high priority. Two more samples were drawn to further illustrate the change:

55 -4 +1 +4 +1 -1 $\bar{x}$ = +.2, R = 8

56 +5 +1 -3 -3 -1 $\bar{x}$ = -.2, R = 8

Both of these ranges are at the upper control limit. Note also the +8 in sample 54, which was out of specifications.

It is desirable to mention that in our experiment, in each case when we changed the bowl, the very first sample gave a signal by a point outside of the control band. This does not always occur. For example, we might just as well have found samples 52 and 53 before finding one like 51. But the stronger the assignable cause is, the more quickly we are likely to obtain a warning signal.

7.4   SOME POPULATION DISTRIBUTIONS
FOR SAMPLING EXPERIMENTS

Sampling experiments, with samples drawn from a bowl containing a
population, can be a great aid in getting the "feel" of sampling
results. This enables the student or reader to experiment as much
as he wishes, until he becomes familiar with the workings of chance
and gains confidence.

We list in Table 7.3 some convenient distributions for sampling experiments. These are the same approximately normal distributions as were extensively used in the famous War Production Board
courses in quality control by statistical methods [2]. Such distributions can be marked on fiber discs or chips. Such pieces can
sometimes be obtained free from stamping holes in fiber sheets.
The smaller the diameter and the thicker the fiber, the more easily
they may be mixed. Different colored inks can be used to distinguish the different distributions, as can circling and underlining.
Or the chips may be dyed different colors. The author uses different colored 1-cm-diameter beads, punched with metal punches and
marked with India ink. Plus and minus signs can be filed out of a
large nail, which is then heated and quenched to harden the metal.

Another available, quite normal distribution is obtainable from
the total showing on a roll of three dice. It is the following
running from 3 to 18

x	3	4	5	6	7	8	9	10	11	12	13	14	15	16	18	Total
f	1	3	6	10	15	21	25	27	27	25	21	15	10	6	1	216

For this, one may find $\mu = 10.5$ by symmetry, and $\sigma = 2.96$. The
average μ may be adjusted by adding on some constant to each x.

Take note that when we have a finite population of numbers and
wish to find the *population standard deviation*, we do not use N - 1
in the denominator as one might expect in looking at (2.4). Instead
we use either of

$$\sigma = \sqrt{\frac{\Sigma(x - \mu)^2 f}{N}} = \frac{\sqrt{N\Sigma\ x^2 f - (\Sigma\ xf)^2}}{N} \qquad (7.11)$$

TABLE 7.3. Approximately "Normal" Population for Sampling Experiments

Number x	Populations						
	A	B	C	D	E	F	G
+11				1			
+10			1	1			
+9			1	1	1		
+8			1	3	3		
+7		1	3	5	10		
+6		3	5	8	23		
+5	1	10	8	12	39		
+4	3	23	12	16	48		
+3	10	39	16	20	39	1	
+2	23	48	20	22	23	3	
+1	39	39	22	23	10	10	1
0	48	23	23	22	3	23	3
-1	39	10	22	20	1	39	10
-2	23	3	20	16		48	23
-3	10	1	16	12		39	39
-4	3		12	8		23	48
-5	1		8	5		10	39
-6			5	3		3	23
-7			3	1		1	10
-8			1	1			3
-9			1	1			1
-10			1				
N	200	200	201	201	200	200	200
μ	0	+2	0	+1	+4	-2	-4
σ	1.715	1.715	3.47	3.47	1.715	1.715	1.715

or we can code as in (2.8) but with the modification as in the last of (7.11).

In Sec. 7.3, the first 50 samples were drawn from Population A of Table 7.3, having $\mu = 0$ and $\sigma = 1.715$. Note that 100% of this population lies within -5 to +5, two of the N = 200, being those extremes. Then for samples 51 through 53, we used Population B having the same σ = 1.715, but with μ shifted up to +2. Finally for samples 54 through 56, we used Population C having μ again centered at 0, but with about double the variability, namely σ = 3.47. When the author obtained the range R of 10, he got interested in how rare an event this was, since in many similar experiments he never saw such a range. Is it not rare to have the *one* +5 and the *one* -5 in the same little sample of n = 5 x's? So he calculated it, finding the probability of such a sample to be about .0005, or about 1 in 2000!

7.5 THE NORMAL DISTRIBUTION

We have been talking about "normal data" and "well-behaved data" and normal distributions. It is now time to be more specific. The so-called "normal distribution" is a theoretical distribution of very wide applicability. That is, a great many observed frequency distributions can be usefully approximated by the normal distribution model. The mathematical properties are used in the mathematical calculations of all of the control chart constants listed in Table A.

The normal distribution is in reality a family of distributions. Any one member of the family is completely determined by specifying μ and σ. Thus for, say, tensile strengths of iron castings we might have a normal distribution with μ = 68,000 and σ = 5,000 psi. Or for "solenoid overtravel" we might have μ = .1400 cm and σ = .0200 cm.

The general shape of a normal distribution is shown in Fig. 7.3. Here the vertical axis represents *relative frequency*, which starts at zero to the left and rises steadily to a maximum then falls steadily back to zero, in a symmetrical pattern. The measurements x could be on the horizontal scale. However, we have chosen to draw a standardized normal curve, for which we use the standardized variable z:

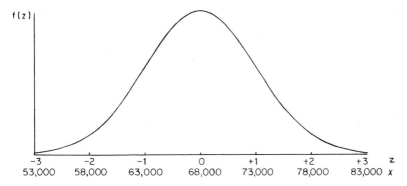

Fig. 7.3. A normal curve shown to two horizontal scales: (1) the standard variable z = (x - μ)/σ, and (2) tensile strength x with μ = 68,000 and σ = 5,000 psi.

$$z = \frac{x - \mu}{\sigma} \tag{7.12}$$

or

$$x = \mu + z\sigma \tag{7.13}$$

If we know μ and σ, then for any x observation x - μ is the observed deviation of x from μ, which tells how far x is from μ. Then we divide this *observed* deviation by the *standard deviation* to see how much larger or smaller the observed deviation is than the standard deviation. For example, if x = 72,500 psi, then the observed deviation is x - μ = 72,500 - 68,000 = +4500 psi. Now dividing by σ = 5000 psi, we find

$$z = +.9$$

We carefully note that because we divided pounds per square inch by pounds per square inch we have no unit on z, that is, it is a pure number. Its interpretation is best seen in (7.13). There we observe that when z = +.9, then x is .9 of a σ above μ.

Similarly, if x = 60,000 psi, the observed deviation is -8000 psi (x being below μ) and we find z = -1.6. Thus x is 1.6σ below μ. In this way, by knowing μ and σ, we may convert any x into z, the number of σ's x lies above or below μ.

Thus z is the dimensionless variable on the horizontal scale. The units on the two scales are so chosen as to make the total area

under the curve to be unity, that is, 100% of the cases. Then the
areas under various portions of the curve can be interpreted as
probabilities. For example, if we draw a vertical line at z = -1
and another at z = +1 we enclose about two-thirds of the cases, or
more precisely a probability of .6826. Therefore, for a normal distribution, we have between μ - 1σ and μ + 1σ, 68.26% of the cases.
As another example, between vertical lines at z = 0 and z = +2 lies
an area of .4772, nearly half. So between μ and μ + 2σ is 47.72%
of the cases. Now where do we obtain such numbers? They are available in any standard normal curve table, such as our Table A. These
tables have been calculated in a variety of forms.

Our Table A gives the cumulative probability from the extreme
negative values (minus infinity, which in practical work is a fic-
tion) up to tabled z values, in .01 steps. The units and tenths
for z are found in the left column and the hundredths in the other
column headings. Thus below z = -2.55 is a cumulative probability
of .0054. Meanwhile below z = +2.55 is a probability .9946. Since
the total probability is one (minus infinity to plus infinity),
this leaves 1.0000 - .9946 = .0054 *above* z = +2.55. This illustrates the symmetry of the normal distribution. There is as much
probability *below* any *negative* z as there is *above* the equal-valued
positive z.

Let us illustrate with a few examples. For a normal distribution of tensile strengths with μ = 68,000 and σ = 5000, find the
probability of a strength below 55,000 psi. First find z by (7.12):

$$z = \frac{55,000 - 68,000}{5,000} = -2.60$$

From Table A, we then find

$$P(x \leq 55,000) = .0047$$

Thus only about 1 in 213 will lie below 55,000. (See Fig. 7.4.)

Next, find the probability of a strength *above* 80,000 psi.
Find z.

$$z = \frac{80,000 - 68,000}{5,000} = +2.40$$

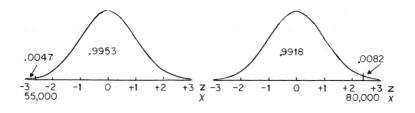

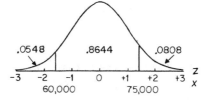

Fig. 7.4. Three figures of the normal curve to illustrate the probabilities x ≤ 55,000, x > 80,000, and 60,000 < x ≤ 75,000 psi, when μ = 68,000 psi and σ = 5,000 psi.

Table A gives

$$P(x \leq 80,000) = .9918$$

so the complementary probability is

$$P(x > 80,000) = 1 - .9918 = .0082$$

again quite small. Next find the probability of a strength lying between 60,000 and 75,000. Convert these strengths to standard z's:

$$z = \frac{60,000 - 68,000}{5,000} = -1.60$$

$$P(x \leq 60,000) = .0548$$

$$z = \frac{75,000 - 68,000}{5,000} = +1.40$$

$$P(x \leq 75,000) = .9192$$

Now the probability of a strength between the two is the difference. (See Fig. 7.4.) Thus

$$P(60,000 < x \leq 75,000) = P(x \leq 75,000) - P(x \leq 60,000)$$
$$= .9192 - .0548 = .8644$$

because the strengths which the event 60,000 < x ≤ 75,000 lacks from those of the event x ≤ 75,000 are precisely those contained in the event $x \leq 60,000$.

Proceeding similarly to the third example, we find the important probabilities:

$$P(\mu - 1\sigma < x \leq \mu + 1\sigma) = P(-1 < z \leq + 1)$$
$$= .8413 - .1587$$
$$= .6826$$
$$P(\mu - 2\sigma < x \leq \mu + 2\sigma) = P(-2 < z \leq + 2)$$
$$= .9772 - .0228$$
$$= .9544$$
$$P(\mu - 3\sigma < x \leq \mu + 3\sigma) = P(-3 < z \leq + 3)$$
$$= .9987 - .0013$$
$$= .9974$$

Thus about two-thirds of the cases lie within 1σ of the mean μ, about 19 out of 20 lie within 2σ of μ, and very nearly all within 3σ of the mean. These are for the normal distribution but it takes quite a bit of nonnormality before they are much in error.

Properties of the standard normal distribution for z:

1. The total area lying between the curve and the horizontal z axis is 1, representing a total probability of 1.

2. The height of the curve represents relative frequency, also called probability density.

3. The relative frequency is greatest at $z = 0$ (or $x = \mu$).

4. Relative frequencies steadily decrease as z moves away from 0 in either direction (or x away from μ).

5. The relative frequency rapidly approaches 0 in both directions.

6. The curve is symmetrical around $z = 0$. Thus, for example, the relative frequency is the same at $z = -.8$ as at $z = +.8$.

7. Desired probabilities for ranges of z may be found from published tables of normal curve areas.

8. The curve has its concave side down between $z = -1$, $z = +1$, and its concave side up outside these limits.

9. The equation for the curve is not often needed because of available tables, but it is

$$f(z) = e^{-z^2/2}/\sqrt{2\pi}, \quad e = 2.7183..., \quad \pi = 3.1415.... \,.$$

10. If we let h be the maximum height (at $z = 0$ or $x = \mu$) we have

Ordinate	h	.969h	.883h	.755h	.607h
z	0	±.25	±.50	±.75	±1.00
x	μ	$\mu \pm .25\sigma$	$\mu \pm .50\sigma$	$\mu \pm .75\sigma$	$\mu \pm 1.00\sigma$
Ordinate	h	.458h	.325h	.216h	.135h
z	0	±1.25	±1.50	±1.75	±2.00
x	μ	$\mu \pm 1.25\sigma$	$\mu \pm 1.50\sigma$	$\mu \pm 1.75\sigma$	$\mu \pm 2.00\sigma$
Ordinate	h	.080h	.044h	.023h	.011h
z	0	±2.25	±2.50	±2.75	±3.00
x	μ	$\mu \pm 2.25\sigma$	$\mu \pm 2.50\sigma$	$\mu \pm 2.75\sigma$	$\mu \pm 3.00\sigma$

These tabled ordinates help one draw a normal curve to any desired scale.

7.6 CONTROL CHARTS FOR x̄ AND R, STANDARDS GIVEN

When a process has shown good control with satisfactory quality performance, then you would usually wish to set standards values of average μ and standard deviation σ. This would be done by use of (7.9) and (7.10). Whenever we have standard values μ and σ, we can use them to set control lines for x̄ and R charts as follows:

$$\mathcal{L}_{\bar{x}} = \mu \qquad\qquad (7.14)$$
$$\text{Limits}_{\bar{x}} = \mu \pm A\sigma \qquad\qquad (7.15)$$
$$\mathcal{L}_R = d_2\sigma \qquad\qquad (7.16)$$
$$\text{Limits}_R = D_1\sigma, \ D_2\sigma \qquad\qquad (7.17)$$

As an example, let us set control lines for x̄ and R for the experimental data of Sec. 7.3, Table 7.2, which were drawn from Population A of Table 7.3. For this $\mu = 0$, $\sigma = 1.715$, $n = 5$. Thus using constants in Table A:

$$\mathcal{L}_{\bar{x}} = 0$$
$$\text{Limits}_{\bar{x}} = 0 \pm 1.342(1.715) = \pm 2.30$$
$$\mathcal{L}_R = 2.326(1.715) = 3.99$$
$$\text{Limits}_R = 0(1.715), \ 4.918(1.715) = \underline{\quad\quad}, \ 8.43$$

The one x̄ point, sample 9, is still out, but now only one range,

R = 10, for sample 42, is out of control.

One advantage in control against standards is that we are ready to interpret each sample as soon as it arises and need not wait for a preliminary run of samples.

The reader may have noticed that appropriate formulas for measurement control charts such as (7.14) through (7.17) are given at the bottom of Table C.

7.7   CONTROL CHARTS FOR STANDARD DEVIATIONS, s

As we have seen, the variability between the measurements x within a sample may be described by either the range R or the standard deviation s. Up to now we have been using range charts. We now briefly cover control charts for values of s.

Why might we wish to use s charts for studying the variabilities within samples? There are several reasons:

1. The standard deviation s, even for small samples, is a slightly more reliable measure than is the range. This is in part because it makes full use of all of the measurements x, instead of just the two extreme x's. Thus s makes more complete use of the information in a sample.

2. As the sample size increases above, say, 10, the range loses out rather rapidly in reliability in comparison with the standard deviation. Hence, above 10 we will always use s rather than R.

3. If a computer is being used we might as well use the "best" measure available. (Actually also it is more difficult to program a computer to find R than to find s.)

Let us therefore give the formulas appropriate (which may also be found at the bottom of Table C).

Case   (Analysis of past data).

$$\pounds_s = \bar{s} \tag{7.18}$$

$$\text{Limits}_s = B_3\bar{s}, \; B_4\bar{s} \tag{7.19}$$

Case   (Standard σ given).

$$\pounds_s = c_4\sigma \tag{7.20}$$

$$\text{Limits}_s = B_5\sigma, \; B_6\sigma \tag{7.21}$$

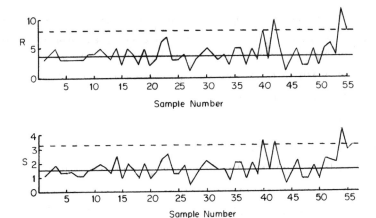

Fig. 7.5. Comparison of range R and standard deviation s charts. Experimental data are from Table 7.2, plus the six other samples for other populations. Note the similarity between the two charts.

For illustration take the first 50 s values of the experiment (Table 7.2, last column). They are plotted in Fig. 7.5. Also plotted above for comparison are the corresponding 50 ranges. Note how they fall and rise in the same way as the standard deviations. We also draw in the control lines from the past data using $\bar{s}$ and $\bar{R}$ for the 50 samples. Thus

$$\bar{s} = \frac{77.78}{50} = 1.556$$

$$\text{Limits}_s = B_3\bar{s}, \ B_4\bar{s}$$

$$= 0(1.556), \ 2.089(1.556)$$

$$= \underline{\quad}, \ 3.25$$

The s values for samples 40 and 42 are out of control, just as the R values were.

Let us also find control limits from the standard value $\sigma = 1.715$, for Population A (Table 7.3) from which the samples were drawn. Using (7.20) and (7.21):

$$\bar{s} = .940(1.715) = 1.61$$

$$\text{Limits}_s = 0(1.715), \ 1.964(1.715)$$

$$= \underline{\quad}, \ 3.37$$

Again the s values for samples 40 and 42 are out of control. So we see that the R and s charts tell the same story as regards within-sample variability.

In analysis of past data, we use $\bar{R}$ in setting limits for $\bar{x}$, via (7.7). Analogous to (7.7) is

$$\text{Limits}_{\bar{x}} = \bar{\bar{x}} \pm A_3\bar{s} \tag{7.22}$$

Using it with $\bar{s} = 1.556$, $\bar{\bar{x}} = +.09$ we obtain

$$\text{Limits}_{\bar{x}} = +.09 \pm 1.427(1.556)$$
$$= +.09 \pm 2.22$$
$$= -2.13, +2.31$$

while using $\bar{R} = 3.80$ in (7.7) gives

$$\text{Limits}_{\bar{x}} = -2.10, +2.28$$

Surely from a practical standpoint these two sets of limits for $\bar{x}$ are virtually identical.

A third use of R *or* s is to estimate σ after reasonable control of variability has been demonstrated. Thus we could use (7.10) or the analogous

$$\frac{\bar{s}}{c_4} \quad \text{estimates } \sigma \text{ if reasonable control shown} \tag{7.23}$$

Using (7.23) we obtain

$$\sigma = \frac{1.556}{.940} = 1.66$$

whereas by (7.10) the estimate for σ was 1.63. Again R and s did a quite similar job in estimating $\sigma = 1.715$.

Comparison of R's and s's from small samples--three uses.

1. The R chart and s chart give practically identical pictures of control of sample variability. When one is in control, the other will be also, and points out of control on one chart will correspond to those out on the other chart.

2. Under analysis of past data, use of $\bar{R}$ and $\bar{s}$ lead to practically the same limits on $\bar{x}$ charts.

3. Estimates of σ, when reasonable control of variability is shown, will be very nearly the same using $\bar{R}$ and $\bar{s}$.

Thus when working with small samples, say up to n = 10, we are at liberty to use either R or s. Take your choice. But, of course, we will not use both R and s. Unless a calculator programmed for s is used, almost everyone uses range charts.

7.8 COMPARISON OF A PROCESS WITH SPECIFICATIONS

In most industrial production, the basic question is whether the pieces or product satisfy requirements. Since it is impossible to produce pieces all exactly alike (except to some very gross measurement scale), we are forced to deal in *distribution of product characteristics*. Is the *distribution* satisfactory? This is the reason why specifications are commonly in terms of maximum and minimum limits for *individual measurements* x. Or, there may be a *minimum limit*, say on strength or the length of life of a part. Or again there may be a *maximum* limit on, say, blowing time of a fuse or root-mean-square finish. But in any case we seek an acceptable distribution. Of course if the distribution of x's is not yet satisfactory, we may well have to sort the parts 100% to remove those outside of limits. But this is expensive and time consuming, possibly throwing the production schedule badly out of balance. The aim should be to *make it right in the first place*. This means to so produce that *all* parts or product lie inside specifications, or else all but some acceptable very small *percentage outside*. These concepts are basic in all production.

Let us now discuss briefly four situations we may well have in a production line. Failure to distinguish among these cases can be devastatingly expensive.

1. Process in control and meeting specifications acceptably
2. Process not in control but meeting specifications
3. Process in control but not meeting specifications
4. Process not in control and not meeting specifications

Situation 1 is of course the desirable one. Control charts are useful in a continuing check or follow-through on maintenance of this desirable condition. Also we may be able to decrease gradually the

frequency of taking samples for charts and/or decrease any audit
inspection. Or we may be able to run closer to a minimum fill-
weight, thereby safely saving on material.

Situation 2 is also desirable, because of meeting specifications. But the lack of control is a danger signal that there are
assignable causes at work, which may suddenly get much worse. At
the very least, a careful watch should be maintained. It may well
be desirable to find out the assignable causes and take appropriate
action, perhaps bringing in some of the advantages mentioned under
situation 1.

Situation 3 may merely require an adjustment of level $\bar{\bar{x}}$, which
in dimensions for example may be easily accomplished sometimes;
also in some chemical processes. But in such things as tensile
strength, hardness, or surface finish, adjustment of the mean level
may not be so easily accomplished. A fundamental change in the process may be required; just tampering will not help.

Situation 4 calls for an aggressive campaign to seek out the
offending assignable cause(s). Often when it or they are found,
and action taken, it will be found that the process improvement
brought about makes the process fully capable of meeting specifications. So seek control in this case.

Note the difference between situations 3 and 4. In the former
we need a fundamental change in the process, or a relaxation of
specifications, and meanwhile must sort by 100% inspection, which
is expensive, time consuming and wasteful. But in situation 4 the
watchword is to seek control.

Conditions when the process is in control. If the process is
not yet in control ($\bar{x}$ chart and/or R chart out of control), we cannot properly talk about *the* distribution of product measurements.
This is because there is *not* just one distribution. Because whenever an assignable cause becomes operative it changes the distribution, possibly drastically from what it was while the assignable
cause was not present. This is true in situations 2 and 4.

What we wish to consider now is various cases when the process is *in control*, especially with the R (or s) chart in control. Then we can obtain a good estimate of σ by (7.10) or (7.23). Consider some of the various conditions as shown in Fig. 7.6. In case (1), the distribution just fits within the specification limits L and U. This was drawn with μ ± 3σ right at the two limits. Very few pieces will be outside, but to prevent pieces outside, the process level μ must be carefully controlled. In case (2), however, the natural process limits μ ± 3σ are comfortably inside L to U. There is some latitude for μ to be permitted to vary, perhaps as in tool wear.

Now look at case (3). The variability is satisfactory, but the level μ needs to be lowered for better centering to cut the

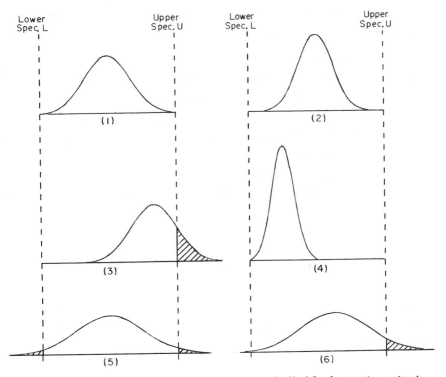

Fig. 7.6. Six cases of distributions of *individual* x's in relation to specification limits L and U for individuals. See text for a description of actions that need to be taken.

percentage above U. This may or may not be easily done. (Of course, another more unlikely possibility would be a drastic cut in σ somehow, while still maintaining the high value of μ.) In case (4), σ is small relative to the tolerance, U - L, and μ may have been purposely set as close as comfortable to L in order to save on material safely.

In case (5), the process standard deviation σ is just too big for the tolerance. But by good centering, the percentage of pieces outside of specifications has been minimized. If the percentage outside is still too big, then steps must be taken to cut σ somehow. In case (6), the variability is the same as for case (5) but the level has been so set as to virtually eliminate pieces below L. This is often done where pieces above U can be reworked as in grinding or machining, while those below L are to be scrapped. This is the expensive case of 100% sorting, rework, and sort again.

A basic point. We must emphasize here that we do *not* compare control limits for *averages*, x̄, with specifications for *individual* x's. Of course if the x̄ control limits lie *outside* x specification limits L and U, then some averages x̄ will not meet x limits, as well as *many* individual x's. But even when x̄ control limits lie between L and U, there may well be individual x's outside. See Fig. 7.7. The x̄ distribution has 3σ limits right on L and U. But the x limits are much wider with quite a few x's outside. This is because of the following fundamental fact:

$$\sigma_{\bar{x}} = \frac{\sigma_x}{\sqrt{n}} \qquad\qquad (7.24)$$

Here σ_x is what we have been calling σ, that is, an only-to-be-expected departure of x's from μ is $\sigma = \sigma_x$. Meanwhile for x̄'s, $\sigma_{\bar{x}}$ is the only-to-be-expected departure of x̄'s from μ. Thus if we have n = 4, then by (7.24)

$$\sigma_{\bar{x}} = \frac{\sigma_x}{\sqrt{4}} = \frac{\sigma_x}{2}$$

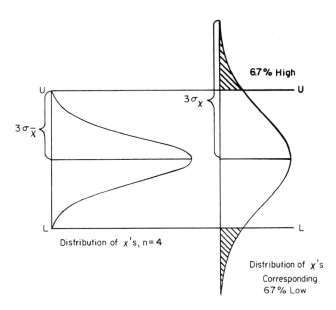

Fig. 7.7. Comparison of x and x̄ distributions with n = 4, where the control limits for x̄ happen to lie right on the specification limits for individual x values, L and U. Virtually no x̄ will lie outside L to U, but this is not important. With n = 4, the $6\sigma_x$ spread for x is twice as great as the $6\sigma_{\bar{x}}$ spread for x̄. And so there are about 13.4% of the individual x's not meeting specification limits L to U.

just half as much scattering for x̄'s as for x's, or again if n = 9, $\sigma_{\bar{x}} = \sigma_x/\sqrt{9} = \sigma_x/3$ so the variation among the x̄'s is only a third as great as that among the x's.

Thus we compare "natural" x limits with specification limits L and U for x's, namely by (7.10)

$$\bar{\bar{x}} \pm \frac{3\bar{R}}{d_2} \quad \text{or} \quad \mu \pm 3\sigma$$

If inside L to U, we are happy. If outside, we may have one of the cases in Fig. 7.6.

In summary, we recall and compare some formulas:

$\bar{x}$ limits: $\mu \pm A\sigma$ All these either give exactly or

$\bar{\bar{x}} \pm A_2\bar{R}$ approximately $\mu \pm 3\sigma_{\bar{x}}$, if there is

$\bar{\bar{x}} \pm A_3\bar{s}$ control. Not to be compared with

L to U (x specifications).

x limits: $\mu \pm 3\sigma$ All these either give exactly or

$\bar{\bar{x}} \pm 3(\bar{R}/d_2)$ approximately $\mu \pm 3\sigma_x$, if there is

$\bar{\bar{x}} \pm 3(\bar{s}/c_4)$ control. It is these limits we com-

pare to L to U (x specifications).

Better reread this section. Unless you clearly understand it, you can hardly be trusted to use measurement control charts in a plant!

7.9 CONTINUING THE CHARTS

In the early stages of a control chart application, we are primarily concerned with letting the process do the talking, telling us how it is doing. That is, we collect the sample data, plot them, calculate the central lines and control limits, draw them in, and interpret the results to all involved. As a consequence of indicated lack of control, we seek out the assignable causes responsible and may possibly find one or more from the preliminary run of data. In any case, we face the problem of continuing the charts. Usually we merely extend the control lines on both the $\bar{x}$ and R (or s) charts. There arises the question, however, as to whether to include the data corresponding to points which lay outside the control band. Data produced while assignable causes were operating should be eliminated, and $\bar{\bar{x}}$ and $\bar{R}$ revised only if *both* of the following are met: (1) The assignable cause for such performance was found, and (2) it was eliminated. Now if the assignable cause behind the unusual performance was not found, or having been found, nothing was done to remove it, then such data are still as typical of the process as any other and should be retained. In fact, whenever any significant change is made in the process, all data produced before the change are no longer typical of the revised process and should be discarded.

In line with this too, we should, in general, be periodically re-
vising the control lines from only fairly recent data.

7.10 WHEN AND HOW TO SET STANDARD VALUES

Standard values of μ and σ are commonly only set after good control
has been achieved and the process is producing a satisfactorily high
percentage of product within the specification limits for x's. In
this desirable situation we set standard values by

$$\bar{\bar{x}} \text{ becomes standard } \mu \qquad\qquad (7.25)$$

$$\frac{\bar{R}}{d_2} \text{ becomes standard } \sigma \qquad\qquad (7.26)$$

or

$$\frac{\bar{s}}{c_4} \text{ becomes standard } \sigma \qquad\qquad (7.27)$$

Such standard μ and σ must be such that the natural x limits

$$\mu \pm 3\sigma \text{ lie on or inside specification limits} \qquad (7.28)$$

Then at least 99.7% of x values will be meeting specifications as
long as control is maintained.

    There is an intermediate situation which is practical. It is
useful in processes where the process mean is relatively easily adjusted. Then if there are two specifications L and U for x, we need
the σ set from (7.26) or (7.27) to be small enough so that

$$6\sigma \leq U - L$$

Thus σ is set from data from the process. But we might set μ, not
from $\bar{\bar{x}}$, but instead at some desirable level. Thus if 6σ is close
to the tolerance U - L, we might well take μ at the nominal (middle
of the specification range) $(U + L)/2$. Or if σ is small enough so
that the natural process spread 6σ is considerably smaller than the
tolerance U - L, then we might wish to place μ at a point 3σ above
L, or at 3σ below U. These latter alternatives may be used also
when there is only one limit, either L or U, but not both. This may
enable saving on material, fill weight, or machining or grinding

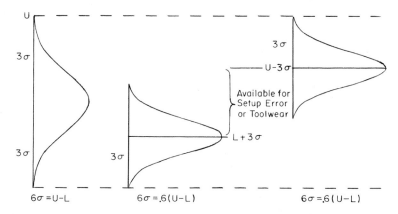

Fig. 7.8. Natural process spread $6\sigma_x$, in relation to tolerance U - L and desired process averages. In the first section, $6\sigma_x$ fills the tolerance, and the process average must be close to the nominal (U + L)/2 in order to be safely away from the specification limits. For the second section $6\sigma_x = .6(U - L)$. The lower distribution has its average at $L + 3\sigma_x$ and thus almost no pieces will lie below L. The upper distribution has its average at $U - 3\sigma_x$, and thus almost no pieces will lie above U. Thus when $6\sigma_x = .6(U - L)$, there is available considerable room for set-up error and/or tool wear, in fact .4(U - L).

time. Or it may be useful as a starting process setting in machining outside diameters, say. Here the aim would be to start a run at a process average level $L + 3\sigma$, and then as the cutting tool wears and the process average increases, end the run when it reaches $U - 3\sigma$. See Fig. 7.8 for illustration of cases.

We now use the standard values of μ and σ to determine control lines against which to compare future sample data $\bar{x}$ and R or s, having confidence that as long as control is maintained, we are meeting specifications.

7.11 EXAMPLES

Example 1. This example is on the *yield* of steel from an open hearth furnace, that is, tons of steel divided by tons of metallics charged. Yield improvement studies frequently disclose evidence of careless practices. In such cases, it is desirable to have operators and supervision have a graphical picture of their progress.

Once such careless practices are eliminated by intensive study and action, the tendency is for vigilance to relax and the gains thus made to be lost. A constant check must be maintained to ensure maximum results and to show when undesirable factors are creeping in.

In the present example, a typical one, period A is a 20-day history period, giving yields for each of the three daily heats of steel, some 200 tons each. Period B gives typical data from a period 3 weeks after control charts became a subject for daily discussion at operating meetings. Period C follows period B by 60 days.

For Period A, the initial history period, we find (see Table 7.4)

$\bar{\bar{x}}$ = 93.8 and $\bar{R}$ = 3.22

and we use n = 3 in Table C to find A_2 = 1.023 and D_3 = 0, D_4 = 2.575 and thus

$$\text{Limits}_{\bar{x}} = 93.8 \pm 1.023(3.22)$$
$$= 93.8 \pm 3.3$$
$$= 90.5, \ 97.1$$
$$\text{Limits}_R = 0(3.22), \ 2.575(3.22)$$
$$= \underline{\quad}, \ 8.29$$

Plotting the $\bar{x}$ and R points and drawing in the control lines we have the first part of Fig. 7.9. Day 6 has an $\bar{x}$ right on the lower control limit, from two yields in the 80's, while the 14th day has an $\bar{x}$ well below the limit, as well as an R out of control. These are a result of one very low yield of 83.0. It would seem that this yield may have acted as a stimulus, because the next 6 days show some improvement.

The 20 days of Period B give

$\bar{\bar{x}}$ = 94.8

R = 2.48

$\text{Limits}_{\bar{x}}$ = 92.3, 97.3

UCL_R = 6.39

Thus there has been an increased yield and more consistency (lower ranges) than in Period A. Day 3 has its $\bar{x}$ practically on the lower limit and day 15 has $\bar{x}$ definitely below. The former contained an

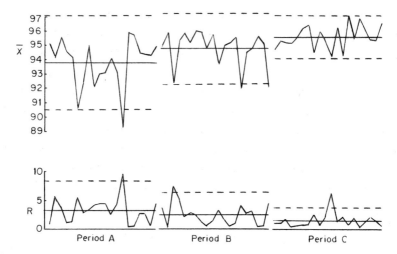

Fig. 7.9. Three periods of percent yield for an open hearth furnace, 3 heats/day. Period A represents the initial history; period B, 3 weeks later; and period C, 60 days later. Note increasing average and decreasing variability over the 3 periods. One percent improvement in yield means a gain of about 2 tons of usable steel per heat. Data in Table 7.4.

88.7 yield, and the latter had three yields all rather low, including a 90.1. Also the range for the third day was high.

For Period C, the 20 days gave

$\bar{\bar{x}} = 95.6$ $\qquad$ Limits$_{\bar{x}}$ = 94.1, 97.1

$\bar{R} = 1.49$ $\qquad$ UCL$_{R}$ = 3.84

Therefore there has been a further increase in yield and more consistency (lower ranges) than in Period A. Day 3 has its $\bar{x}$ practically on the lower limit, and day 15 has $\bar{x}$ definitely below. The former contained an 88.7 yield and the latter had three yields all rather low, including a 90.1. Also the range for the third day was high.

For Period C, the 20 days gave

$\bar{\bar{x}} = 95.6$ $\qquad$ Limits$_{\bar{x}}$ = 94.1, 97.1

$\bar{R} = 1.49$ $\qquad$ UCL$_{R}$ = 3.84

TABLE 7.4. Three Periods of a Yield Study on an Open Hearth Furnace, Making Steel, Three Heats per Day. Period A, Initial History; Period B, 3 Weeks Later; Period C, 60 Days Later

	% yield, 3 Heats			Period A		Period B		Period C	
Day	x_1	x_2	x_3	Average, $\bar{x}$	Range, R	Average, $\bar{x}$	Range, R	Average, $\bar{x}$	Range, R
1	95.6	94.8	94.8	95.1	.8	95.0	3.8	94.7	1.0
2	94.5	91.0	96.8	94.1	5.8	95.9	.3	95.3	1.0
3	96.1	97.3	93.4	95.6	3.9	92.4	7.4	95.2	1.9
4	95.0	94.0	94.7	94.6	1.0	95.3	5.2	95.2	.3
5	94.0	94.8	93.7	94.2	1.1	95.9	2.1	95.6	.6
6	88.5	94.0	89.2	90.6	5.5	95.2	2.7	96.2	.6
7	91.3	94.2	91.6	92.4	2.9	96.0	2.4	96.4	.8
8	96.7	94.8	93.4	95.0	3.3	95.9	1.2	94.4	2.6
9	90.5	91.1	94.7	92.1	4.2	94.8	.5	96.0	.7
10	90.3	94.0	94.6	93.0	4.3	95.8	1.4	95.2	2.1
11	91.7	91.6	95.9	93.1	4.3	93.7	3.4	94.2	6.4
12	93.2	95.7	93.3	94.1	2.5	95.0	1.9	96.3	1.3
13	90.7	95.0	93.7	93.1	4.3	95.2	.4	94.3	2.1
14	92.4	92.6	83.0	89.3	9.6	95.6	1.0	97.1	.8
15	96.0	95.9	95.7	95.9	.3	92.0	4.1	95.5	2.0
16	95.9	95.7	95.5	95.7	.4	94.5	2.9	96.9	.2
17	96.2	93.6	93.6	94.5	2.6	94.8	3.2	96.1	1.2
18	95.6	94.6	93.0	94.4	2.6	95.6	.4	95.4	2.2
19	94.3	94.0	94.6	94.3	.6	95.0	.5	95.3	1.5
20	96.8	92.3	95.5	94.9	4.5	92.1	4.7	96.6	.5
				1876.0	64.5	1895.7	49.5	1911.9	29.8
				$\bar{\bar{x}}$ = 93.8	$\bar{R}$ = 3.22	$\bar{\bar{x}}$ = 94.8	$\bar{R}$ = 2.48	$\bar{\bar{x}}$ = 95.6	$\bar{R}$ = 1.49
			UCL	97.1	8.29	97.3	6.39	94.1	3.84
			LCL	90.5	--	92.3	--	97.1	--

Therefore, there has been a further increase in yield, and an improvement again in consistency. There are two x̄'s practically on the lower control limit and a high range on day 11 mostly because of a low yield of 91.1.

The three periods give a most encouraging picture of improvement. The gain from A to C is 1.8%, or 3.6 tons more steel for each heat, on the average, from the same amount of metallics used. Continuing occasional evidence of lack of control gives hope of still further improvement. Continuation of the charts can help this and also act as a follow-through monitor.

Example 2. In Table 7.5, there are given measurements x, average x̄, and ranges R for two periods of a machining process. Let us consider the first period, with points as plotted in Fig. 7.10.

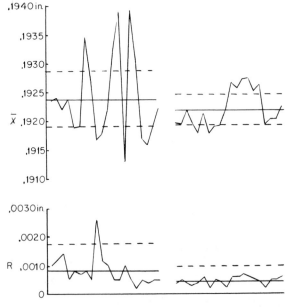

Fig. 7.10. Control charts for averages x̄ and ranges R for the outside diameter of a low-speed plug. Data are from Table 7.5. Specifications are .1900 to .1940 in. for periods before and after action taken. In second period, x̄ values are still out of control, but variability is much improved, so that now specifications are being safely met.

TABLE 7.5. Production Data on Outside Diameter of a Brass Low-Speed Plug, Blue Print Specifications .1900, .1940 in.

	x_1	x_2	x_3	x_4	x_5	$\bar{x}$	R	Comment
October 7								
11:30 AM	.1920	.1930	.1921	.1923	.1923	.19234	.0010	
12:45 PM	.1916	.1925	.1926	.1928	.1925	.19240	.0012	
1:30 PM	.1927	.1927	.1923	.1913	.1920	.19220	.0014	(a)
1:45 PM	.1924	.1922	.1922	.1927	.1924	.19238	.0005	
2:15 PM	.1919	.1917	.1918	.1924	.1916	.19188	.0008	(b)
2:55 PM	.1918	.1917	.1924	.1919	.1917	.19190	.0007	
3:15 PM	.1936	.1937	.1932	.1930	.1938	.19346	.0008	(c)
4:23 PM	.1925	.1930	.1930	.1925	.1925	.19270	.0005	
4:35 PM	.1930	.1927	.1915	.1902	.1910	.19168	.0028	
5:08 PM	.1920	.1922	.1920	.1917	.1910	.19178	.0012	
5:25 PM	.1925	.1925	.1920	.1925	.1915	.19220	.0010	
5:45 PM	.1930	.1930	.1935	.1930	.1935	.19320	.0005	
6:05 PM	.1940	.19?5	.1940	.1940	.1940	.19390	.0005	(b)
7:40 PM	.1915	.1905	.1915	.1915	.1915	.19130	.0010	(c)
8:13 PM	.1940	.1940	.1940	.1935	.1940	.19390	.0005	(b)
8:35 PM	.1930	.1930	.1930	.1931	.1929	.19300	.0002	
9:02 PM	.1915	.1915	.1920	.1915	.1920	.19170	.0005	
9:25 PM	.1915	.1915	.1915	.1915	.1919	.19158	.0004	
9:50 PM	.1920	.1915	.1920	.1920	.1920	.19190	.0005	
10:16 PM	.1925	.1921	.1920	.1925	.1920	.19222	.0005	
						3.84762	.0165	

$\bar{\bar{x}}$ = .19238; $\bar{R}$ = .000825; Limits$_{\bar{x}}$ = .19190, .19286; UCL$_R$ = .00174

	x_1	x_2	x_3	x_4	x_5	$\bar{x}$	R	Comment
December 10								
11:30 AM	.1920	.1921	.1917	.1920	.1920	.19196	.0004	
12:20 PM	.1921	.1916	.1921	.1918	.1920	.19192	.0005	
12:45 PM	.1921	.1922	.1920	.1924	.1922	.19218	.0004	
1.05 PM	.1920	.1920	.1919	.1921	.1918	.19196	.0003	
1:20 PM	.1920	.1916	.1920	.1917	.1916	.19178	.0004	
1:43 PM	.1917	.1923	.1921	.1923	.1922	.19212	.0006	
2:10 PM	.1917	.1919	.1918	.1918	.1917	.19178	.0002	
2:30 PM	.1918	.1922	.1920	.1917	.1918	.19190	.0005	
2:48 PM	.1918	.1922	.1918	.1918	.1920	.19192	.0004	
3:05 PM	.1922	.1921	.1922	.1923	.1921	.19218	.0002	
2:28 AM	.1923	.1926	.1926	.1929	.1929	.19266	.0006	
2:50 AM	.1922	.1926	.1924	.1928	.1928	.19256	.0006	
3:05 AM	.1924	.1927	.1926	.1931	.1928	.19272	.0007	
3:25 AM	.1928	.1927	.1928	.1924	.1930	.19274	.0006	
3:45 AM	.1923	.1928	.1925	.1926	.1923	.19250	.0005	
4:10 AM	.1924	.1924	.1928	.1927	.1928	.19262	.0004	
4:30 AM	.1920	.1918	.1920	.1919	.1920	.19194	.0002	
4:48 AM	.1923	.1920	.1921	.1918	.1919	.19202	.0005	
5:05 AM	.1923	.1918	.1919	.1920	.1921	.19202	.0005	
5:25 AM	.1921	.1919	.1924	.1924	.1925	.19226	.0006	
						3.84374	.0091	

$\bar{\bar{x}}$ = .19219; $\bar{R}$ = .000455; Limits$_{\bar{x}}$ = .19193, .19245; UCL$_R$ = .00096

Comments: (a) Out of stock; (b) adjustment made; (c) sharpened tool.

For the 20 samples, we have

$$\bar{\bar{x}} = .19238$$

$$\bar{R} = .000825$$

$$\text{Limits}_{\bar{x}} = .19238 \pm .577(.000825) = .19190, .19286$$

$$\text{UCL}_R = 2.115(.000825) = .00174 \text{ (all in inches)}$$

Drawing in the control lines, we find only one *range* out of control, in fact way outside. But the $\bar{x}$ chart is in very poor control, although shift one is in much better control than shift two. Obviously assignable causes are at work, affecting the level. This might have been a result of poor set-up, or set-up changes made on insufficient evidence. But there were also causes affecting the process level μ.

The question arises as to whether specifications were being met. A frequency tabulation of the 100 diameters shows none outside of the specifications limits of .1900 to .1940 in. However, measurements tended to concentrate on whole and half thousandths, such as, .1930, .1935, and .1940, and thus the measuring was apparently not done fully to the nearest .0001 in. Moreover, there were eight at .1940 in., some of which might well have been above .1940 in. if more accurately measured. Nevertheless, we might regard this as a case of situation 2: specifications are met fairly well but the process is out of control. So effort was made to improve the process, for greater security and assurance.

Actions taken were (1) developing a new cam and tool layout, (2) changing the sequence of operations, and (3) adding a steady rest and skiving tool and holder.

The second set of data was taken 3 months later. First note the range chart. $\bar{R}$ is about one-half of that for the earlier period showing much more uniformity around the current level, whatever it is. Moreover, the R chart shows perfect control. The $\bar{x}$ points for December 10 would have been in control relative to the $\bar{x}$ control limits for October 7. But because $\bar{R}$ is so much smaller, the $\bar{x}$ limits from the present data show much lack of control. The first shift is well centered on the nominal .1920 in., but some $\bar{x}$'s are below the lower limit because $\bar{\bar{x}}$ includes the high $\bar{x}$'s of the second shift. The operator on the second shift must have changed the process setting on insufficient evidence, in distrust of the first shift operation. One often sees such intershift rivalry.

Now for maximum security against the specifications, the plant could have used the nominal of .1920 in. as the average and limits, around this standard of $\pm A_2\bar{R}$. That is

$$\mu \pm A_2\bar{R} = .1920 \pm .577(.000455)$$
$$= .19200 \pm .00026$$
$$= .19174, .19226$$

Here we have used a desired *standard* μ and used *past data* to set the *width* of the control band, a mixture of the two control chart purposes. All of the $\bar{x}$'s for shift one lie inside these limits showing good adherence to the nominal. The first 6 $\bar{x}$'s of the second shift are all outside this control band; the last 4 $\bar{x}$'s are inside. This is a problem of set-ups rather than an assignable cause affecting the mechanics of the process.

Now is this so very bad? How about specifications? Is the process meeting specifications, even when at the high level of the first 6 points of the second shift? Let us see. Whenever the process average is at some level μ, within what limits will the individual x's lie? This requires us to know or else estimate σ_x for the x's, as we have seen in Sec. 7.8. Here, using $d_2 = 2.326$ from Table C:

$$\sigma_x = \frac{\bar{R}}{d_2} = \frac{.000455}{2.326} = .000196$$

So the natural limits of individual x's around any process average μ are

$$\mu \pm 3\sigma_x = \mu \pm 3(.000196) = \mu \pm .00059$$

Now does μ have to be right at the nominal .1920 in. in order to avoid trouble with the specification limits .1900, .1940 in.? No. (See Fig. 7.11.) How low and how high could μ be and still have us meeting the limits? If μ is as high as

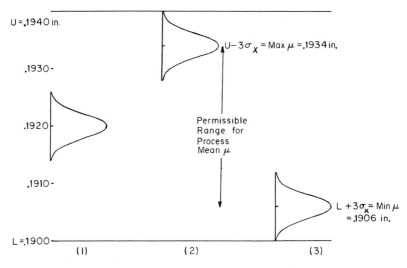

Fig. 7.11. Distributions for individual x's for diameter of low-speed plug. (1) Distribution centered, the ideal. (2) The highest safe distribution. (3) The lowest safe distribution. Shows the permissible variation for the process mean μ. σ_x was estimated from $\bar{R}$ by $\bar{R}/d_2$.

$$\text{Max } \mu = U - 3\sigma_x = .19400 - .00059 = .19341 \text{ in.}$$

or as low as

$$\text{Min } \mu = L + 3\sigma_x = .19000 + .00059 = .19059 \text{ in.}$$

we are still meeting specification limits. Even for the first 6 points of the second shift, the process average μ seems to be only about .19265 in., and so we are safely meeting the specification limits for x's, since the R chart shows perfect control. (If it were not in control, we could not be so sure.) Thus if desirable, we could simply aim to try to maintain the true process average somewhere between .19059 and .19341. Such an approach does give up, however, on the program of looking for assignable causes and merely concerns itself with checking the process setting. It is up to management as to whether to permit μ to vary between such limits, or to maintain somewhat tighter control.

7.12 SOME BACKGROUND OF CONTROL CHARTS

In all of the control charts described in Chapter 6 and the present
chapter, we have been making use of theory of sampling from distri-
butional models. These models are the binomial distribution for
defectives d and fraction defective p, the Poisson distribution for
defects c and defects per unit u, and the normal distribution for
measurements x. To find the behavior of sample characteristics such
as d, c, $\bar{x}$, R, or s, draw a large number of random samples from the
assumed population distribution, which might have been placed on
chips or beads in a bowl. We have seen a *small* run of such sample
results from each of the 3 distribution models. By taking a *large*
number of samples and tabulating the results, we would have a good
picture of the way the sample characteristic in question behaves,
that is, the sampling distribution.

However, there are obvious limitations to such an experimental
approach, even if we use an electronic computer. Fortunately, available theory enables us to determine the exact sampling distributions.
In fact for d and c, they are, respectively, the binomial and Poisson
distributions for the samples. For drawings from a normal distribution, theory provides the exact sampling distributions of $\bar{x}$, R, and s.

Now in all these cases, the exact sampling distributions are
fully determined by given standards p', c', or μ and σ, and, of
course, the sample size n. Each sampling distribution has an average and a standard deviation. In all cases the limits

average ± 3·(standard deviation)

for the characteristic will contain virtually all of them. Hence
if we find a sample value outside such 3σ limits we conclude that
the population has somehow changed (rather than that a rare accident
of sampling has occurred). Thus we conclude that there has been a
change in p', c', μ, or σ from the standard value, and we seek the
assignable cause for the change in the process.

Now if no standard was given, we need a preliminary run of samples upon which to base our analysis. Provisionally assuming that

the process is in control (sample results behaving like drawings from one bowl), we proceed to estimate the standard by

$\bar{p}$ for p'

$\bar{c}$ for c'

$\bar{\bar{x}}$ for μ

$\dfrac{\bar{R}}{d_2}$ or $\dfrac{\bar{s}}{c_4}$ for σ

Then these estimates are used as though they were the standard values.

The control chart constants and formulas given in Table C all come from theory on the normal distribution. Moreover, even if the actual distribution of x departs quite a little from the symmetrical normal distribution, the control chart constants still apply well. Thus we use

Standard given: μ, σ	Analysis of past data
$\ell_{\bar{x}}$: μ	$\bar{\bar{x}}$
Limits$_{\bar{x}}$: $\mu \pm A\sigma$	$\bar{\bar{x}} \pm A_2\bar{R}$ or $\bar{\bar{x}} \pm A_3\bar{s}$
ℓ_R: $d_2\sigma$	$\bar{R}$
Limits$_R$: $D_1\sigma$, $D_2\sigma$	$D_3\bar{R}$, $D_4\bar{R}$
ℓ_s: $c_4\sigma$	$\bar{s}$
Limits$_s$: $B_5\sigma$, $B_6\sigma$	$B_3\bar{s}$, $B_4\bar{s}$
ℓ_x: μ	$\bar{\bar{x}}$
Limits$_x$: $\mu \pm 3\sigma$	$\bar{\bar{x}} \pm \dfrac{3\bar{R}}{d_2}$ or $\bar{\bar{x}} \pm \dfrac{3\bar{s}}{c_4}$

The center line and limits for individual x's are used in determining the "natural process limits," but only when the $\bar{x}$ and R (or s) charts show good control.

Again let us emphasize that limits for $\bar{x}$'s and x's must be carefully distinguished. Limits for $\bar{x}$'s are always narrower than those for x's, by a factor of $\sqrt{n}$. Thus

Limits$_x$: $\mu \pm 3\sigma_x$

Limits$_{\bar{x}}$: $\mu \pm 3\sigma_{\bar{x}} = \mu \pm \dfrac{3\sigma_x}{\sqrt{n}}$

It is the x limits which we must compare with specification limits
for x's.

7.13 SUMMARY

In this chapter, we have studied control charts for measurements x,
for the two characteristics of sample data, namely, level by $\bar{x}$ and
variability by R or s. We thus use two charts for measurement data.
It seems unnecessary to list again the appropriate formulas which
have been scattered throughout the chapter and collected in the preceding section, and also in Table C.

These methods may be used to analyze practically any series of
varying results or data. It is however desirable to collect the
data in *small* samples so that conditions have little chance to vary
while the n x values for the sample are taken. In this way all of
the variation between these x's can be attributed to chance causes.
Between this sample and the next, there will be a time interval, and
conditions might vary. Thus we aim at within-sample variation being
by chance causes alone, whereas sample-to-sample variation may also
contain assignable causes as well as chance.

Typical types of analyzable data are all kinds of dimensions
such as diameters and lengths, electrical characteristics, timing
such as relays and blowing time of fuses, production and unit cost
figures, weights of contents or pieces, placement of holes, thicknesses of paper or rubber, chemical compositions of all kinds, percent impurities, lengths of life, bending, tensile or breaking
strengths, thickness of coating, mechanical properties such as horsepower or torque, bounce of balls, color and taste properties, lengths
of yard goods measured versus ordered, temperature controls, and resistance to corrosion.

More examples of applications and some of the theory are in
Ref. 3. Many good case histories are given in Ref. 4. Another useful reference is Ref. 5, with many numerical examples.

PROBLEMS

7.1 Percentage moisture content of an oat cereal were subject to a
maximum specification limit of 4.0. Final checks were made to the
nearest .1%. Samples 1, 2, and 20 follow: 4.0, 4.2, 3.6, 3.8, 3.6;
3.3, 3.7, 3.4, 3.6, 3.7; 4.2, 4.9, 3.6, 3.5, 4.0. The 20 $\bar{x}$'s and
R's were:

Sample	1	2	3	4	5	
Average, $\bar{x}$	3.84	3.54	3.96	3.58	3.50	
Range, R	.6	.4	1.1	.3	.9	
Sample	6	7	8	9	10	
Average, $\bar{x}$	3.34	3.76	3.74	3.56	3.58	
Range, R	.6	.5	.9	.5	1.0	
Sample	11	12	13	14	15	
Average, $\bar{x}$	3.70	3.64	3.54	3.72	3.58	
Range, R	.9	.7	.6	.6	1.1	
Sample	16	17	18	19	20	Total
Average, $\bar{x}$	4.04	3.42	3.86	3.94	4.04	73.88
Range, R	.6	.9	1.3	.5	1.4	15.4

(a) Make $\bar{x}$ and R charts and comment. (b) Is the process meeting the
specification limit? (c) If justified set the natural limits for
the x's. (d) What action would you suggest? A frequency tabulation
gives the following:

x	3.0	3.1	3.2	3.3	3.4	3.5	3.6	3.7	3.8	3.9
f	5	3	4	5	6	10	13	11	9	10

x	4.0	4.1	4.2	4.3	4.4	4.5	4.6	4.7	4.8	4.9
f	11	2	5	4	1	0	0	0	0	1

(e) Comment on this distribution.

7.2 This problem considers wire, cold drawn to size, then galvanized and fabricated into field fence. The present data are for size
12-1/2, carrying specifications on diameter of .096, .102 in. Wire
is produced by the ton, coating is applied by ounces, and the final

fence is sold by the rod. It is obvious that holding to the low side of the diameter specification will produce more miles of finished wire per ton of raw stock, and hence more rods of final fence. The following x and R values are for diameter, with n = 7. (a) Draw $\bar{x}$ and R charts, check control, comment. (b) Can you tell from the data what the precision of measurement was? (c) What can you say about meeting specifications for the diameter? Decimal points omitted for the diameters in inches:

Sample	1	2	3	4	5	6	7	8	9	10
$\bar{x}$	0985	0980	0970	0985	0973	0972	0980	0965	0970	0975
R	0030	0025	0010	0030	0025	0005	0040	0010	0010	0010

Sample	11	12	13	14	15	16	17	18	19	20
$\bar{x}$	0980	0980	0975	0980	0984	0985	0980	0970	0968	0975
R	0025	0015	0010	0040	0030	0030	0040	0010	0010	0010

Sample	21	22	23	24	25	26	27	28	29	30
$\bar{x}$	0972	0970	0980	0978	0985	0970	0985	0974	0972	0970
R	0010	0010	0010	0010	0010	0020	0010	0020	0015	0040

Total
2.9288
.0570

7.3 This problem concerns the same wire discussed in Prob. 7.2. The data are on the weight of coating in ounces per square foot of area. There is a *minimum* specification of .250 oz/ft^2. It is obvious that holding as closely as possible to the minimum coating weight, and as uniformly as possible, will save considerable coating materials. Values of $\bar{x}$ and R are given below on coating weight with n = 2. (a) Make $\bar{x}$ and R control charts, check control and comment. (b) What can you say about meeting the minimum specification on weight of coating? (c) What action would you suggest? (d) In the first sample of 2, $\bar{x}$ = .255, R = .010, can you then determine the two x's? Decimal points omitted for the coating weights in ounces per square foot area:

Sample	1	2	3	4	5	6	7	8	9	10
$\bar{x}$	255	330	280	215	220	223	255	235	222	315
R	010	020	100	010	040	000	020	030	020	030

Sample	11	12	13	14	15	16	17	18	19	20
$\bar{x}$	380	280	260	270	410	310	250	275	275	280
R	010	040	020	050	150	100	040	030	030	040

Sample	21	22	23	24	25	26	27	28	29	30
$\bar{x}$	275	300	360	270	290	300	260	280	285	295
R	030	040	000	010	020	010	010	020	010	010

$\Sigma \bar{x} = 8.455 \qquad \Sigma R = .950$

7.4 The following data give results for n = 10 measurements of corrosion resistance for each of 12 minor additions of an element to
alpha aluminum bronze in $mg/dm^2/day$ [6]. Twelve elements were chosen
randomly from the 22 in the study. $\bar{x}$'s and s's are given below.
(a) Make $\bar{x}$ and s charts. Is there control? (b) Would you expect
control? Would you want it?

Element	$\bar{x}$	s
Antimony	29.50	2.13
Arsenic	29.89	1.51
Cadmium	28.94	2.05
Beryllium	25.87	3.89
Chromium	32.13	1.20
Zinc	29.14	.65
Zirconium	26.38	1.73
Phosphorous	28.64	1.42
Silicon	30.10	2.24
Calcium	28.63	1.41
Lead	31.67	.97
Tin	29.48	1.29
	350.37	20.49

7.5 The table below gives information on a machining process, three
phases of actual data. Treating each set of 20 samples as a run of

past data, draw the $\bar{x}$ and R charts, plotting the 60 points ($\bar{x}$ and R) in succession and drawing the respective control lines. Comment on the three phases with respect to control, meeting of specifications, and need for action.

UNIT OF MEASUREMENT: .0001 in. from 1.1760 in.

SPECIFICATIONS: 1.1760 in. - 1.1765 in.

Comments: Samples 1-20   First stage--experimental data

Samples 21-40   Second stage--after some action was taken

Samples 41-60   Third stage--after more assignable causes were eliminated

Sample	x_1	x_2	x_3	x_4	x_5	$\bar{x}$	R	Notes
1	+3	0	-3	+4	+4	+1.6	7	
2	0	+3	+3	+2	+2	+2.0	3	
3	+5	+3	+5	+6	+5	+4.8	3	
4	+3	+2	+3	+2	+1	+2.2	2	
5	+3	+2	+2	+1	+1	+1.8	2	
6	+1	+1	0	-3	+3	+ .4	6	
7	+2	-4	+1	-2	+2	- .2	6	
8	+2	+4	+3	+3	+2	+2.8	2	
9	+3	+2	+4	+4	+3	+3.2	2	
10	+3	+4	+3	+1	+3	+2.8	3	
11	+3	+2	+4	+4	+5	+3.6	3	
12	+2	+4	+2	+1	+5	+2.8	4	
13	+1	+3	+3	+2	+3	+2.4	2	
14	-5	-6	0	+4	+4	- .6	10	
15	+1	0	+4	+4	+5	+2.8	5	
16	+4	+4	+3	+5	+1	+3.4	4	
17	+6	+6	+6	+8	+4	+6.0	4	
18	+3	+8	-2	+4	+3	+3.2	10	
19	-5	+3	-3	+1	+2	- .4	8	
20	+1	+7	+8	-2	0	+2.8	10	
						+47.4	96	

Sample	x_1	x_2	x_3	x_4	x_5	$\bar{x}$	R	Notes
21	+4	+3	+3	+2	+1	+2.6	3	
22	+1	-1	+5	+3	+4	+2.4	6	
23	+1	+2	-1	0	+2	+ .8	3	
24	+1	+4	+4	+2	+2	+2.6	3	
25	+4	+4	+2	+4	0	+2.8	4	
26	0	+7	+4	+6	+3	+4.0	7	
27	+5	+4	+6	+5	+4	+4.8	2	
28	+4	+1	+3	0	+1	+1.8	4	
29	0	+1	0	+4	0	+1.0	4	
30	+1	+3	+4	0	0	+1.6	4	
31	+2	+5	+2	+2	+4	+3.0	3	
32	+7	+3	+3	+5	+4	+4.4	4	
33	+4	+3	+4	+4	+1	+3.2	3	
34	-2	+4	+3	-4	+5	+1.2	9	
35	+4	+5	-2	+7	+4	+3.6	9	
36	0	+4	+4	0	+1	+1.8	4	
37	+3	+1	+1	+3	+1	+1.8	2	
38	+2	+3	+2	+6	+5	+3.6	4	
39	+3	+3	+4	+5	+2	+3.4	3	
40	+4	+4	+1	+5	+4	+3.6	4	
						+53.0	85	
41	+3	+4	+3	0	+3	+2.6	4	
42	+4	+1	+1	+3	+3	+2.4	3	
43	+3	+2	+3	+3	+3	+2.8	1	Attention focused
44	+3	+2	+2	+2	+2	+2.2	1	on process
45	+3	+3	+3	+3	+1	+2.6	2	Dirt found on holding device cleaned off
46	+2	+2	+3	+2	+3	+2.4	1	
47	+4	+3	+3	+4	+3	+3.4	1	
48	+3	+2	+3	+3	+2	+2.6	1	
49	+3	+3	+2	+3	+3	+2.8	1	
50	+3	+2	+3	+3	+2	+2.6	1	

Sample	x_1	x_2	x_3	x_4	x_5	$\bar{x}$	R	Notes
51	+3	+3	+2	+3	+2	+2.6	1	Operator dressed
52	+3	+3	+2	+3	+2	+2.6	1	wheel
53	+2	+3	+3	+2	+3	+2.6	1	
54	+2	+2	+2	+3	+3	+2.4	1	
55	+2	+3	+3	+2	+3	+2.6	1	
56	+3	+3	+2	+2	+3	+2.6	1	
57	+3	+3	+3	+2	+2	+2.6	1	
58	+3	+3	+3	+3	+3	+3.0	0	
59	+4	+3	+3	+3	+3	+3.2	1	
60	+3	+3	+3	+3	+3	+3.0	0	
						+53.6	24	

7.6 The minimum distance between rubber gasket and the top of a metal cap for glass jars to contain food is an important characteristic. If too small, the cap may come off; if too large, the cap may leak and lose the vacuum. Given here are 24 values of $\bar{x}$ and R for this distance, which has specification limits of .117 to .133 in. Make $\bar{x}$ and R charts, and comment on control and the meeting of specifications. What action would you suggest? $\bar{x}$ and R are in .001 in. units:

Sample	1	2	3	4	5	6	7	8
$\bar{x}$	120	122	120	118	120	120	124	113
R	20	10	20	20	20	30	15	15

Sample	9	10	11	12	13	14	15	16
$\bar{x}$	120	108	117	118	122	118	127	124
R	0	20	10	20	20	20	10	15

Sample	17	18	19	20	21	22	23	24
$\bar{x}$	127	122	123	127	118	122	118	122
R	10	15	10	10	10	10	20	20

The chuck was adjusted after sample 10.

7.7 The following data are shown on a data sheet form of the kind
which is useful in collecting relevant information on a study.
Complete information including notes is of great importance. A
form can easily be made up for your own company. Samples 1 through
20 were taken as preliminary data. After considerable work seeking
assignable causes and taking appropriate action, samples 21 through
40 were taken. Treating each as a separate run of past data, construct the two sets of $\bar{x}$ and R charts in succession on the same
scales. Comment on control and meeting of specifications (±5 in
coded units).

DATA SHEET

PART NAME: Transmission Main Shaft Bearing Retainer

PART NUMBER: 6577-D

OPERATION NUMBER: 27   MACHINE NUMBER: TC 3677

   DEPARTMENT NUMBER: 64

CHARACTERISTIC MEASURED: Inside Diameter

METHOD OF MEASUREMENT: Dial Indicator

UNIT OF MEASUREMENT: .0001 in. from 2.8346 in.

SPECIFICATIONS: 2.8341 - 2.8351 in.

DATA RECORDED BY: Davis   DATE:

Sample number	Measurements of each of five items in sample					Average for sample	Range for sample	Sample number	Average for sample	Range for sample
	a	b	c	d	e					
1	+4	+4	+6	+3	0	+3.4	6	11	-5.0	5
2	+2	+3	+2	0	0	+1.4	3	12	-10.2	9
3	+20	-4	-1	-3	+16	+5.6	24	13	+2.2	7
4	-3	-5	-5	+2	-2	-2.6	7	14	+.8	8
5	0	-2	+1	+4	+3	+1.2	6	15	+1.0	2
6	-3	-3	-2	-2	0	-2.0	3	16	+1.6	4
7	-4	-7	-8	-11	-11	-8.2	7	17	-7.8	3
8	-2	-4	-7	-8	-8	-5.8	6	18	-3.8	7
9	-9	-7	-8	-7	-8	-7.8	2	19	-2.0	2
10	-4	-7	-1	-6	-2	-4.0	6	20	0	0

Sample number	Average for sample	Range for sample	Sample number	Average for sample	Range for sample
21	+1.2	2	31	-4.0	2
22	+.8	3	32	-4.4	8
23	+.6	3	33	+3.0	4
24	-.4	1	34	-.8	3
25	-.2	1	35	-.4	4
26	+.6	1	36	+.4	1
27	+.2	2	37	+1.2	2
28	+.2	1	38	+.8	2
29	0	2	39	-1.0	3
30	-.2	2	40	-1.0	2

7.8 The data given below are on thickness of wax on the inside of cartons (in .001 in.). The desired thickness is 5, with specification limits of 3 and 7. Write a brief report on your findings. Sample size was n = 3.

Date	Time	$\bar{x}$	R	Date	Time	$\bar{x}$	R
2/1	7:10 AM	5.23	.9	2/3	7:40 AM	5.90	1.0
	1:40 PM	5.03	.5		9:40 AM	5.53	.1
	5:30 PM	4.20	.9		11:40 AM	4.90	.5
	9:45 PM	5.20	.6		1:30 PM	4.80	.4
	1:15 AM	5.40	1.0		7:30 PM	5.20	.5
	2:10 AM	4.63	.8		10:50 PM	5.00	.9
	4:50 AM	5.10	.5		2:10 AM	5.70	1.0
	6:45 AM	4.70	1.0		4:10 AM	5.70	1.0
2/2	8:05 AM	3.97	.9	2/4	9:05 AM	5.03	.7
	10:00 AM	4.13	.9		11:40 AM	4.90	.2
	1:30 PM	4.50	.7		1:30 PM	3.93	.4
	5:00 PM	5.33	.2		5:30 PM	3.43	.9
	7:30 PM	5.00	.8		7:40 PM	2.93	.4
	12:05 AM	4.73	.1		11:30 PM	4.87	.3
	2:10 AM	5.23	.8		2:00 AM	5.30	.5
	5:20 AM	5.73	.9		4:20 AM	4.90	.3

7.9   Three heats of steel are made each day. Chemical analysis
and tapping and pouring temperatures must be carefully controlled.
Values of $\bar{x}$ and R for samples of the n = 3 heats per day, for manganese in 1045 steel are given here. The data are in hundredths of
a percent. Specifications are 70 to 90 in .01%.

Day	$\bar{x}$	R	Day	$\bar{x}$	R	Day	$\bar{x}$	R
1	76.3	5	10	77.7	6	19	79.0	8
2	73.0	15	11	77.3	6	20	74.3	4
3	77.0	5	12	79.3	2	21	85.3	23
4	80.0	16	13	75.7	11	22	81.3	15
5	72.0	18	14	78.3	4	23	80.0	20
6	72.7	11.	15	79.3	9	24	79.0	22
7	78.7	5	16	78.0	8	25	84.0	8
8	80.3	6	17	83.3	9	26	75.0	10
9	77.7	17	18	80.7	18	Sum	2035.2	281

(a) Check for control by $\bar{x}$ and R charts and comment. (b) What can
you say about meeting specifications? (c) Do you have any recommendations? (d) Can you set natural limits for x for the process? If
so, set them?

7.10   A drill depth dimension on a heater flange carried specification limits of 1.217, 1.222 in. Trouble with shallow holes had been
encountered for years on this part. Control charts were run on $\bar{x}$
and R, the latter being in control. Control limits for $\bar{x}$ were
1.215, 1.221 in. for samples of 5. (a) What should this information
immediately tell you? But now how about control? Of the first 25
samples, all but three $\bar{x}$ values were on or below the central line.
But of samples 28 through 34, all but one were outside of the upper
control limit. So control was poor giving even more trouble meeting
specifications. In samples 1 through 25, holes more shallow than
1.217 were very common requiring 100% inspection and reworking.
Some of the deep-hole pieces had to be scrapped.

     As a result of this control chart study, the quality control
manager made a number of recommendations which were adopted.

A subsequent run of 38 samples had $\bar{\bar{x}}$ = 1.2194 in. and $\bar{R}$ = .0026.
The R chart was in good control. (b) Therefore, estimate σ_x for
the process. The nominal (middle of specification range) is 1.2195
in. Thus the process is now well centered. (c) If the $\bar{x}$'s are in
control, what are the natural limits for x? (d) How do these compare with the specifications? But actually the $\bar{x}$'s were not in
control, out of 38 $\bar{x}$ values there being 5 above the upper control
limit and 2 at or below the lower control limit. The trouble would
seem to have been poor adjustment of the process particularly in
the form of overcompensating on insufficient evidence. If the $\bar{x}$'s
are brought into full control, with the same $\bar{R}$, then use of σ_x from
(b) and μ = 1.2194 in. along with the normal curve, Table A shows
that the percent outside specification limits is about 2.6%:

$$z = \frac{1.2220 - 1.2194}{.00112} = +2.32 \qquad .0102 \text{ above}$$

$$z = \frac{1.2170 - 1.2194}{.00112} = -2.14 \qquad \underline{.0162 \text{ below}}$$

$$.0264 \text{ outside}$$

This was a very profitable and successful project.

7.11   What would you say to a shop man or union steward who con-
tends that your $\bar{x}$ limits, which must be well inside the specification limits U and L for x, mean that you are cutting the tolerance
and making the job harder and that therefore the pay should be
greater?

7.12   Which is likely to be more difficult to correct: out-of-
control conditions on $\bar{x}$ or on the R chart?

REFERENCES

1. American Society for Quality Control Standard (B3) or American
    National Standards Institute Standard (Z1.3-1969), *Control Chart
    Method of Controlling Quality During Production*, ASQC, 161 W.
    Wisconsin Ave., Milwaukee, Wisc. 53203.

2. H. Working and E. G. Olds, *Manual for an Introduction to Sta-
    tistical Methods of Quality Control in Industry, Outline of a
    Course of Lectures and Exercises*. Office of Production Research
    and Development, War Production Board, Washington, D. C., 1944.

3. I. W. Burr, *Statistical Quality Control Methods*, Dekker, New York, 1976.

4. E. L. Grant and R. S. Leavenworth, *Statistical Quality Control*. McGraw-Hill, New York, 1972.

5. American Society for Testing Materials Manual, *Quality Control of Materials*, ASTM, Philadelphia, Penna.

6. R. H. Hoefs, "The Effect of Minor Alloy Additions on the Corrosion Resistance of Alpha Aluminum Bronze." Ph.D. Thesis, Purdue Univ., Lafayette, Ind., 1953.

Chapter Eight

FURTHER TOPICS IN CONTROL CHARTS AND APPLICATIONS

We now take up some additional concepts and techniques of control charts. They are useful in appropriate situations, and of interest in their own right too, in consolidating your understanding. The chapter concludes with a listing of applications.

8.1 TYPES OF SAMPLING

This section considers several different ways of drawing samples of pieces for measurement or inspection and analyzing the results. Success in quality control depends to a large extent upon how samples for analysis are drawn.

8.1.1 *The Rational Sample or Subgroup*

A rational sample is one in which, insofar as possible, all pieces or units within the sample were produced *and* measured under the same conditions. In this way all of the variation among pieces *within* the sample is due to chance causes alone. The usual way to insure that conditions are substantially constant is to take *small samples* on pieces produced close together in time. By the time the next sample is taken and measured, some of the conditions may have changed. In this way we can say that when we use *rational* samples, all of the variation *within* each sample is a result of chance causes only, whereas variation *from sample to sample* is attributable not only to chance causes, but possibly to assignable causes as well.

If we let an assignable cause creep into our sample, then the range between the measurements will not reflect chance alone but will be enlarged by the assignable cause. *Don't inflate the range by assignable cause variation.*

The three experimental examples for np, c, $\bar{x}$, and R charts in which all samples were drawn from the same population, in each case, were rational samples. (And further, of course, there were no assignable causes at work in the preliminary run of data, to cause assignable variation from sample to sample.)

The concept and importance of rational sampling will be clarified further as we discuss other types of sampling.

8.1.2 The Stratified Sample

In production we frequently have pieces produced from multispindle machines, multicavity molds, multiorifices, and so on, where a number of pieces are produced more or less at the same time. For example, a five-spindle automatic machine produces a piece from spindle one, then spindle two, and finally spindle five, after which spindle one makes the next piece, always in this rotation. It may well seem quite logical to take one piece from each spindle as a sample. Such a sample is a "stratified sample," because the pieces come from "strata" or layers. Taking a series of such samples does do an excellent job of estimating the *average level* of all of the pieces produced, because it gives an equal opportunity for each spindle to contribute to the grand average. If specifications are being consistently met, then such a collection of samples is probably satisfactory.

If, however, specifications are not being met, or process improvement is desired, then it may well be desirable to take a small sample of pieces all from spindle one, another sample of pieces from spindle two, and so on. We thus have a *rational* sample from each spindle. Then later on another set of samples can be taken. Such an approach has two distinct advantages over stratified samples: (1) the variation within such samples is entirely due to chance causes only, whereas the variation within a stratified sample contains not only chance variation, but also any assignable cause variation occasioned by spindle-to-spindle differences, (2) the rational samples enable us to determine how each spindle is doing by itself, that is, to measure spindle-to-spindle differences, whereas this is not possible if stratified samples are taken.

Let us now illustrate these points by an experimental collection of data. Consider the results in Table 8.1. First, we have the stratified samples *in rows*: x_1 from spindle one, x_2 from spindle two, and x_3 from spindle three. The averages and ranges for the stratified samples are listed, and they are plotted as shown in Fig. 8.1, the upper pair of charts. Look first at the $\bar{x}$ chart. The $\bar{x}$'s seem to be running randomly enough. But now note the 3σ control limits calculated from the 36 row-samples. Does not control seem to be too good? There is not one $\bar{x}$ as much as one $\sigma_{\bar{x}}$ away from the central line, let alone three $\sigma_{\bar{x}}$. (Of 36 $\bar{x}$'s, if control were perfect, the probabilities for a normal distribution of $\bar{x}$'s would say that about a third of the 36 $\bar{x}$'s should be more than one $\sigma_{\bar{x}}$ away from $\bar{\bar{x}}$.) Something is obviously wrong. To only a somewhat lesser extent, the same thing is true for the R chart. Only three of the 36 R values are more than one σ_R from the central line to the limit. Before further discussion, let us set down the calculations (with $n = 3$).

$$\mathcal{E}_{\bar{x}} = \bar{\bar{x}} = \frac{+.9}{36} = .02$$

$$\text{Limits}_{\bar{x}} = \bar{\bar{x}} \pm A_2 \bar{R} = .02 \pm 1.023(8.47)$$
$$= -8.64, +8.68$$

$$\mathcal{E}_R = \bar{R} = \frac{305}{36} = 8.47$$

$$\text{Limits}_R = D_3\bar{R}, D_4\bar{R} = 0(8.47), 2.575(8.47)$$
$$= \underline{\quad}, 21.8$$

The trouble lies with the make-up of the stratified (row) samples. In each sample, there is one piece from each spindle. Thus such a sample's range will contain not only chance variation (or causes) but also any mean differences between spindles (assignable causes). By looking at the x's in each sample, we see that, with very few exceptions, spindle one contributes the maximum x and spindle three the minimum x for R = max x - min x. Thus $\bar{R}$ has been *inflated by spindle-to-spindle differences.*

TABLE 8.1. Data Illustrative of Sampling a Three-Spindle Automatic, with Specifications on Outside Diameter of .2010 ± .0012 in. Recorded in .0001 in. from .2010 in. Stratified Samples in Rows, One from Each Spindle. Rational Samples in Columns.

Sample number	x_1	x_2	x_3	$\bar{x}$	R	Sample number	x_1	x_2	x_3	$\bar{x}$	R
1	+5	+2	-3	+1.3	8	19	+4	+1	-6	-.3	10
2	+5	-2	-1	+.7	7	20	+1	0	-3	-.7	4
3	+3	-4	-3	-1.3	7	21	+5	0	-7	-.7	12
$\bar{x}$	+4.3	-1.3	-2.3			$\bar{x}$	+3.3	+.3	-5.3		
R	2	6	2			R	4	1	4		
4	+7	0	-5	+.7	12	22	+5	+1	-4	+.7	9
5	+5	+1	-6	0	11	23	+3	0	-3	0	6
6	+7	0	-4	+1.0	11	24	+4	-1	-4	-.3	8
$\bar{x}$	+6.3	+.3	-5.0			$\bar{x}$	+4.0	0	-3.7		
R	2	1	2			R	2	2	1		
7	+4	-1	-2	+.3	6	25	+5	+2	-5	+.7	10
8	+3	0	-3	0	6	26	+2	0	-4	-.7	6
9	+5	-3	-8	-2.0	13	27	+6	-3	-2	+.3	9
$\bar{x}$	+4.0	-1.3	-4.3			$\bar{x}$	+4.3	-.3	-3.7		
R	2	3	6			R	4	5	3		
10	+3	-1	-4	-.7	7	28	-1	+1	-4	-1.3	5
11	+3	+2	-5	0	8	29	+2	+3	-5	0	8
12	+9	0	-5	+1.3	14	30	+6	+1	-3	+1.3	9
$\bar{x}$	+5.0	+.3	-4.7			$\bar{x}$	+2.3	+1.7	-4.0		
R	6	3	1			R	7	2	2		
13	+1	+1	-4	-.7	5	31	+6	-1	-4	+.3	10
14	+5	+2	-3	+1.3	8	32	+3	-1	-2	0	5
15	+5	+3	-5	+1.0	10	33	+4	-2	-5	-1.0	9
$\bar{x}$	+3.7	+2.0	-4.0			$\bar{x}$	+4.3	-1.3	-3.7		
R	4	2	2			R	3	1	3		
16	+2	-2	-3	-1.0	5	34	+5	-2	-4	-.3	9
17	+2	+3	-3	+.7	6	35	+6	0	-4	+.7	10
18	+3	+3	-8	-.7	11	36	+6	0	-5	+.3	11
$\bar{x}$	+2.3	+1.3	-4.7			$\bar{x}$	+5.7	-.7	-4.3		
R	1	5	5			R	1	2	1		
						$\Sigma\bar{x}$	+49.5	+1.0	-49.7	+.9	
						ΣR	38	33	32	103	305

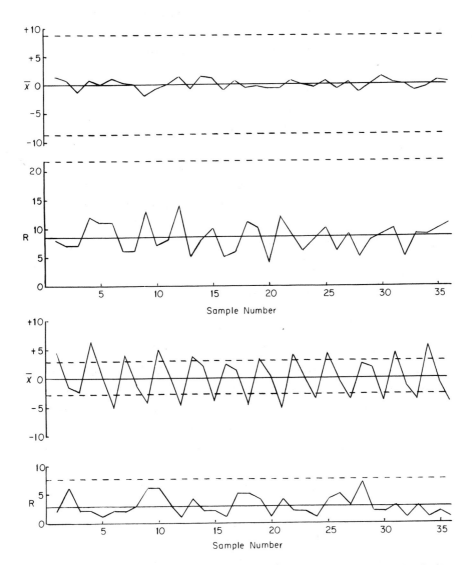

Fig. 8.1. Control charts for outside diameters from a three-spindle automatic. Upper x̄ and R charts are for stratified samples: one piece from each spindle. Lower x̄ and R charts have n = 3 from spindle 1 plotted as sample 1, then three from spindle 2, and so on, in rotation. These are two different ways to plot the basic data of Table 8.1, giving radically different pictures.

This gives the unreasonably wide control bands we see in the upper part of Fig. 8.1.

Now consider another way of analyzing these same results. Taking the first three x_1 values as a sample to represent spindle one, the first three x_2 values to represent spindle two, and so on, we have the *column* samples as shown in Table 8.1. These are plotted in rotation: spindles one, two, and three, then the second sample of spindle one, and so on. When plotted, this gives the lower $\bar{x}$ chart of Fig. 8.1. We immediately note the regular cycles with all $\bar{x}_1$ values high, all $\bar{x}_3$ values low, and the $\bar{x}_2$ values intermediate. Meanwhile the R chart looks like a reasonable random control chart showing perfect control, but not *suspiciously* good control. The calculations for control lines follow:

$$\ell_{\bar{x}} = \bar{\bar{x}} = \frac{.8}{36} = .02$$

$$\text{Limits}_{\bar{x}} = .02 \pm 1.023(2.86) = -2.91, +2.95$$

$$\ell_R = \bar{R} = \frac{103}{36} = 2.86$$

$$\text{Limits}_R = 0(2.86), 2.575(2.86) = \underline{\quad}, 7.36$$

Although the two $\bar{\bar{x}}$ values are identical and must be except for round-off errors, the new $\bar{R}$ is only about a third of that for the stratified samples. This is because the rational samples contained only chance variation within a single spindle, whereas the stratified samples also contained spindle-to-spindle differences. Thus the R's were inflated in the stratified samples.

Again look at the $\bar{x}$ chart for the rational samples. All but two of the 12 $\bar{x}_1$ values are above the upper control limit, and all but one of the $\bar{x}_3$ values are below the lower control limit. Yet $\bar{\bar{x}}$ is practically "on the beam" at .02. Clearly spindles one and three need adjustment. By how much? For this we find

$$\bar{\bar{x}}_1 = \frac{+4.95}{12} = +4.12 \qquad \bar{\bar{x}}_3 = \frac{-49.7}{12} = -4.14$$

so the two spindles need to be adjusted down and up, respectively, by about .0004 in. Once this is accomplished, what specification

limits can be met? Proceeding as usual for limits for individual
x's, since the R values are in control:

$$\sigma_x = \frac{\bar{R}}{d_2} = \frac{2.86}{1.693} = 1.69$$

$$\text{Limits}_x = .02 \pm 3(1.69) = .02 \pm 5.07 = -5.05, +5.09$$

Thus by proper adjustment of the spindles the process can be made
to meet specifications of ±5, that is, .2005, .2015 in.

Now what really were the populations being used for this experiment? They were from Table 7.3, respectively, Populations E, A,
and G with averages μ: +4, 0, -4, but each with $\sigma = 1.715$. These
rather radically different populations were used to make the above
concepts perfectly clear. We could also have used, say, B, A, and
F; or, say, three of A, one each of B and F.

Another possible way to plot the rational samples for the spindle is to plot the 12 samples for spindle one in separate $\bar{x}$ and R
charts, then those for spindle two, and finally for spindle three.
Then control lines can be figured for each spindle *separately* and
compared.

Still another plotting method is called the "group control
chart." It is especially useful when we have several sources, perhaps many, and do not wish to plot all values of $\bar{x}$ and R. We may
then take a small sample of n = 2, over some short time period, from
each spindle or *source*, figuring $\bar{x}$ and R for the sample for each
source. Then instead of plotting all $\bar{x}$'s, only the maximum $\bar{x}$ and
the minimum $\bar{x}$ are plotted for each time period. Then a small number indicating the source is placed by the maximum $\bar{x}$ point and the
number of the source contributing the minimum $\bar{x}$ by its point; similarly, for the R's at each time period. Then after, say, 20 time
periods, one can look along the high points to see whether any one
source is contributing disproportionately to the high $\bar{x}$'s or the
low ones. Also the same thing is done for the R's. Control limits
are set from *all* of the $\bar{x}$'s and R's, and, of course, out-of-control
points carry their usual meaning.

A friend of the author had decided to study a 14 station machine

doing 14 different jobs simultaneously on castings, such as, drill-
ing holes and facing. Then the machine would move each casting in
place for the next job. He was gathering data on one job, a facing.
After completing about four rows of the 14 station measurements in
a table, the foreman took notice. After explaining what he was doing, the foreman looked over the data, and at once said "Station 6
has to be jacked up and 11 lowered." Such adjustments were clearly
apparent even on such meager data. This illustrates the power of
gathering data in a rational manner, even without the statistical
analysis.

8.1.3 *Charts from Mixed Product*

Product from multispindle automatic machines, multicavity molds, or
other multiple sources is often dumped onto a conveyor or in a tote
pan. Thus the pieces from the various spindles or cavities get more
or less completely mixed up, and when we draw a sample, we can no
longer expect one from each source in the sample. Moreover, we cannot trace back the source of each piece. What happens then? Let us
see. Again we run an experiment. In order to tie in this type of
sampling with the previous experiment, we took one set of each of
Populations A, E, and G and thoroughly mixed the 600 different-col-
ored numbered beads. Then samples, each of n = 3, were drawn giving
the data shown in Table 8.2. Care was taken to not look in the box
while a sample of 3 was drawn, in order to avoid conscious or uncon-
scious bias. Note that we cannot in industrial practice trace the
pieces back to their source, unless some distinguishing mark is made
on the piece by the process. Also it is worth mentioning again that
we cannot expect one from each of the three spindles in a sample.
(In fact the probability for such a sample here turns out to be
only .223.)

Thirty-six samples of mixed product are given in Table 8.2,
and $\bar{X}$ and R are plotted in Fig. 8.2. The calculation of control
lines with n = 3 follow as usual.

$$\pounds_{\bar{x}} = \bar{\bar{x}} = \frac{-14.9}{36} = -.41$$

TABLE 8.2. Data Illustrative of Sampling Mixed Product from a Three-Spindle Automatic, with Specifications on Outside Diameter of .2010 ± .0012 in. Recorded in .0001 in. from .2010 in.

Sample	x_1	x_2	x_3	$\bar{x}$	R	Sample	x_1	x_2	x_3	$\bar{x}$	R
1	-8	+1	-4	-3.7	9	19	-3	+5	0	+.7	8
2	+4	-3	+3	+1.3	7	20	+3	-4	-4	-1.7	7
3	0	+3	0	+1.0	3	21	-3	+3	-6	-2.0	9
4	-5	-2	-1	-2.7	4	22	+5	-1	+5	+3.0	6
5	+3	-5	-1	-1.0	8	23	-2	+2	0	0	4
6	+4	+2	-4	+.7	8	24	+2	+2	-6	-.7	8
7	-2	+1	-5	-2.0	6	25	-2	+3	-3	-.7	6
8	+2	-1	-3	-.7	5	26	+3	-4	-2	-1.0	7
9	+3	+3	-2	+1.3	5	27	+2	-6	+2	-.7	8
10	-1	-5	+3	-1.0	8	28	+2	+4	-7	-.3	11
11	+5	-5	-2	-.7	10	29	+1	-5	+7	+1.0	12
12	-3	+3	-6	-2.0	9	30	+5	-5	+4	+1.3	10
13	-1	+7	-5	+.3	12	31	-4	+4	+1	+.3	8
14	-6	-3	-1	-3.3	5	32	-1	0	-4	-1.7	4
15	-5	+1	+1	-1.0	6	33	+1	-4	-4	-2.3	5
16	-4	+3	+4	+1.0	8	34	-5	0	+1	-1.3	6
17	+5	+4	+1	+3.3	4	35	-5	+1	-6	-3.3	7
18	+1	-1	+6	+2.0	7	36	+3	-3	+5	+1.7	8
										-14.9	258

$$\text{Limits}_{\bar{x}} = -.41 \pm 1.023(7.17) = -7.74, +6.92$$

$$\mathcal{E}_R = \bar{R} = \frac{258}{36} = 7.17$$

$$\text{Limits}_R = 0(7.17), \ 2.575(7.17) = \underline{}, \ 18.5$$

Plotting these control lines, we find both charts in perfect control. And this is despite the fact that the pieces from which the samples were drawn came from three quite distinct populations. The reason is that we were drawing in reality from just one population,

namely, the compound population made up of the three separate populations (one population from each spindle).

Now what do these control charts of Fig. 8.2 tell us about
meeting specifications of ±12 (in the coded units)? As usual, when
we have good control, we set limits for the individual x's:

$$\sigma_x = \frac{\bar{R}}{d_2} = \frac{7.17}{1.693} = 4.24$$

$$\text{Limits}_x = \bar{\bar{x}} \pm 3\sigma_x = -.41 \pm 3(4.24) = -13.13, +12.31$$

Since these limits for x lie approximately on the specification limits ±12, we are indeed meeting specifications quite well. Thus with
such lenient limits, it would not be necessary to examine more deeply for spindle-to-spindle differences.

The population from which we drew the samples of Table 8.2
was the following composite of Populations A, E, and G of Table 7.3:

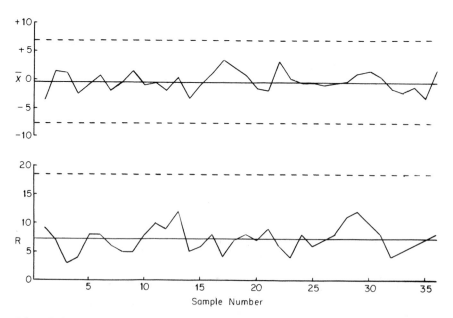

Fig. 8.2. Samples from mixed product from three spindles. Outside
diameter to meet specifications of .2010 ± .0012 in. Data from
Table 8.2 and recorded in .0001 in. from .2010 in.

x	-9	-8	-7	-6	-5	-4	-3	-2	-1	0
f	1	3	10	23	40	51	49	46	50	54

x	+1	+2	+3	+4	+5	+6	+7	+8	+9	
f	50	46	49	51	40	23	10	3	1	600

This is not a "normal" population, being (irregularly) flat-topped, but it is the population we drew from. Its nonnormality caused the x limits calculated to be wider than the actual population limits of ±9.

But now suppose that specifications were not ±12, but instead ±5. Then we would examine the process in more detail for differences between spindles by taking rational samples *separately* from each spindle. Then, as we have seen, the process could meet limits of ±5, provided each spindle is set up properly to have an average close to the nominal 0 (.2010 in.).

Let us give a couple of examples to touch up this subsection. One was a problem worked on by Holbrook Working, who was one of those who developed and taught the World War II short courses in statistical quality control. In a plant visit, the foreman said that they were having trouble with a production process. Almost everyone blamed the raw material (an all-too-common tendency). Working examined the operation and then said, "It can't be the raw material." His reason was that the way this material came into the shop was such that it was very well mixed. Thus any charts would show good control because of the mixing, even from quite an out-of-control process producing the raw material. But they *were* finding charts out of control on the product, following the two supposedly identical production machines. Then the foreman admitted that they did have some misgivings on one of them. Additional carefully taken samples from each machine separately enabled them to find the trouble. The whole affair was a matter of an hour or so, whereas the trouble had been occurring for quite some time.

In another case, a man came back from a short course and made a lot of control charts in a sheet and tin mill associated with a steel plant. His charts all showed good control! (It is very rare

in an initial study to find good control immediately.) The super-
intendent, foreman, and others were very skeptical. About this time,
the man was taken in the draft! Examining the charts and data indicated that the man had taken the cards for the process *in random
order*. This of course causes the charts to show control, despite
noncontrolled production conditions.

8.2 TOOL WEAR, SLANTING LIMITS

In this section, we describe a variation of control charting which
can prove of much use when the tolerance T = U - L for individual x
values is considerably greater than $6\sigma_x$, where σ_x is the short-term
standard deviation of pieces or product produced over a short time
interval. Under these conditions, there is some latitude within
which we may let the process average μ vary and still be meeting the
specifications. The techniques we are about to describe are especially useful when μ tends to shift gradually. For example, we may
find that because of tool wear, the average *outside* diameter of the
parts gradually increases, or the average *inside* diameter gradually
decreases. Other examples of drifts are (1) grinder wear, (2) pol-
isher wear, (3) viscosity, (4) temperature as of room conditions,
(5) human fatigue, (6) moisture in lumber, (7) surface finish, and
(8) die wear.

Let us consider for an example the data on an outside diameter
of a spacer as given in Table 8.3. The quality control man was considerably confused by this operation when he applied the usual analysis of a preliminary run to the data, just preceding those in Table
8.3. The x̄ chart showed very bad control, but there seemed to be
cycles, and the spacers *were* meeting specifications. Then he began
to realize that the process mean could be permitted to drift somewhat.

Suppose we analyze the data of Table 8.3. Look first at the R
chart.

$$\bar{R} = \frac{133}{44} = 3.02$$

$$UCL_R = D_4\bar{R} = 2.115(3.02) = 6.39$$

TABLE 8.3. Department: Automatic; Sampling interval: 30 min; Sample size: 5; Part: Spacer; Specifications: Outside diameter = .1250 +.0000, -.0015 in.; Recorded in .0001 in. from .1250 in.

Sample number	x_1	x_2	x_3	x_4	x_5	$\bar{x}$	R
1	-10	-3	-9	-2	-3	-5.4	8
2	-12	-13	-13	-11	-10	-11.8	3
3	-12	-8	-13	-13	-11	-11.4	5
4	-11	-10	-8	-12	-11	-10.4	4
5	-11	-10	-13	-10	-12	-11.2	3
6	-11	-9	-13	-11	-13	-11.4	4
7	-10	-12	-11	-10	-9	-10.4	3
8	-12	-12	-11	-8	-11	-10.8	4
9	-8	-10	-9	-10	-11	-9.6	3
10	-11	-10	-8	-9	-8	-9.2	3
11	-10	-8	-10	-9	-9	-9.2	2
12	-9	-10	-9	-8	-10	-9.2	2
13	-10	-10	-8	-9	-8	-9.0	2
14	-8	-7	-9	-9	-8	-8.2	2
15	-7	-7	-6	-8	-10	-7.6	4
16	-7	-6	-9	-8	-7	-7.4	3
17	-6	-6	-7	-7	-8	-6.8	2
18	-4	-5	-6	-7	-4	-5.2	3
19	-4	-2	-4	-3	-3	-3.2	2
20	-11	-2	-10	-9	-10	-8.4	9
21	-10	-9	-8	-11	-9	-9.4	3
22	-11	-10	-11	-9	-9	-10.0	2
23	-9	-7	-11	-11	-9	-9.4	4
24	-9	-7	-9	-10	-8	-8.6	3
25	-8	-10	-9	-8	-7	-8.4	3
26	-10	-11	-8	-8	-7	-8.8	4
27	-7	-8	-8	-7	-6	-7.2	2
28	-6	-8	-7	-9	-9	-7.8	3
29	-6	-7	-7	-8	-8	-7.2	2
30	-7	-9	-5	-7	-7	-7.0	4
31	-5	-7	-7	-6	-5	-6.0	2
32	-5	-8	-7	-6	-7	-6.6	3
33	-7	-6	-8	-8	-7	-7.2	2
34	-8	-8	-7	-9	-7	-7.8	2
35	-7	-8	-7	-6	-7	-7.0	2
36	-6	-5	-6	-5	-5	-5.4	1
37	-5	-8	-5	-6	-4	-5.6	4
38	-5	-6	-4	-6	-6	-5.4	2
39	-4	-4	-5	-5	-4	-4.4	1
40	-11	-12	-10	-11	-12	-11.2	2
41	-9	-8	-11	-10	-10	-9.6	3
42	-9	-9	-8	-10	-8	-8.8	2
43	-7	-8	-9	-10	-8	-8.4	3
44	-8	-9	-8	-7	-10	-8.4	3
						-361.4	133

Drawing in these control lines shows two points clearly out of control. What is the cause? Looking at the make-up of these two samples of five x values, we can see the presence of two quite distinct
levels of x. Looking now also at the $\bar{x}$ graph, it is seen that the
high ranges occur at or just after a peak on the $\bar{x}$'s. This then
looks like a case of resetting the process average downward because
it is getting too close to the upper specification. Moreover, the
high-diameter spacers produced before the resetting and the smaller
diameter pieces produced after the resetting were both in the tote
pan, and some of each were included in each of these two samples.
This will give a high range inflated by the adjustment. Such a
range does not measure chance variation only. The appropriate action is obvious, namely, to arrange to have kept separate the pieces
produced before a resetting from those produced afterward. If this
action is carried out, then the two high ranges become no longer
typical of the process, and can be discarded. Hence we now revise $\bar{R}$

$$\bar{R} = \frac{133 - 8 - 9}{42} = 2.76$$

$$UCL_R = 2.115(2.76) = 5.84$$

Now the remaining 42 R values are all below this limit, and thus
control is perfect. We are, therefore, in a good position to estimate σ_x for the process by (7.10).

$$\sigma_x = \frac{\bar{R}}{d_2} = \frac{2.76}{2.326} = 1.19$$

Comparing this with the tolerance = U - L = 15 we find that the
tolerance is 15/1.19 = 12.6 standard deviations. Thus, there is a
considerable amount of latitude or permissible variation within
which we can permit the process average μ to vary.

Now *suppose* we were to set the *usual* $\bar{x}$ limits. We have (omitting the two $\bar{x}$ values for samples with the high R's):

$$\bar{\bar{x}} = \frac{-361.4 + 5.4 + 8.4}{42} = -8.28$$

$$Limits_{\bar{x}} = -8.28 \pm .577(2.76) = -9.87, -6.69$$

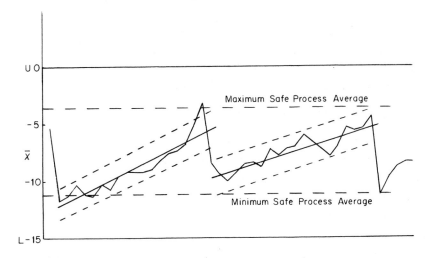

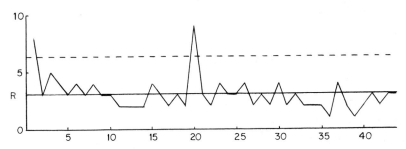

Fig. 8.3. Charts for $\bar{x}$ and R from data on spacers in Table 8.3.
The $\bar{x}$'s show tool wear and were analyzed by trend lines for μ and
parallel control limit lines. Also shown are maximum and minimum
safe process average lines for guidance as to where to start and
end a run.

A great many $\bar{x}$'s are outside of these limits, occurring at the beginning of a run upwards and also at the end. Of course, the acid
test is whether spacers outside of the specifications have been or
are being produced.

A good way to analyze the process is to draw by eye a "trend
line" following the $\bar{x}$ points. That is, we try to picture the way
the true process average grows when we disregard the chance variations of the $\bar{x}$'s around this trend. We have drawn in two such trend
lines for μ in Fig. 8.3. (This was done by eye, but there are

objective mathematical methods. Also the trend could be a curve.)

Now how much variation of the $\bar{x}$'s around this trend for μ is only to be expected? As usual we use $3\sigma_{\bar{x}}$ limits, *measured vertically* from the estimated trend line. This is found by

$$\pm A_2\bar{R} = \pm .577(2.76) = \pm 1.59$$

We therefore draw slanting control limit lines, above and below the trend lines, by a *vertical* distance of 1.59. In the first run, the last point indicates an assignable cause of rapid rise in μ and should be investigated. It was noticed, however, and resetting was done promptly. The tool wear in the second run was in control, and at a slightly slower rate than for the first run.

We now ask the useful questions as to what level μ we should aim at when beginning a run, and to what level we should let μ rise before resetting. For these, we find the following two levels from our estimated standard deviation σ_x for individual x's.

Maximum safe process average = U - $3\sigma_x$ (8.1)

Minimum safe process average = L + $3\sigma_x$ (8.2)

(Use of $3\sigma_x$ is quite conservative, since only about .1% would be out at these levels. We could use $2.5\sigma_x$ if we can permit .5% to be out for a short time.) In the example since σ_x = 1.19, these give

Maximum safe process average = 0 - 3(1.19) = -3.57

Minimum safe process average = -15 + 3(1.19) = -11.43

Then our objective, in order to maximize the length of a run, is to aim the initial setting at -11.4, and then to try to reset when the *trend line* for μ hits -3.6. Note carefully that when μ is at -11.4, a $\bar{x}$ point can well be below -11.4, in fact half of them will be. And similarly when μ is at -3.6, half of the $\bar{x}$'s will lie above -3.6. Thus we watch the *trend* for μ, much more than the separate $\bar{x}$ points. Reset when it reaches -3.6.

Using these as guide lines, the first run seems to have been started about right or perhaps a little low. But for the out-of-control run-out point, the run might have been permitted to continue longer. The second run seems to have been started high and ended too soon.

In one division of a corporation manufacturing farm implements, the management said that die life had been doubled by use of trend charts. In one example cited, the time between downtimes was increased from 4 to 11 days, by investigating points out of the slanting control lines.

Commonly the trend line for μ is not available until several $\bar{x}$ points are plotted. Then the line can be sketched in. However, if the slopes of the trends appear to be quite constant, it may become possible to draw in a tentative trend and slanting control limits quite early.

8.3 CHARTS FOR INDIVIDUAL x's AND MOVING RANGES

Normally when we wish to analyze a process by control charts for measurements, we take a preliminary run of some 100 x values divided into rational samples, within which conditions are supposed to be basically identical. But between samples, conditions may vary. The 100 x values may yield 20 samples of n = 5, or 25 of n = 4. However, if the individual measurements occur at some sizable time gap, perhaps 1 hr, 4 hr, a shift, or a day, then it is more hazardous to assume that conditions have remained constant, and moreover it may take a very long time to accumulate 100 measurements. Another point too is that the measurements may be quite costly, and we cannot afford to take very many. It then becomes desirable to obtain as much information as we can from rather meager data in a time or serial order of production. The methods of this section provide one approach to analyzing such data. This is the method of control charts for individual x's and for "moving ranges."

Let us consider the data of Table 8.4. These are daily percent of steel rejections in a plate mill. They are daily figures for the first 27 days of June, at which time the mill was shut down for general repairs. The moving ranges are the numerical differences between each set of two *consecutive* x's. Thus, using vertical bars

TABLE 8.4. Percent Steel Rejections in a Plate Mill for
27 Days in June. Measurements x and Moving Ranges Shown

Date	x	R	Date	x	R
1	3.69		14	2.30	
		.16			.79
2	3.53		15	3.09	
		.92			.63
3	2.61		16	3.72	
		.25			.71
4	2.36		17	4.43	
		2.00			1.75
5	4.36		18	2.68	
		2.78			.04
6	1.58		19	2.64	
		2.45			2.19
7	4.03		20	4.83	
		.99			.47
8	3.04		21	4.36	
		1.09			1.25
9	4.13		22	5.61	
		2.31			.46
10	1.82		23	6.07	
		.75			4.02
11	1.07		24	2.05	
		2.08			3.14
12	3.15		25	5.19	
		.95			1.39
13	4.10		26	3.80	
		1.80			3.85
			27	7.65	
				97.89	39.22

for the absolute value,

$$R_1 = |x_1 - x_2|, \quad R_2 = |x_2 - x_3|, \quad R_3 = |x_3 - x_4|, \quad \ldots,$$
$$R_{n-1} = |x_{n-1} - x_n| \tag{8.3}$$

This yields only n - 1 moving ranges out of n x values. In Table
8.4,

$$R_1 = |3.69 - 3.53| = .16, \quad \ldots, \quad R_{26} = |3.80 - 7.65| = 3.85$$

One must be very careful in the subtraction. Also note that the
ranges are recorded between the x's from which they were found.

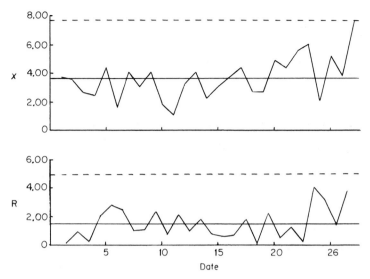

Fig. 8.4. Control charts for individual x values and moving ranges R for daily percent steel rejections. Data from Table 8.4. Moving ranges plotted between the two associated x values.

We next plot the 27 x values as shown in Fig. 8.4, one for each date. But in plotting the moving ranges, we plot R_1 half-way between June 1 and 2, because, in time, it is related to both dates. This gives the 26 R points as shown in the figure.

For central lines we use as usual the averages

$$\bar{x} = \frac{\Sigma\ x}{n} = \frac{97.89}{27} = 3.63$$

$$\bar{R} = \frac{\Sigma\ R}{n - 1} = \frac{39.22}{26} = 1.51$$

For the moving range chart

$$UCL_R = D_4\bar{R} = 3.267(1.51) = 4.93$$

All ranges lie below this limit, and so no assignable cause is in-dicated. (Such an assignable cause if found on the moving range chart would be for an exceptionally large jump or drop between x's.)

Next the limits for the individual x's are taken from σ_x, as estimated by $\bar{R}/d_2$. [See (7.10).] Here using d_2 for n = 2

$$\sigma_x = \frac{\bar{R}}{d_2} = \frac{1.51}{1.128} = 1.34$$

Then 3σ control limits for x's are

$$\bar{x} \pm 3\sigma_x = \bar{x} \pm \frac{3\bar{R}}{d_2} \qquad (3\sigma \text{ x limits}) \qquad\qquad (8.4)$$

In our case we obtain from this

$$\text{Limits}_x = 3.63 \pm 3(1.34) = \underline{\quad}, \; 7.65$$

the lower limit being negative, and thus meaningless for percent rejection. The last x is on the control limit. This high point is
probably associated with the mill being shut down after the 27th.

Notice particularly that we did not use $\bar{x} \pm A_2\bar{R}$, which are
limits for $\bar{x}$ of n = 2 each, which we do not have here.

Now there is one variation some workers use, the author being
one of them. This variation uses 2σ limits for x. The reason is
that with such meager data available we must try to get "something"
out of it. So we deliberately increase the *risk* of thinking an
assignable cause is present when *none* is, so as to *decrease* the
chance of missing a signal when an assignable cause *is* present.
Then we have

$$\bar{x} \pm 2\sigma_x = \bar{x} \pm \frac{2\bar{R}}{d_2} \qquad (2\sigma \text{ x limits}) \qquad\qquad (8.5)$$

and in our example this gives

$$\text{Limits}_x = 3.63 \pm 2(1.34) = .95, \; 6.31$$

It happens here that these limits do not exclude any additional x
points.

8.4   PERCENT DEFECTIVE OF BULK PRODUCT

There are some types of percent or fraction defective data which
should not be analyzed by a fraction defective chart. For example,
consider percent spoilage of bulk product, say, paint. We might
have

Batch	Volume (gal)	Spoilage	Percent
1	20,000	400	2.00
2	18,000	390	2.17
3	20,000	420	2.10
4	21,000	380	1.81

Now why cannot the last column be analyzed as ordinary percent defective data (or converted to fraction defective data)? The reason
is that we have not taken 20,000 1-gal cans of paint and declared
each one either good or defective. If this were the case, then n
would be 20,000. But instead the defective paint might be skimmed
off the top of a vat or strained out. (The author is no paint manufacturer.) Thus there is no natural sample size. Furthermore, if
we used pints instead of gallons, the first sample size would seem
to be 160,000 instead of 20,000. Thus we cannot use a p chart for
analysis.

But we can treat the series of percents of spoilage as measurements x. Then, if we have quite a few in relatively short time
periods, we can use $\bar{x}$ and R charts. Or if the percents available
are few in number and/or relatively far apart in time, we can use
x and moving range charts, as in the preceding section.

Another case where we do not use the p chart approach on fraction defective data is that in which we have overall data on total
production. Thus for an organization making piston rings, a week's
production may be 4,000,000. Suppose that at final inspection the
average fraction defective is about .02. Then what would be the
standard control limits?

$$\bar{p} \pm 3\sqrt{\frac{\bar{p}(1 - \bar{p})}{n}} = .02 \pm 3\sqrt{\frac{.02(.98)}{4,000,000}} = .02 \pm .0002$$

$$= .0198, .0202$$

Probably *no* weekly figure will lie between such exceedingly narrow
limits! The reason is that so many things can happen over all lines
and types of rings in a week that fluctuations in p will be much
greater. In this case again, we can treat such fractions as

measurements x, and use an x chart.

One other case where fraction defective data are not appropriately analyzed by a p chart is that in which we make moldings,
stampings, plastic parts, rubber molded parts, or stacks of small
metal castings. The reason is that the occurrence of defective
pieces is not independent. Thus if one piece is defective, it is
more likely that others near it are defective too, because of defect-producing conditions. Or if a good one occurs, this tends to
mean that conditions were adequate and we have a bit *lower* probability of defectives near it. The binomial distribution is not
applicable. Again we can take such a fraction defective and treat
as an x measurement.

8.5* AVERAGE RUN LENGTH FOR A POINT OUT

An important technique for comparing control charts as to their efficiency is to determine the *average run length* of pieces measured
or gauged until a point outside of the limits is found. In other
words, we are interested in the number of samples, *on the average*,
which we must observe before a point goes outside of the control
limits. There are two sides to this matter. On the one hand, if
the process remains in control relative to the standard distribution assumed, we would like to have a long average run length (ARL);
the longer the better. But if the standard distribution changes by
a significant amount, we would like to have a quick warning of this
change, that is, a short ARL. To be truly comparable, two charts
should have the same ARL when the process is in control. Then we
can meaningfully compare the ARL's when the standard distribution
changes by some amount of interest, that is, goes out of control by
some amount. The shorter the ARL the better, under the changed
condition.

To determine the ARL for a chart, we need to know the probability of a point, representing a sample of n, to be outside of the
control band. Then calling this probability P, we have

$$\text{ARL} = \frac{n}{P} \qquad\qquad (8.6)$$

for the average *number of pieces* until a warning occurs.

Thus for example if we have a c chart with c' = 3, the control limits are

$$c' \pm 3\sqrt{c'} = 3 \pm 3\sqrt{3} = \underline{\quad}, \quad 8.2$$

Now in Table B, given c' = 3, P(8 or less) = .996, so that P(point out) = P(9 or more) = .004. Then by (8.6)

$$ARL = \frac{1}{.004} = 250$$

So, *on the average*, there will be 250 samples of one piece before a false alarm occurs by a point outside. This is quite a comfortable ARL.

Now suppose that we continue to use the upper control limit of 8.2 but suddenly production conditions change from the standard c' = 3 to c' = 6. Again using Table B with c' = 6, we find

$$\begin{aligned} P(\text{point out}) &= P(9 \text{ or more}) \\ &= 1 - P(8 \text{ or less}) \\ &= 1 - .847 \\ &= .153 \end{aligned}$$

Then using (8.6)

$$ARL = \frac{1}{.153} = 6.5$$

This may or may not be short enough to suit us. If not, then we might try taking three pieces of the product, upon which we will count the total defects. Now, however, c' = 9 (for the count on three), which give control limits of $9 \pm 3\sqrt{9} = 0, 18$. Given c' = 9,

$$\begin{aligned} P(\text{point out}) &= P(0) + 1 - P(17 \text{ or less}) \\ &= .000 + 1 - .995 = .005 \end{aligned}$$

$$ARL = \frac{3}{.005} = 600$$

Comparable to a jump to c' = 6 on *one* piece is a jump to 18 defects on *three*.

Given c' = 18,

$$P(\text{point out}) = P(0) + 1 - P(17 \text{ or less})$$
$$= .000 + 1 - .469$$
$$= .531$$

$$\text{ARL} = \frac{3}{.531} = 5.6$$

Comparing, we see that we have considerably improved the length until a false alarm when in control, but have not obtained much improvement of an ARL when the one-piece c' goes from 3 to 6. To obtain a more comparable ARL under control, we can try an upper control limit, for the count on three pieces, of 17 or 16. (These would not be 3σ limits, but might be desirable in balancing risks.)

Moreover, we can consider the option of $\bar{x}$ and R charts versus a p chart. The former requires numerical measurements, whereas the latter merely requires a determination as to whether the measurement for the piece does, or does not, lie between the specification limits. For example, this can be done by a "go-not-go" gauge. Thus, suppose that a production process is just comfortably able to meet the limits L to U with a normal distribution. That is, $L = \mu - 3\sigma_x$, $U = \mu + 3\sigma_x$. Then by Table A, since L corresponds to $z = -3$, and U to $z = +3$, the fraction defective when in control is .0026. Now for 3σ limits on the $\bar{x}$ chart, the probability for a false alarm of a point out while still in control is .0026 for all sample sizes. Thus the ARL is

$$\text{ARL} = \frac{n}{.0026} = 385n$$

Now suppose that the process mean should jump up to $\mu + 1\sigma_x$. Then the z values for L and U for individual x's are

$$z_L = \frac{\mu - 3\sigma - (\mu + \sigma)}{\sigma} = -4$$

$$z_U = \frac{\mu + 3\sigma - (\mu + \sigma)}{\sigma} = +2$$

Using Table A,

$$P(-4 < z \leq +2) = P(z \leq +2) - P(z \leq -4)$$
$$= .9772 - .0000 = .9772$$

and so the fraction defective under the changed condition is
1 - .9772 = .0228.

We may now try a sample size for $\bar{x}$ and a sample size for a p
chart. For the latter, given n, we can use p' = .0026 and find the
probability of a false-alarm point out. Then do the same when p'
jumps to .0228. These will give ARL's by (8.6).

Likewise with a chosen n for $\bar{x}$ values, the ARL for a false
alarm is 385n, as given above. And for the ARL when μ jumps up by
σ_x we have the control limits $\mu \pm (3\sigma_x/\sqrt{n})$

$$z = \frac{\mu + (3\sigma/\sqrt{n}) - (\mu + \sigma)}{\sigma/\sqrt{n}} = \frac{3/\sqrt{n} - 1}{1/\sqrt{n}} = 3 - \sqrt{n}$$

$$z = \frac{\mu - (3\sigma/\sqrt{n}) - (\mu + \sigma)}{\sigma/\sqrt{n}} = \frac{-3/\sqrt{n} - 1}{1/\sqrt{n}} = -3 - \sqrt{n}$$

If n = 4, then z = 1, z = -5. Neglecting z = -5, the probability
above z = 1 for the upper control limit for $\bar{x}$ is .1587 from Table A.
This is for a warning when μ jumps up by $1\sigma_x$. Then (8.6) gives

$$\frac{4}{.1587} = 25$$

So for $\bar{x}$ (n = 4), the two desired ARL's are 385(4) = 1540 and 25.

Now compare with a p chart and n = 250. When in control, p'
= .0026 and np' = 250(.0026) = .65 for the central line, and by (6.7)

$$\text{Limits}_{np} = .65 \pm 3\sqrt{.65(.9974)}$$
$$= .65 \pm 2.42$$
$$= \underline{}, 3.07$$

So 4 is out of control.

Now if the process is still in control, we have np' = .65 and
use Table B for

$$P(\text{point out}) = P(4 \text{ or more, given } .65)$$
$$= 1 - P(3 \text{ or less})$$
$$= 1 - .996 = .004$$

This gives

$$\text{ARL} = \frac{250}{.004} = 62,500$$

Now if p' jumps to .0228, by the assumed process change, we have
np' = 250(.0228) = 5.7. Now when we *expect* 5.7, we have

$$P(\text{point out}) = P(4 \text{ or more, given } 5.7)$$
$$= 1 - P(3 \text{ or less})$$
$$= 1 - .180 = .820$$

Thus by (8.6)

$$ARL = \frac{250}{.820} = 305$$

The working out of specific plans to maximize efficiency at
some particular out-of-control condition does not seem to the author
to be very fruitful, because we cannot expect a process to be either
in control, or else out by the one specified amount. Nevertheless,
the concept of average run length is of much importance.

One example of p charts versus measurement charts was one where
the head-to-shoulder length of cartridge cases was the characteristic in question. Originally snap gauges were used giving an attribute decision on each piece: either within or outside of limits.
Samples of n = 10,000 were used because even in this many, only 1 or
2 would be rejects. But even this few was undesirable. Little progress was made because too much could happen while 10,000 were being
produced. So a method of *measurement* was developed. Then with samples of n = 4, $\bar{x}$ and R charts permitted much better control, and
causes for off-length were found and eliminated. And fewer defectives were *produced* than formerly had been slipping past 100%
inspection.

8.6* CHART FOR DEMERITS, RATING QUALITY

A method of rating product quality for more or less complicated
product, such as electronic assemblies, was developed by Dodge [1].
(See also Ref. 2.) The method can, however, be used for any product
which may contain defects of varying degrees of seriousness.

The plan involves giving demerit points for each defect observed in a unit of product. All possible defects are listed in

classes, according to seriousness, and assigned demerit values appropriate. In the original operation of the plan in the Bell System, four classes were used:

    1--Very serious (demerit value 100 points)

    2--Serious (demerit value 50 points)

    3--Moderately serious (demerit value 10 points)

    4--Not serious (demerit value 1 point)

Defects included in each class must be carefully defined. Then inspection of a unit of product is done, yielding counts c_1, c_2, c_3, and c_4 of defects within each class, where the c's may be 0, 1, 2, ... Now calling the weights w_1, w_2, w_3, and w_4, the demerit points D for a unit of product is defined by

$$D = c_1 w_1 + c_2 w_2 + c_3 w_3 + c_4 w_4 \qquad (8.7)$$

This D score provides a demerit rating for each unit of product.

    A control chart can be run for such individual D values. But probably the more commonly used application is for describing the quality for a shift, day, week, or a month. Then for one such time period we will have, say, n units and thus demerit ratings D_1, D_2, ..., D_n. Thus the total demerits for the time period is

$$\Sigma D_i = w_1 \Sigma c_1 + w_2 \Sigma c_2 + w_3 \Sigma c_3 + w_4 \Sigma c_4 \qquad (8.8)$$

Such an aggregate of demerit points is of some use, but since the number of units may vary a bit, it is usually desirable to use as the measure, the *demerits per unit*, called U. Thus for n units, we define

$$U = \frac{\Sigma D_i}{n}$$

$$= \frac{w_1 \Sigma c_1 + w_2 \Sigma c_2 + w_3 \Sigma c_3 + w_4 \Sigma c_4}{n}$$

$$= w_1 \bar{c}_1 + w_2 \bar{c}_2 + w_3 \bar{c}_3 + w_4 \bar{c}_4 \qquad (8.9)$$

Such a U value then describes the quality over the time period in terms of average demerits per unit. As such, it is quite readily interpreted.

For a series of k time periods, we may wish to run a control chart, and thus we need a central line and control limits. We will have k U values, such as in (8.9). This will provide

$$\bar{U} = \frac{\Sigma U_i}{k} \qquad (8.10)$$

Or to find this we could average the last expression of (8.9) over k time periods finding

$$U = w_1 \bar{\bar{c}}_1 + w_2 \bar{\bar{c}}_2 + w_3 \bar{\bar{c}}_3 + w_4 \bar{\bar{c}}_4 \quad \text{(central line for U's)} \qquad (8.11)$$

the $\bar{\bar{c}}$ values being grand averages over all nk units in question. We then have the formulas for analysis of past data:

$$\sigma_U = \sqrt{\frac{w_1^2 \bar{\bar{c}}_1 + w_2^2 \bar{\bar{c}}_2 + w_3^2 \bar{\bar{c}}_3 + w_4^2 \bar{\bar{c}}_4}{n}} \qquad (8.12)$$

$$\text{Limits}_U = \bar{U} \pm 3\sigma_U \qquad (8.13)$$

If standard values c_1', c_2', c_3', and c_4' had been decided upon, they would be used to replace the respective $\bar{\bar{c}}$'s in (8.11) and (8.12).

Trends are especially to be looked for. Also when a U value goes above the upper control limit, we look in the records for the particular defect or defects which were responsible, and seek the causes so as to control the trouble.

The assignment of demerit points is somewhat arbitrary. For example, Hill [3] uses demerit points 50, 20, 5, and 1. It might also be mentioned in passing that we could instead run a c or u chart for the defects in *each* class, thus giving four c charts instead of one U chart.

8.7 SOME TYPICAL APPLICATIONS

1. Saving time in looking for assignable causes when none are present. In one case, a process was producing about 40% of the pieces defective. One thing after another was tried and sometimes improvement seemed to be obtained, but then a reversal would be noted. But the sample sizes were quite small. The quality control

man said, "I don't think any of our experimenting has had any effect.
I think I can duplicate these results out of one box." So he got
beads of two colors, 40% being red, and repeatedly scooped up the
sample size being used. Wild fluctuations in the observed p values
occurred from the one box, much like the variations noted during
the trying of different remedies. So after "brainstorming" they
agreed to try something entirely new. This dropped $\bar{p}$ to 5%. Then
another new idea dropped it to 1%. The point was that during the
early experimentation, the process had remained nearly in control
but with a very poor $\bar{p}$.

An extremely widespread evil in manufacturing is to reset processes on insufficient evidence. When a process is in reasonable
control (even though not meeting specification limits), then tampering with the setting when a sample gives a seemingly off-result, but
which is not out of control, only serves to *make matters worse*, by
increasing the variability above that natural to the process. In
one actual case, a process was not meeting specifications well. It
was reset 68 times in an 8-hr shift. Two quality control men got
permission to run it for a whole shift, without adjustment. They
had the set-up man do a careful job, and then let it run. The result was much fewer pieces out of limits. Such excessive tampering
with the setting is called "hunting."

*2. Help in finding causes of trouble by telling when to look,
and by rational sampling where.* Many cases could be cited for this
basic aim of a control chart program. In one plant making vacuum
tubes, excellent results were obtained in refining and improving the
manufacture and assembly of the tubes. In one instance, an extremely difficult but important tube was developed. The engineers said,
"We are only getting five percent good tubes, and you will probably
never get over a ten percent *yield*. But don't worry. We need these
tubes." Manufacturing with help from quality control people using
charts to control all component characteristics worked up to a 95%
yield.

In another instance in wool processing, trouble occurred, and

experimentation for months was unsuccessful. Then someone ran a
control chart on all the wool sample results and found the first
point, 6 weeks *before* trouble really hit. Checking the records,
they found a process change at that time. Restoring the earlier
condition solved the problem.

3. *Knowing when a machine or process is doing the best that
can be expected of it.* If a process is not in control, whether or
not meeting specifications, we really do not know what it is capable of. Only when the process is gotten into control, can we say
what it can do. In one case a man just returning from a course
asked permission to experiment with a couple of old machine tools
which were not being used, because everyone said they would not
hold useful tolerances. Armed with control charts and with help,
he was able to find quite a few assignable causes. After getting
good control at last, they found that these machine tools could do
almost as close work as any others (although at a slower rate than
the newest ones). So they were able to save the two, and also
learned a lot.

It is very useful to obtain good control of all one's manufacturing processes and thus to know the *process capabilities*. Then
it is possible to efficiently allocate jobs to the various lines,
as well as to maintain their known capabilities.

4. *Determining the repeatability of a measuring technique and
its error.* Dimensional or weight measurements and chemical analyses
are of course subject to variability. That is, when the same material or piece is repeatedly measured the results will differ at least
slightly, if we try to work closely. A series of repeated measurements made on identical (as nearly as possible) material, without
the measurer knowing it is identical, should show good control on $\bar{x}$
and R, or x and moving R charts. If not, steps should be taken to
secure control. Then $\bar{R}/d_2$ estimates the repeatability standard deviation. If there is no bias in the measurement, this is the standard deviation measurement error. But if $\bar{\bar{x}}$ or $\bar{x}$ is substantially
off from a known true measurement or analysis μ, then there is a

bias error. Steps should be taken to eliminate the bias or else to obtain a calibration curve.

Sometimes a proposed measurement technique or gauge proves incapable of controlled results and is, therefore, to be abandoned. Moreover, such, unfortunately not infrequently, proves to be the case with some time-honored measuring techniques.

5. *Decreasing product variability.* The very act of eliminating undesirable assignable causes commonly reduces the process variability, often drastically, to perhaps a half or less.

6. *Saving on scrap and rework costs.* In an actual case in automobile manufacture, the division management of the lower-priced automobile had been urging the management of the top quality automobile division to give statistical quality control a try. Finally they agreed and asked that someone be sent over. They asked that he look into the crankshaft production. They said to themselves, "No bad crankshaft has gone out in a car in two years. Let's see what he can do with that record." The visitor got out the blue-prints, borrowed gauges and followed the processes through. Within a month he found that he had an ideal spot for the methods. For one thing 14% of the crankshafts went out the back door "grade A scrap iron." Then too there was a huge amount of unnecessary rework and excess machining. His charts clearly showed that in the first rough work very little attention was paid to specified diameters. Sometimes too much metal was removed, necessitating scrapping. At others, too little was removed calling for excessive metal removal on subsequent operations. Presenting his case within about 6 weeks, action was taken reducing scrappage to 1% and greatly reducing machining and grinding time by an unknown amount!

Often when processes are incapable of meeting limits and reworking is necessary, it is possible to run at a level so as to balance scrap and rework costs for a minimum loss overall.

7. *Increasing tool and die life.* This is a "natural" for the methods discussed in Sec. 8.2, using slanting control limits. The idea is to aim at starting the process as close to one specification

limit as is safe and let the average increase or decrease watching
control till approaching as near as is still safe to the other limit.
Thus maximum length runs are obtainable. Some chemical processes
can use the same approach.

8. *Decreasing inspection for processes in control at a satisfactory level.* When the appropriate charts show the process to
be in control at a satisfactory quality performance, we can in general decrease the frequency of taking control chart samples. Or
possibly, the size n may be cut. In one famous case for a molded
part in World War II, after showing the process to be in control
but incapable of meeting specification limits, the latter were examined and parts out of limits assembled. They worked satisfactorily. So the tolerance was much increased. No longer was 100% sorting necessary. Samples of n = 5 for an $\bar{x}$ and R chart set-up were
taken only every hour. Then only one sample each 4 hr and finally
only one in 8 hr.

Note that some follow-through of an occasional sample was still
maintained. This is highly recommended, for otherwise production
may become careless. Moreover, a record of some surveillance is desirable in case of some legal action. One must be able to defend
his methods.

9. *Safer guaranteeing of product, reducing customer complaints.*
The basis of sound guarantees and customer satisfaction is controlled
processes at satisfactory level. If control is uncertain, we are in
a much weaker position. Sales managers need to know our process
capabilities, what we *can* do, and what we *cannot*. Then they are in
a position to promise only what can be delivered.

10. *Improving production-inspection relations.* Control charting of production can make a real contribution by objective treatment,
taking account of variation. It is the variability in results which
can cause much argument, unless allowance is made for discrepancies
explainable by chance. Differing results may or may not be
compatible.

11. *Improving producer-consumer relations.* Many problems in this relationship are similar to those in 10. In addition there are problems of sampling-acceptance results. See Chapters 9 and 10. One commonly occurring problem is that of compatibility of measuring technique, such as gauges, and of definitions of defects or nonconformances. All such problems can be aided by statistical methods.

12. *Setting specifications more realistically, sounder relation between engineering and production.* The objective is to obtain good control of processes so as to know the production *capabilities*. Often by obtaining full control of processes enough decrease in variability is obtained to meet the specified tolerances. But also not infrequently, the process, even when in control, is incapable of meeting specifications. It then becomes a choice between (1) making a fundamental change in the process, (2) sorting 100% to the specification limits, or (3) widening the specification limits. Solution (2) is very expensive and time consuming. It also requires scrapping and reworking. Solution (1) may be expensive too, perhaps extremely so. Solution (3) is often feasible. If engineering can be convinced that production is getting all it can out of the process, then engineering may well relax the tolerance, if the parts made can be shown to work satisfactorily.

In one case, 30-in. rubber tubes (extruded) were cured on mandrils, then each tube was cut into 300 gaskets for assembly in metal caps for food jars. Thickness of rubber was important for proper sealing. Specified limits for doubled thickness of rubber from single gaskets were so tight that no one tube could yield gaskets all within limits, let alone a whole lot of tubes. The reason was that in curing on a mandril the rubber at each end of a tube would pull together a bit, giving thicker gaskets at the ends than in the middle. Thus some of the 300 gaskets from *single* tubes would always lie outside *both* limits. When this was clearly shown, engineering was willing to set much more realistic limits, within the process capabilities.

13. *Decreasing defects or nonconformances on subassemblies.*
As we have seen, c and u charts are potent tools for attacking such
problems. Posting results can be a strong inducement to improvement. In one case a very large manufacturer of auto radiators decreased the number of leaks at the initial test following the assembly of two sides, from seven to less than two, in a few months.

14. *Comparison of several inspectors, machines, or processes.*
Taking rational samples from each as in Sec. 8.1.2 and putting on
one control chart permits comparison with reasonable allowance for
natural variability. If the various samples are all in control,
then there is no reliable evidence of any real differences. But if
one or more points are out we do have reliable evidence of real differences in quality performance. Charts could be on measurements or
attributes.

15. *Stabilization of chemical or metallurgical processes.* By
running control charts on chemical additions, temperatures, pressures,
timing, throughput, and then on the resulting chemical compositions
and physical characteristics of the final product, the whole operation may be improved and stabilized. For example, in a large production of cast manganese steel tank shoes, spectacular results were
obtained, the percentage of nonconforming castings dropping from
13.9 to 2.0% in 10 months. And this was at a benefit-to-cost ratio
of 20 to 1.

16. *Justifying a "pat on the back," fostering quality mindedness.* If carefully approached and sold to personnel, control charts
can easily become a source of pride, even friendly competition among
workers. It is a help to be complimented by a foreman, when an objective chart shows that a fine job is being done by a workman.
Moreover, under some incentive systems statistical quality control
methods may actually put more money into a worker's pocket.

17. *Determining the stability and quality level of a producer.*
Control charts run on samples from a producer form an excellent
record of his performance. If in control at a satisfactory level,

receiving inspection may well be decreased, perhaps drastically.
And if not in control and/or the quality level is inadequate, appropriate steps can be taken objectively. Producers may also be objectively compared. Regular reports to producers are facilitated.
Moreover, if the producer uses statistical control charting, he may
submit his results to the consumer, which may facilitate little or
no receiving inspection. This can be the basis of "vendor
certification."

   18. Convenient and meaningful records. Control charts form a
compact and objective record of product performance throughout the
production process. Such records can be vital in the case of customer complaints, or in the extreme case of a liability law suit.
In such a situation our methods are likely to be under close scru-
tiny, and we must be able to show that due care was exercised, along
with objective methods. Even with such sound methods, judges and
juries may be difficult to convince because of ignorance of probability and statistics.

   19. Saving on weight control. There are enormous savings
available in all types of manufacturing. The first thing you may
think of is the package or container content weight. The general
approach is to secure better and better control of weight of fill,
in two respects (1) freedom from assignable causes of erratic weights
and (2) decreasing the variability. As improvement is made in both
respects, it becomes possible to run the process average closer to
the specified average or the specified minimum, whichever is applicable. But this subject is much broader than container weights. It
literally applies to nearly all products. Furthermore it applies to
factors of safety, which some, not without reason, call "coefficients
of ignorance."

   20. Interpretation of the many management figures. One often
sees in manager's offices a series of charts with figures plotted,
such as, production, unit costs, late shipments, absenteeism, man-
hours per ton, power consumption, and so on. But unfortunately such
running records frequently have no central line nor control limits.

The placement of control lines can aid decision making, so that matters which deserve attention receive it, and those apparently off-
figures, which are in reality within control, do not receive unjust-
ified attention and concern. Blaming people for what is really only
a random fluctuation with nothing different being done, does not improve morale. Quite the contrary. Moreover action may be taken
which is harmful rather than beneficial.

   21. Improving visual inspection. The first step, of course,
is to obtain clear, objective definitions of each defect. A set of
"limit samples" helps, that is, a collection of pieces with the bor-
derline size of defect: Larger ones are to be regarded as "defects,"
smaller ones are to be disregarded. Such can be a great help to
inspectors. But this may not altogether solve the problem. If no
trouble seems to be coming along, inspectors may become lax, and
then if trouble hits, the foreman may get "tough" and clamp down.
Then inspectors may start rejecting pieces with defects much smaller
than the marginal size.

   In a piston-ring corporation, this precise problem was encountered on holes in completed rings. For each inspector *two* audit
charts of the np type were run. On an ordinary vertical scale, 0,
1, 2, ... was plotted for the day, all of the *good* rings *rejected*
by the inspector, found among all the rings he rejected. These were
found by a lead inspector auditing the rejected rings. Then immediately beneath the foregoing chart was another np chart, with scale
0, 1, 2, ..., *increasing downward* (back to back with the other one).
On this chart was plotted the number of *defective* rings found by the
lead inspector from among a sample of 200 rings *accepted* by the inspector. Now if both points were higher (many good ones rejected,
few bad ones missed), this tended to mean the inspector was too
tough. But if both points were low (few good ones rejected, many
bad ones missed), this tended to mean that the inspector was too
lenient. Out-of-control points on either chart were, of course,
most significant. The savings on good rings, which were no longer
being rejected, amounted to 25% of the corporation's net income one
year.

22.   *Decreasing the number of clerical errors.*   The c chart is
a natural approach to cutting the number of clerical errors.   Clerical work can be audited by a top clerk or inspector.   All sorts of
clerical jobs may thus be controlled, including typographical errors
and accuracy of proofreading.   A mail order company achieved excellent results in this way.   So many applications have been made that
there is an Administrative Applications Division in the American
Society for Quality Control.

23.   *Decreasing accidents.*   One company found a relation between
hospital calls and serious or lost-time accidents.   By watching a
chart on hospital calls and putting on a more intensive program of
accident prevention when the calls significantly increased, the
accident rate was cut.

24.   *Classifying product to be processed.*   One highly lucrative
application was made in a hardwood veneer company.   Lumber to be
dried is measured with respect to moisture content.   Prior to the
application of a control chart method, lumber of many different
moisture contents would be put in a kiln and dried as the wettest
lumber.   The following approach was devised by a man returning from
a short course in quality control.   For each stack of lumber he would
take six readings x, find $\bar{x}$ and $\hat{R}$.   Then he would find $\bar{x} + (3R/d_2)$
as an estimate of the wettest lumber in the stack, which he called
"moisture number."   He chose categories for moisture numbers, plac-
ing the data on different colored cards.   Then he would have all,
say, blue card stacks dried together.   They would be dried to the
estimated wettest lumber using $\bar{\bar{x}} + (2\bar{R}/d_2)$ from the various blue
card stacks to be dried.   In this way all the lumber would be dried
at about the right rate, thus avoiding spoilage ("honeycombing").
And millions more board-feet of lumber were able to be dried per
kiln-month, because some wet lumber would not be holding up each
load to be dried.

25.   *Improving packaging.*   In one case of packaging of cleaning
powder, the metal ends were not always well crimped onto the card-
board cylinder.   Rigorous tests on finished packages in connection

with np charts got at the causes of trouble, which were eliminated.
The quality control people were called "the boys with the iron
thumbs" by production. Control charts can be used in a wide variety
of ways in packaging.

26. *Facilitating random assembly*. In a corporation, piston
ring castings were being ground three or four times for edge width
(thickness) by disc grinders. An efficient materials-handling method
put all the rings in order onto stakes. After the final fine grind,
the rings would be made up into pots of 15 or 20 for machining
grooves and the like. Unfortunately, as a result of poor control,
shims had to be used in making up the pots. A complete graphing of
6,000 consecutive rings gauged for edge width led to the discovery
that the thicknesses showed runs of thicker, then thinner rings. By
the materials-handling method, these runs would be accentuated in
the next grinds. Finally, it was decided to dump the rings from
each grind helter-skelter into a box. This broke up the cycling.
Also it facilitated much better control of edge width so that 100%
sorting after each grind was eliminated in favor of sampling inspection, decreasing the cost in the grinding department by 80%. Another
big gain was that with a random order of rings and better control,
the pots of rings for machining could be assembled randomly without
resorting to the use of shims of various thicknesses.

27. *Getting at the causes of trouble rather than expecting the
nonconforming pieces to be sorted out and not missed by inspection.*
This is the basic aim in all process control and is greatly aided by
the control chart approach.

If this list does not give you any ideas to work on, then you
must either lack knowledge of your processes and product, or you
lack imagination. A bibliography of applicational references in
journals is given in Ref. 4.

PROBLEMS

8.1 Obtain 30 stratified samples, each of three x values, consisting of the total points in a throw of two dice, then of three dice,
and finally of four dice. Tabulate the three x values always in
the same order, as in Table 8.1. (a) Analyze the stratified samples as in rows: using the usual control chart formulas and drawing the charts. Comment. (b) Similarly analyze the rational samples in turn. Comment. (For the totals on two dice $\mu = 7$, $\sigma = 2.42$;
three dice $\mu = 10.5$, $\sigma = 2.96$; four dice $\mu = 14$, $\sigma = 3.42$; and in
general for k dice $\mu = 3.5k$, $\sigma = \sqrt{35k/12}$.)

8.2 Simulate a three-spindle automatic being run for production at
specification limits of ±5 (coded). Use 30 samples of n = 3, each
consisting of one from each of populations A, B, and F of Table 7.3.
Always take in same order as in Table 8.1. (a) Analyze the stratified samples in rows: using the usual control chart formulas and
drawing the charts. Comment. (b) Similarly analyze the rational
samples in turn. Comment.

8.3 This problem involves measurements on the outside diameter at
base of the stem of an exhaust-valve bridge. Specifications were
1.1550 to 1.1560 in. All pieces from the machining operation were
measured, and samples of n = 5 in succession were formed. Three
vendors were supplying forgings. There were constant tool changes
and tool resets. As a result of these changes, tool life was short,
and pieces were often out of specifications. It was the general
opinion that differences among vendors were responsible for pieces
being out of specifications. However, a study was made on each vendor's pieces, and there was no significant difference in the trends
of any of them. A chart was placed on the machine and the operator
instructed as to where to start a run and at what level to reset.
The chart indicated that approximately 75 pieces could be produced
before resetting. By using this type of trend chart, the process
was made to run without defects, and 100% inspection was reduced to
five pieces every 30 min. Tool life was considerably increased.

The data given below were in .0001 in. above 1.1550 in., with n = 5.

Sample	1	2	3	4	5	6	7	8	9	10
$\bar{x}$	3.0	4.0	4.4	4.6	5.0	5.6	6.8	7.0	7.6	7.8
R	2	2	2	2	2	2	1	2	1	2

Sample	11	12	13	14	15	16	17	18	19	20
$\bar{x}$	8.0	8.6	9.0	1.8	2.4	2.6	2.8	3.0	3.4	3.6
R	2	1	2	2	1	2	1	2	1	2

(a) Plot the points for $\bar{x}$ and R. Check the R chart for control. Estimate σ if justified. (b) Draw by eye the trend lines for μ, and show on the chart the minimum and maximum safe process averages. Also draw slanting control limits around the tre id lines. (c) Were the runs started and ended at about the right levels? Was the tool wear running in control?

8.4 The data shown below constituted the original run, preceding that of Table 8.3. As in that table, specifications were .1235 to .1250 in., with data in .0001 in. units below .1250 in. n = 5.

Time	July 27	8:00	8:30	9:00	9:30	10:00	11:00	12:30
$\bar{x}$		-8.0	-7.4	-6.8	-5.4	-5.6	-4.0	-3.4
R		2	3	2	2	3	4	3

Time	July 27	1:15	2:00	2:45	3:15	4:00	4:30
$\bar{x}$		-10.8	-11.6	-10.4	-9.6	-9.8	-9.4
R		2	3	2	3	2	4

Time	July 30	9:00	9:30	10:00	11:00	12:15	12:45	1:15
$\bar{x}$		-7.4	-7.8	-7.4	-5.4	-7.0	-5.8	-3.8
R		4	4	2	3	2	3	2

Time	July 30	1:45	2:15	2:45	3:30	4:00	4:30
$\bar{x}$		-4.6	-4.2	-2.8	-6.2	-5.6	-6.2
R		4	4	3	5	4	2

Time	July 31	8:00
$\bar{x}$		-9.4
R		4

(a) Plot the $\bar{x}$ and R points, check the R chart for control. Estimate σ if justified. (b) Would the $\bar{x}$ chart with ordinary limits be in control? (c) Draw by eye the trend lines for μ. Show the minimum and maximum safe process averages, and the slanting control limits for $\bar{x}$. (d) Were the trend lines for μ started and ended at about the right level?

8.5 The following analyses x give the carbon content of 20 consecutive heats of 1045 steel. Three heats are made in an open hearth furnace per day. Check the control by an x chart (with 2σ limits) and a moving range chart. Specified limits are .45 to .50%. The following are given in .001% units: 480, 470, 470, 455, 515, 495, 460, 465, 520, 530, 485, 495, 455, 475, 515, 455, 500, 470, 480, 505. (a) How is control? (b) Can you estimate the proportion of heats meeting specifications?

8.6 A steel plant was producing hot-rolled bars which were shipped to a customer who cold drew them. After cold drawing, the bars were inspected and certain ones rejected for defects purporting to originate in the steel plant. Rejections were totaled for each month and expressed as a percentage of the total drawn. Twenty-four monthly figures in percent rejection follow: 4.4, 2.9, 4.2, 5.4, 6.6, 3.6, 1.0, 1.3, 1.2, 1.2, 1.0, 2.8, 2.0, 3.0, 1.8, 2.3, 1.3, 5.2, 4.1, 4.3, 5.3, 5.3, .6, .8. Make an x chart (2σ limits) and a moving range chart and comment.

8.7 Give two examples from your own experience of data on fraction defective which cannot appropriately be analyzed by a p chart. State why not in each case.

8.8 Find the average run lengths in terms of individual pieces, when using an upper control limit of 17 (17 defects is out of control) for the total defects on three units: (a) when c' for one is 3, and (b) when c' for one is 6.

8.9 Suppose that on 100 units of equipment, $\Sigma c_1 = 1$, $\Sigma c_2 = 3$, $\Sigma c_3 = 32$, $\Sigma c_4 = 350$, with corresponding weights $w_1 = 100$, $w_2 = 50$, $w_3 = 10$, $w_4 = 1$. Standard values are $c_1' = .001$, $c_2' = .005$, $c_3' = .3$,

$c_4' = 3$. Find U, and compare with the control lines.

REFERENCES

1. H. F. Dodge, A method of rating manufactured product. *Bell System Tech. J.*, *7*, 350-368 (1928). Bell Telephone Laboratories Reprint B315.

2. H. F. Dodge and M. N. Torrey, A check inspection and demerit rating plan. *Indust. Quality Control*, *13* (No. 1), 5-12 (1956).

3. D. A. Hill, Control of complicated product. *Indust. Quality Control*, *8* (No. 4), 18-22 (1952).

4. I. W. Burr, *Statistical Quality Control Methods*, Dekker, New York, 1976.

Chapter Nine

ACCEPTANCE SAMPLING FOR ATTRIBUTES

For the moment, let us consider the general problem of deciding
whether to accept a lot from a vendor, subcontractor, or another
department or line. This can of course be done by inspecting or
testing *every* item or piece in the lot. But this can be expensive
and time consuming. Moreover, it is quite often unnecessary, as we
shall see. Sound decisions can often be made from a sample from
the lot. The criteria of quality of pieces and of the lot may be
on the basis of measurements or the presence or absence of defects,
that is, by attributes. It is the latter class of criteria which
we will take up in this chapter and the next. Sampling acceptance
by measurements is discussed in Chapter 11.

9.1 WHY USE A SAMPLE FOR A DECISION ON A LOT?

There are several reasons why we may wish to use a sample from a
lot or process for decision making as a basis for action, rather
than inspecting or testing all pieces:

1. *To save money and time.* Unless the characteristic in ques-
    tion is extremely critical, sampling acceptance plans are
    often satisfactory for a decision. They can be set up so
    as to *reliably* distinguish between "good" lots and "reject-
    able" lots. The degree of reliability of such discrimina-
    tion can be specified in advance, and to any desired degree
    of confidence.

2. Very often the inspection of a sample of pieces can be and
    is done much more carefully than 100% inspection of the
    entire lot. This is because of fatigue and psychological
    factors. It may even be possible to find out the lot qual-
    ity more accurately from a sample than from inspection of
    the entire lot! One hundred percent inspection can be

notoriously poor. Some mechanical inspection can, of
course, be highly accurate.

3. When the test is *destructive*, 100% testing of pieces is
    impossible, for then there would be no pieces left to use
    even if the lot were shown to have been perfect. In this
    case the only possibility is to base the decision on the
    lot upon a sample.

9.2 LEVELS OF INSPECTING OR TESTING A LOT

Consider the problem of assuring that a lot of pieces is of adequate
quality in all characteristics. How many characteristics are there
on one piece? The author once saw a rather small piece part for the
electronics industry that had some 60 dimensions of varying importance! In another case, there were 20 pages of specifications on
insulated wire. It is simply not possible to give full attention to
all characteristics. As a consequence, it can be said that inspection and testing of the characteristics will run from (1) none, (2)
spot-checking, (3) sampling inspection, (4) 100% inspection, to (5)
several hundred percent inspection. Which level to use will depend
upon (1) the practical importance of the characteristic in the product, (2) the ease of controlling the characteristic, and (3) the
producer's record or history on it.

In sampling inspection, as in this chapter and the next two,
we are assuming that it is feasible and desirable to base the decision about a lot upon a sample. Because of the great flexibility in
the setting up of sampling plans, such acceptance sampling can be
used for incidental defects of little significance, minor defects,
major defects, and even quite critical defects. We can set the
risks or probabilities of wrong decisions at whatever levels we
wish, keeping in mind that the smaller the risks are which we specify, the larger the sample size must be. One way to cover a mul-
tiplicity of different characteristics is to include them in classes
of defects. Thus "major defects" may include, say, 12 defects (some
measurable, some visual).

You can have almost any desired reliability built into your
acceptance sampling plan, provided you are willing to pay the price

of the required sample size, and provided you will take steps to
see that samples are randomly chosen and that inspection is done
soundly.

9.3 THE OPERATING CHARACTERISTIC OF A PLAN

Probably the most important characteristic of a sampling plan is its
operating characteristic (OC). This is the probability that a lot
or process will be approved, called Pa for "probability of acceptance." Now such a probability for any given sampling plan is a function of the submitted lot or process quality, that is, the better
this quality, the higher the probability of acceptance is. Of course,
Pa also depends upon the criteria of the sampling plan itself, such
as the sample size or sizes and the acceptance numbers given in the
plan, as we shall be seeing. But always keep in mind the "if/then"
character of any plan: *If* the lot is of such and such quality, *then*
Pa is thus and so. In mathematical terms, Pa is a conditional probability, conditioned upon the submitted lot quality.

Furthermore in practice, Pa, as calculated, will assume random
sampling and accurate inspecting or testing. If one samples only
out of one corner of a lot where good pieces happen to be, but there
are many defective ones in other parts of the lot, it is possible to
accept a very bad lot.

9.4 ATTRIBUTE SAMPLING INSPECTION

Up to now what we have said applies to both sampling inspection by
attributes and by variables (or measurements). We will now concentrate on the former, namely, where the lot quality of interest is
concerned with attributes. As we have seen in Chapter 6, there are
two ways to inspect:

1. By counting *defectives* wherein each piece is "good," that
    is, has no defects, or else it is a defective, that is, it
    possesses one or more defects.
2. By counting the number of *defects* on one or more pieces.

For the moment we shall be primarily concerned with the case of
defectives.

Let us again emphasize that although the word "defective" has
a "bad" meaning to many and is undesirable to all, there are defects
of all sorts of severity and importance, from quite unimportant discrepancies all the way to very critical defects. It may be safer in
some situations (especially in a court room) to speak of "nonconforming" pieces, rather than "defective" pieces. But we shall continue
to use the term, a "defective."

9.5 CHARACTERISTICS OF SINGLE SAMPLING PLANS

A "single sampling" plan is one in which we take a random sample of
n pieces from the lot or process, inspect or test them, at the end
of which a decision is made to either accept or reject the lot (or
process). On the other hand for "double sampling," we select a random sample of n_1 pieces and inspect or test them. The results lead
to a decision to accept or to reject, or to request that another
sample of n_2 pieces be drawn randomly from the remainder of the lot.
These are then inspected or tested, after which a firm decision of
acceptance or rejection is always reached. We may also have "multiple sampling," in which several samples may be required before reaching a decision.

Let us now set down a few notations for acceptance sampling
based upon defectives.

N = number of pieces in the lot	(9.1)
n = number of pieces in the sample	(9.2)
Ac = c = acceptance number for defective pieces in sample	(9.3)
Re = c + 1 = rejection number for defective pieces in sample	(9.4)
D = number of defective pieces in lot	(9.5)
d = number of defective pieces in sample	(9.6)
$p' = \dfrac{D}{N}$ = fraction defective in the lot	(9.7)
$p = \dfrac{d}{n}$ = fraction defective in the sample	(9.8)

Pa = probability that lot will be accepted by plan

= Pa(given p')

= probability of acceptance, if p' in lot (9.9)

The acceptance-rejection decision process for single sampling may be diagrammed as follows:

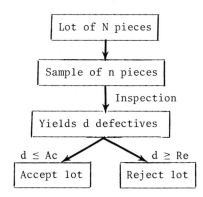

This is simple and exactly what you the reader would come up with, given a little thought.

Thus a single sampling plan consists of three numbers: N for lot size, n for sample size, and Ac for acceptance number. Now how do we find out what any given plan does for us, for example, how well does it distinguish between lots of various qualities or p' values?

9.5.1 The Operating Characteristic Curve

It is basic to know for lots of any given quality p' how often they stand to be accepted by the sampling plan. Naturally if p' is relatively small, then Pa will be high, that is, close to 1. And the larger p' becomes, the lower we can expect Pa to be. How do we obtain specific answers? This is accomplished through use of the appropriate attribute distribution, from Chapter 4. We have the following two cases:

1. *Type A OC curve.* Probability of accepting a lot of N pieces, where account is taken of the lot size N. The

exact calculation involves using the "hypergeometric distribution" of Sec. 4.4, which you may not have studied.
(But do not worry, it can be readily approximated!)

2. *Type B OC curve.* Probability of acceptance of a lot of N
chosen at random from a *process* with true fraction defective p'. The exact calculation involves using the binomial distribution of Sec. 4.2. But again, we can in general
obtain entirely useful approximations by using the Poisson
distribution of Sec. 4.3 and thus Table B.

Type A OC curves are correct when we have a series of lots of
N pieces, in each of which there are *exactly* D defectives. This
seems to be and is quite artificial. But if we should consider that
we have just one lot of N, containing D defectives, we likely would
wish to know the probability of such an isolated lot being accepted
by our sampling plan.

On the other hand, a Type B OC curve is in effect, if we think
of a series of lots of N chosen from a process with fraction defective p'. This is usually what we are interested in if we have a
series of lots. Given a process at p', what proportion of lots will
be accepted in the long run? Here the binomial distribution is exactly correct (and the size of the lot, N, is immaterial). If n is,
say, at least 20 and p' is .05 or less, the Poisson distribution is
an excellent approximation to the binomial, by letting c' = np'.
Even when n = 10 and p' = .1, the approximation is quite good.
Therefore, we will use the Poisson distribution.

Now for an example, let us find the Type B OC curve for the
sampling plan

N = 1000   n = 80   Ac = 3

In the first column of Table 9.1 are listed some average lot or process fraction defectives of interest, and in the second column, the
expected or theoretical average number of defectives *per sample,*
that is, np' from (4.11). Here since n = 80, this is 80p'. Then
using this to set the row in Table B, we use the column for c = 3
to give P(d ≤ 3) = P(d ≤ Ac) = Pa. Thus the entries in Table B give
directly the probabilities of acceptance. Fortunately no interpola-
tion for entries was needed here. The last column gives the Pa's as

TABLE 9.1. Type B OC Curve Calculations for Plan: N = 1000, n = 80, Ac = 3, Re = 4.

Process, p'	Expected defectives in sample, 80p'	Poisson approximation for Pa from Table B	Exact binomial for Pa from [1]
.00	.0	1.000	1.000
.01	.8	.991	.991
.02	1.6	.921	.923
.03	2.4	.779	.781
.04	3.2	.603	.602
.05	4.0	.433	.428
.06	4.8	.294	.286
.07	5.6	.191	.181
.08	6.4	.119	.109
.09	7.2	.072	.063
.10	8.0	.042	.035

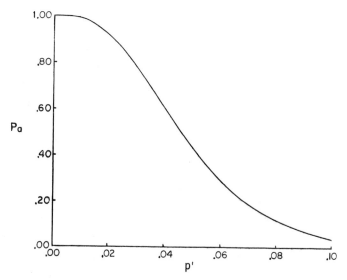

Fig. 9.1. The type B OC curve for the single plan: N = 1000, n = 80, Ac = 3, Re = 4.

found by the formally correct binomial distribution using an available table. Comparing the last two columns of Table 9.1, we can see that the Poisson distribution provided an excellent approximation to the binomial.

Figure 9.1 shows the OC curve for the plan. If the full grid were shown, we could easily read off Pa for any given quality p'. Or we can reverse the question and ask for what p' is Pa .95? This appears on the graph to be about .017. Of course, we can answer such questions from Table B. In this instance, we look in the $c = 3$ column of Table B for P(3 or less) to be .950. We obtain

np'	P(3 or less)
1.3	957
	7
	950
	4
1.4	946

The desired np' is interpolated for, by $1.3 + (7/11)(.1) = 1.364$. This being np' we have but to divide by $n = 80$ for the desired p', that is, .017. Such a p' being accepted .95 of the time is called p_{95}'.

9.5.2 *The Average Sample Number Curve*

For a single sampling plan, we always inspect (or test) precisely n pieces in order to reach a decision. Thus the average sample number (or ASN) for a single sampling plan is n. The graph of the ASN is merely a horizontal line at a height of n, for the whole collection of p's. However as we shall see in Sec. 9.6, the ASN curve for a double sampling plan really is a curve. Thus in order to reach a decision on a lot which contains no defectives, we will always accept on n_1 pieces. But as p' increases above zero, we will sometimes require a second sample of n_2 additional pieces, and therefore it takes $n_1 + n_2$ pieces for a decision. However if p' continues to increase, the chance of *rejecting* the lot on the first sample comes into play, and the n_2 additional pieces may not be needed for a decision. So the ASN begins to decrease toward n_1 again. The ASN is a *cost* characteristic of a sampling plan.

9.5.3 *The Average Outgoing Quality Curve*

This characteristic of a sampling plan is another index to the *protection* it supplies, the OC curve being the first such protection characteristic which we considered. The average outgoing quality is the average fraction defective for *all* lots submitted, after those lots which were rejected by the sampling plan have been sorted 100% and cleared of defectives. Thus, for example, suppose a series of lots comes in at about 2%, that is, p' = .02, and we sample inspect each lot. Some lots will be accepted as they stand, still at about 2%. Meanwhile other lots, even though they also had p' = .02, will be rejected and screened of defectives so that for them p' has become zero. Therefore, the average outgoing quality is a weighted average of p' = .02 and p' = .00. In particular, suppose Pa = .80. This means that 80% of the lots are passed and remain at p' = .02, whereas 20% of the lots are rejected and rectified by screening so that p' = .00. Therefore, using the respective weights .80 and .20 (which add to one), we have

$$AOQ = .80(.02) + .20(.00) = .016$$

In general, it can be seen that we have the following formula, since the second term, (1 - Pa)(.00), always drops out:

$$AOQ = (Pa)p' \qquad (9.10)$$

Now there are in practice various ways of handling the defectives found in the samples and the sorted remainder of the lots. That is, we may consider that we replace by good pieces each defective piece found in the sampling and/or in the sorted remainder. If replaced, this gives more good pieces in the outgoing lots than if not replaced. Also there is the question as to whether we are using Type A calculations (all lots at exactly p') or Type B (lots chosen randomly from a process at p'). However, for nearly all situations in practice, it makes comparatively little difference on the AOQ which assumptions we use. Thus in general we use (9.10) for AOQ's. For our plan of n = 80, Ac = 3, we have, using the Poisson column of Table 9.1:

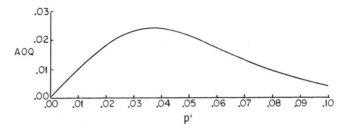

Fig. 9.2. Average outgoing quality curve for the single plan
N = 1000, n = 80, Ac = 3, Re = 4, AOQL = .024 (maximum AOQ).

p'	.00	.01	.02	.03	.04	.05
AOQ	.0000	.0099	.0184	.0234	.0241	.0217

p'	.06	.07	.08	.09	.10
AOQ	.0176	.0134	.0095	.0065	.0042

The AOQ curve is drawn in Fig. 9.2. We see there that the curve
rises initially along the 45° line AOQ = p', then begins to fall
away from this line, reaching a maximum AOQ of about .0242 when p'
is about .036.

Definition. The maximum of the AOQ's for all p' values is
called the *average outgoing quality limit* (AOQL) of the plan.

Thus the AOQL tells us what is the worst long-run *average* fraction
defective we will have to work with, no matter what comes in. But
p' has to be at just one level for the average outgoing quality to
approach the AOQL. Thus for the sampling plan being analyzed, p'
must be close to .036 in order to have an AOQ close to the limit
AOQL of .0242. See the curve. For p' on either side of .036, the
AOQ will be less.

9.5.4 The Average Total Inspection Curve

The fourth characteristic of a sampling plan considered here is also
a cost curve, as was the ASN curve.

Definition. The *average total inspection* for a sampling plan
is the average number of pieces inspected per lot for a series of
lots, including those pieces in samples and the pieces inspected

in the remainder of rejected lots.

Thus the ATI measures the total inspection load to maintain an AOQ for given p'. The ATI depends upon the lot size, the sampling plan, and, of course, the incoming fraction defective p'.

In order to calculate the ATI for any p', we note that for the accepted lots, we only inspect n pieces. A proportion, Pa, of all the lots are accepted. Meanwhile for the rejected lots, we look at n pieces for the decision, and, having rejected the lot we now must look at N - n *additional* pieces. Among the lots the proportion of rejected lots is 1 - Pa. Since we look at n pieces in all lots, we have as an average inspection load

$$ATI = n + (N - n)(1 - Pa) \quad \text{(single sampling)} \quad (9.11)$$

The ATI for a single sampling plan starts at n for p' = 0, since then the Pa = 1 and the second term on the right-hand side of (9.11) drops out. As p' increases, Pa starts to decrease, and the second term begins to increase. The ATI rises above n and eventually reaches the lot size N, when the probability of acceptance Pa becomes zero.

The ATI curve is shown in Fig. 9.3. The calculations are simple. For our example we have n = 80, N - n = 920, and we use Pa from the Poisson approximation column of Table 9.1. Thus we have the following if p' = .04: Pa = .603, 1 - Pa = .397,

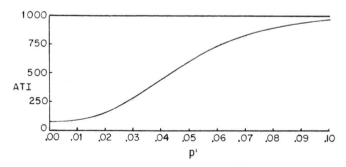

Fig. 9.3. The average total inspection curve for the single sampling plan N = 1000, n = 80, Ac = 3, Re = 4.

ATI = 80 + 920(.397) = 80 + 365 = 445. Continuing we obtain

p'	.00	.01	.02	.03	.04	.05
1 - Pa	.000	.009	.079	.221	.397	.567
(N - n)(1 - Pa)	0	8	73	203	365	522
ATI	80	88	153	283	445	602

p'	.06	.07	.08	.09	.10
1 - Pa	.706	.809	.881	.928	.958
(N - n)(1 - Pa)	650	744	811	854	881
ATI	730	824	891	934	961

The four curves, two protection--OC and AOQ--and two cost--ASN and ATI--give a wealth of information about a sampling plan and how it operates. In passing, we may mention that none of these curves is of much practical importance beyond a p' value where Pa is .6. If such a p' quality level should continue, there would be so many rejected lots that sampling inspection would be discontinued and action taken to have the producer drastically improve his process. In our example, this would mean that if p' goes much above .04, we would drop sampling inspection and begin 100% sorting each lot until the quality level improves.

9.6 DOUBLE SAMPLING PLANS AND THEIR CHARACTERISTICS

We now take up the somewhat more complicated but highly useful subject of double sampling. The objective of double sampling is to obtain a more favorable (lower) ASN curve while still supplying the same power of discrimination, that is, the same OC curve. Two other advantages of double sampling are (1) the psychological appeal to giving a lot a second chance and (2) the fact that no lot is ever rejected on less than two defectives, under double sampling.

The following notations will be used for double sampling, acceptance-rejection criteria:

n_1 = number of pieces in first sample   (9.12)

n_2 = number of pieces in second sample, when required   (9.13)

d_1 = number of defectives in first sample   (9.14)

d_2 = number of defectives in second sample $\qquad$ (9.15)

Ac_1 = acceptance number for d_1 $\qquad$ (9.16)

Re_1 = rejection number for d_1 $\qquad$ (9.17)

Ac_2 = acceptance number for $d_1 + d_2$ $\qquad$ (9.18)

Re_2 = rejection number for $d_1 + d_2$ $\qquad$ (9.19)

$Re_2 = Ac_2 + 1$ $\qquad$ (9.20)

These notations can perhaps better be described by an example than
by a flow chart. As our example we shall take a double sampling
plan, which has an OC curve quite similar to that of the single plan
we have already analyzed. This is the plan N = 1000, n_1 = 50,
n_2 = 50, Ac_1 = 1, Re_1 = 4, Ac_2 = 4, and Re_2 = 5.

For the operation of this plan, we take a random sample of 50
from the lot of 1000 and inspect the pieces, finding d_1 defectives.
Now if d_1 is 0 or 1, that is, less than or equal to Ac_1, we accept
the lot at once. Or if d_1 is 4 or more we reject the lot at once
because of Re_1 being 4. But if d_1 is 2 or 3, we are in between the
acceptance and rejection numbers for the first sample, and cannot
make a decision until after we have drawn a second sample of 50,
randomly, from the 950 pieces as yet uninspected in the lot. Then
after this inspection yields d_2 additional defectives, we find the
total defectives $d_1 + d_2$ in the two samples. Now if $d_1 + d_2$ is less
than or equal to Ac_2 = 4, we accept the lot, but if $d_1 + d_2$ is equal
to or greater than Re_2 = 5, we reject the lot. Since there is no
gap between Ac_2 and Re_2, there is always a firm decision after completing inspection of the second sample.

Let us summarize the foregoing by a general flow chart.

Flow Chart, Double Sampling

$$\boxed{\begin{array}{c} n_1 \text{ from lot yielding} \\ d_1 \text{ defectives} \end{array}}$$

Accept lot $\leftarrow d_1 \leq Ac_1 \qquad Ac_1 < d_1 < Re_1 \qquad d_1 \geq Re_1 \rightarrow$ Reject lot

$$\boxed{\begin{array}{c} n_2 \text{ more from lot yielding} \\ d_2 \text{ defectives} \end{array}}$$

Accept lot $\leftarrow d_1 + d_2 \leq Ac_2 \qquad\qquad d_1 + d_2 \geq Re_2 \rightarrow$ Reject lot

9.6.1 The OC Curve, Double Sampling

For the OC curve, we need to find the probability for a lot of given
incoming fraction defective, p', to be accepted by a given double
sampling plan. This probability of acceptance, Pa, is made up of
several distinct ways in which acceptance can occur. Thus for the
double sampling plan which we are using as an example, these distinct routes for acceptance are

On first sample of 50: $d_1 = 0$, $d_1 = 1$

On second sample with total pieces inspected of 100, the
acceptable combinations are

$d_1 = 2$ $d_2 = 0$
$d_1 = 2$ $d_2 = 1$
$d_1 = 2$ $d_2 = 2$
$d_1 = 3$ $d_2 = 0$
$d_1 = 3$ $d_2 = 1$

These may be grouped as follows in order to use Table B--the Poisson
distribution:

On first sample: $d_1 \leq 1$
On second sample: $d_1 = 2$ with $d_2 \leq 2$
                    $d_1 = 3$ with $d_2 \leq 1$

Then we have only to calculate the probabilities for each way and
add.

We recall that Table B--the Poisson distribution--has cumulative or additive probabilities such as $P(d \leq 3)$ for given np' or c'.
Therefore, to find $P(d = 2)$, we use $P(d \leq 2) - P(d \leq 1)$. We now
work out Pa for our sampling plan, for the case where p' = .04. We
can do this calculation in a simple table form as follows. A little
explanation may be useful. On the first sample, as explained in
general terms, $P(2) = P(2$ or less$) - P(1$ or less$) = .677 - .406$
$= .271$. Similarly $P(3) = .857 - .677 = .180$. Now suppose that we
had 1000 lots coming in from a process with p' = .04. Of these, we
would expect (average) to have 406 accepted on the first sample by
having only 0 or 1 defective. Then also we would expect to have

$p' = .04$ $n_1 = 50$ $n_2 = 50$ Contrib.
Expectations: $n_1 p' = 2.0$ $n_2 p' = 2.0$ to Pa
   P(1 or less) = .406 406
       P(2) = .271   P(2 or less) = .677   .677(.271) = .183
   P(2 or less) = .677
       P(3) = .180   P(1 or less) = .406   .406(.180) = .073
   P(3 or less) = .857
                                                  Pa = .662

271 lots go into a second sample with 2 defectives against them.
Of these, .677 will be passed by the second sample yielding 2 or
less defectives (total 4 or less). Thus of the original 1000 lots,
we expect to pass .677 of the 271 lots yielding $d_1 = 2$, or .677(.271)
as a probability. Similarly, we expect to have 180 lots of the 1000
with 3 defectives against them from the first sample. To pass they
can afford not over 1 defective on the second sample. The probability of this latter event is .406. So the probability for this route
to acceptance is .406(.180) or .073. Adding the probabilities for
the three distinct routes to acceptance yields Pa = .662. Let us do
one more for comparison:

$p' = .08$ $n_1 = 50$ $n_2 = 50$ Contrib.
Expectations: $n_1 p' = 4.0$ $n_2 p' = 4.0$ to Pa
   P(1 or less) = .092 092
       P(2) = .146   P(2 or less) = .238   .238(.146) = .035
   P(2 or less) = .238
       P(3) = .195   P(1 or less) = .092   .092(.195) = .018
   P(3 or less) = .433
                                                  Pa = .145

In these calculations we have used the Poisson distribution
(Table B) as an approximation. The correct distribution for a series
of lots chosen from a process at p' is the binomial. As an illustration of the accuracy, the author used binomial tables [1] on the
first calculation. This yielded Pa = .66116 instead of .662.

Doing other similar calculations to those we have shown yields

p'	.00	.01	.02	.03	.04	.05	.06	.07	.08	.09	.10
Pa	1.000	.996	.950	.831	.662	.488	.339	.224	.145	.091	.057

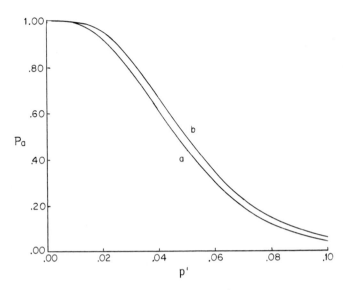

Fig. 9.4. Operating characteristic curves (Type B) for (a) the single plan n = 80, Ac = 3, Re = 4 and (b) the double plan $n_1 = n_2 = 50$, $Ac_1 = 1$, $Re_1 = 4$, $Ac_2 = 4$, $Re_2 = 5$.

OC curves are shown in Fig. 9.4 for our single and double sample plan examples. It is easily seen that the double sampling plan is everywhere more lenient than the single plan, but that they give rather similar protection against making wrong decisions on lots. Thus lots at p' = .00 to .02 are accepted the great majority of the time, and those at p' = .08 or more seldom are accepted whenever offered.

9.6.2 *The Average Sample Number Curve, Double Sampling*

Let us now consider the determination of the average sample number (ASN) curve, that is, the *average* number of pieces to be inspected per lot to reach a decision. For any double sampling plan we always inspect a first sample of n_1 pieces, but then go on to inspect a second sample of n_2 pieces only some of the time. How often does this occur, that is, what is its probability? Now if

$$d_1 \leq Ac_1 \quad \text{or} \quad d_1 \geq Re_1$$

we have a decision on the first sample, respectively, acceptance
and rejection. But if d_1 lies *between* Ac_1 and Re_1 then a second
sample is required, that is, when $Ac_1 < d_1 < Re_1$. What is the
probability? Let us see by studying the calculation table for Pa
when p' was .04. There we see that when d_1 = 2 or 3 a second sample is needed, so we want P(2) + P(3) = .271 + .180 = .451. Another
way to obtain this is by P(3 or less) - P(1 or less). Hence at
p' = .04

$$ASN = 50 + .451(50) = 72.6$$

since of 1000 lots at .04, all need 50 pieces inspected, and we expect that 451 lots of the 1000 will require a second sample for a
decision. So, on the *average*, the number inspected is as given.
Generalizing

$$ASN = n_1 + n_2 P(Ac_1 < d_1 < Re_1) \quad \text{(doubling sampling)} \quad (9.21)$$

Notice especially that there are no equals signs in the probability,
only inequality signs.

Similarly for p' = .08, we have P(1 < d_1 < 4) = P(3 or less)
- P(1 or less) = .433 - .092 = .341. Therefore

$$ASN = 50 + 50(.341) = 67.0$$

Proceeding we may find

p'	.00	.01	.02	.03	.04	.05
ASN	50.0	54.4	62.2	68.8	72.6	73.6

p'	.06	.07	.08	.09	.10
ASN	72.4	70.0	67.0	64.0	61.2

These results are plotted in Fig. 9.5, giving a curve starting at
ASN = 50, increasing to a maximum of around 73.6, then dropping back
gradually toward 50. Also shown is the ASN graph for the single
plan, a straight line. The ASN for the double plan lies everywhere
below that for the single plan. Remember, however, that for the
double plan the actual number of pieces to be inspected in a lot
will be either 50 or 100, depending upon whether or not a second
sample is required for a decision.

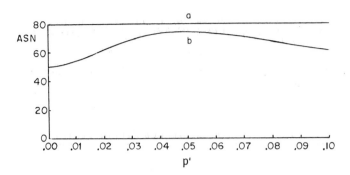

Fig. 9.5. Average sample number curves for two quite comparable sampling plans: (a) single, n = 80, Ac = 3, Re = 4; (b) double, $n_1 = n_2 = 50$, $Ac_1 = 1$, $Re_1 = 4$, $Ac_2 = 4$, $Re_2 = 5$.

9.6.3 The Average Outgoing Quality Curve

As we have seen, this characteristic of a sampling plan gives the average fraction defective, outgoing, when lots come in at fraction defective p'. The average outgoing quality (AOQ) is made up of the accepted lots, still at about p', and the rejected lots which have been screened of defectives and are now substantially free of them, giving a zero fraction defective. We could have either of two cases (1) all lots exactly at some fraction defective p' = D/N, and (2) lots of N chosen at random from a process in control at p'. Then too, there are cases according to whether we replace with good pieces all defective pieces found in the samples and/or the sorting. However, unless the lot size is quite small, these various cases all give quite closely the same AOQ. Thus as a practical matter we may again use the simple formula (9.10), that is,

$$AOQ = (Pa)p' \qquad\qquad (9.10)$$

Using this formula we find the following AOQ's for the double sampling plan being studied.

p'	.00	.01	.02	.03	.04	.05
AOQ	.0000	.0100	.0190	.0249	.0265	.0244

p'	.06	.07	.08	.09	.10
AOQ	.0203	.0157	.0116	.0082	.0057

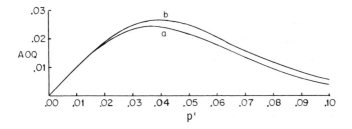

Fig. 9.6. Average outgoing quality curves for the two sampling plans (a) single, n = 80, Ac = 3, Re = 4; (b) double, n_1 = n_2 = 50, Ac_1 = 1, Re_1 = 4, Ac_2 = 4, Re_2 = 5.

These are plotted in Fig. 9.6, along with the AOQ curve for the single sampling plan for comparison. Since Pa for the double sampling plan was everywhere above that for the single plan (except at p' = 0), by (9.10) the AOQ curve for the double sampling plan lies everywhere above that for the single plan (except at p' = 0).

For the double sampling plan, the maximum AOQ, that is the AOQL, is about .0265, whereas for the single plan it is about .0242. Thus, no matter what the incoming lot fractions defective may be, the *average* outgoing fractions defective will be no worse than these respective AOQL values. In practice, the actual outgoing fraction defective is much less than the AOQL because to reach this level, p' would have to remain at the level giving a peak on the AOQ curve. And this occurs when Pa is low, perhaps .60. Should p' persist at such a level, sampling acceptance would soon be abandoned in favor of 100% sorting.

9.6.4 *The Average Total Inspection (ATI) Curve, Double Sampling*

The total amount of inspection per lot, made up of sampled pieces for a decision and of 100% screening of pieces in rejected lots, together make up the ATI. There are many possible formulas for ATI in double sampling. But it seems easiest to understand by using three categories: (1) lots *accepted* on the first samples, (2) lots *accepted* after a second sample, and (3) lots *rejected*. The respective inspection loads are (1) n_1, (2) n_1 + n_2, and (3) N. Meanwhile

the probabilities of such cases are, respectively, (1) $P(d_1 \leq Ac_1)$,
(2) $Pa - P(d_1 \leq Ac_1)$, and (3) $1 - Pa = P(\text{reject})$. The first is for
immediate acceptance on the first sample, while the second covers
the remainder of the *accepted* lots. Then the third covers all of
those lots which are *rejected*. Using the probabilities as weights
of the respective inspection amounts we have

$$ATI = n_1 \cdot P(d_1 \leq Ac_1) + (n_1 + n_2) \cdot [Pa - P(d \leq Ac_1)]$$

$$+ N \cdot [1 - Pa] \quad \text{(double sampling)} \tag{9.22}$$

The calculation of such an ATI looks complicated to one not
accustomed to using formulas. But it is really quite easy to do,
using the tables made in finding Pa for a double sampling plan.
Let us see.

For our double sampling example, we have at $p' = .04$, $Pa = .662$,
and also $P(d_1 \leq Ac_1) = P(d_1 \leq 1) = .406$. Therefore, using (9.22),

$$ASN = 50(.406) + 100(.662 - .406) + 1000(1 - .662)$$
$$= 20 + 26 + 338 = 384$$

At $p' = .08$, we have $Pa = .145$, $P(d_1 \leq 1) = .092$ and hence

$$ASN = 50(.092) + 100(.145 - .092) + 1000(1 - .145)$$
$$= 5 + 5 + 855 = 865$$

Proceeding in this way we obtain

p'	.00	.01	.02	.03	.04	.05	.06	.07	.08	.09	.10
ATI	50	59	109	224	384	546	685	792	865	915	947

This ATI curve is shown in Fig. 9.7, along with the ATI curve for
the single sampling plan. The latter is above the former, in part
because it has a uniformly higher ASN curve.

9.6.5 Truncated Inspection and the ASN Curve

Truncated or curtailed inspection is quite often used in practice.
What this means is that for decision making it may be possible to
stop sampling inspection on a lot before completing the sample size(s)
called for in the plan. For example, in our single sampling plan

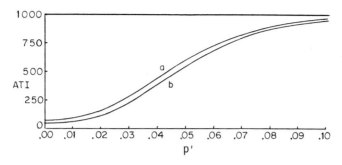

Fig. 9.7. Average total inspection curves for lots of N = 1000, for (a) single plan, n = 80, Ac = 3, Re = 4, and (b) double plan, $n_1 = n_2 = 50$, $Ac_1 = 1$, $Re_1 = 4$, $Ac_2 = 4$, $Re_2 = 5$.

n = 80, Ac = 3, Re = 4, it could happen that the fourth defective occurs on the 37th piece inspected. It has now become certain that the lot will be rejected, because even if the remaining 43 pieces to complete n = 80 were all good ones we would still have d ≥ Re = 4. This could be quite a saving in the case of a rejection. On the other hand, in the case of lot acceptance, much smaller savings might occur. The maximum saving would be if all of the first 77 pieces inspected were good. Then, even if all of the remaining three pieces should be defectives, we would still have d < Re = 4. Hence we would know what we could accept. But 77 is the smallest number to which we could ever truncate under *acceptance*. Thus with such small potential saving *under acceptance*, we do not bother with truncation.

But even with lot *rejection* under *single* sampling, we ordinarily do not truncate, but instead complete the inspection of the sample for the record, so as to have the same sample size for each lot, and thus can give each lot the same weight in determining p̄. In fact, this also makes easier the construction of a control chart record of performance, p or np, which is an excellent thing to maintain in acceptance sampling.

Now with *double sampling* the situation is a bit different. In general, we will always complete the *first* sample for the record. Since many lots do not call for a second sample, only the results

of the first samples will be used for the control chart record.
Thus we are free to truncate the second sample. But for the reasons given above, it only pays to truncate under early rejection.
But this does provide substantial savings and gives a lower ASN
curve than such a one as in Fig. 9.5, especially at relatively high
p'. (See Ref. 2 for an explanation of how to make the calculations.)

9.6.6 *Use of Calculations by Others*

Sections 9.5 and 9.6 on the characteristics of single and double
sampling plans have been given so that you may be able to understand
the meanings of them and how the curves are developed. That is,
this material provides a broad orientation to acceptance sampling
plans. But you may never need to calculate any plans yourself. The
reason is that there are so many available which have been calculated by others. (See the published plans in Chapter 10.) Then too,
if you do need some curves calculated, there is likely to be available in your organization calculators and programmers who can do the
job for you. But you should know what is going on and what the results mean.

9.7 ACCEPTANCE SAMPLING FOR DEFECTS

Up to now in this chapter we have been studying plans where the
acceptance-rejection criteria were in terms of counts of *defective
pieces* in samples, and determining how the characteristics depended
upon lot or process quality p'. Let us now consider criteria in
terms of counts of *defects*. We are quite likely to use such an approach to quality when we are concerned with a variety of possible
defects, any one of which makes a piece defective. Moreover, two
or three defects on a single piece still gives only one defective.
Now if c' is so small that we have on the average only five *defects*
per 100 pieces, it is unlikely that two or more of the five occur
on a single piece thus giving less than five *defectives* per 100.
But if c' is great enough to yield 10 defects per 100 pieces, then
duplication becomes more likely, and a c' of 10 per 100 is about
equivalent to a p' of .09. So with such a c' or a higher one, we

tend to use criteria in terms of counts of defects rather then defectives, although either could be used. Then instead of criteria
such as d ≤ Ac or d ≥ Re, we merely change to counts, c, on samples
of n pieces and use c ≤ Ac, c ≥ Re. And interest centers on Pa,
AOQ, ASN, or ATI versus lot or process *defects per 100 units* rather
than defectives per 100 pieces, that is, percent defective. The
basic distribution for such calculations on defects is the Poisson
(Table B) rather than the binomial.

9.8 FINDING A SINGLE SAMPLING PLAN TO MATCH
TWO POINTS ON THE OC CURVE

An approach to acceptance sampling plans which has rather wide
appeal is to set two fractions defective p' and appropriate risks
or probabilities of wrong decisions. This makes especially good
sense when we are concerned with a single isolated lot, or with
lots coming along at quite infrequent intervals. Then we want to
be given good assurance that we will accept a lot if p' = D/N is
at some acceptably low p', and also good assurance that we will reject a lot if p' is at some rejectably high p'. For example, we
might ask for Pa to be .95, if p' = .01. Thus the producer of a
lot with p' = .01 is subjected to a risk of only .05 of his lot being rejected. On the other hand, a lot at p' = .04 might be considered quite undesirable, and we might be willing to accept such
a lot only 10% of the times offered, that is, Pa = .10. Thus we
desire to find a plan with the following two points on the OC curve:

$$p_1' = .01, \quad Pa = .95 \quad \text{and} \quad p_2' = .04, \quad Pa = .10$$

Now with such a plan what are the odds on a lot at say p' = .02?
This can be found from the complete OC curve for the plan which was
found. But obviously its Pa would lie between .95 and .10.

Here let us say a word about "risks." At p' = .01 = p_1', we
might be willing to accept all such lots. But decision making is
always subject to some risks of wrong decisions. Here the risk of
rejecting a lot at p' = .01 is set to be .05. Thus about one time
in 20 such an *acceptable* lot will be rejected. This risk or

probability is often called the Producer's Risk, PR, and is sometimes known by the small Greek alpha, α. On the other hand, the consumer regards lots at p' = .04 as undesirable and would like to never accept such lots. But in order to avoid having to sort every lot 100%, he must be willing to take some risk of erroneously *accepting* such a *rejectable* lot, should one be offered. This risk was set at .10 and is called the Consumer's Risk, CR, and is sometimes known by the small Greek beta, β. Thus the consumer is willing to erroneously accept such lots 10% of the times they are offered. The .05 value for PR and .10 value for CR have somehow gotten quite well entrenched in the literature, but they do make quite good sense.

Now if you are willing to use a 5% PR and a 10% CR, then the problem of finding a plan for defectives which matches as well as possible the two points on the OC curve

$$p_1', \ Pa = .95 \qquad \text{and} \qquad p_2', \ Pa = .10$$

becomes very easy using Table 9.2. This follows the plan in Ref. 3.

Let us use Table 9.2 on our problem. We first take the desired fractions defective $p_2' = .04$, $p_1' = .01$ and divide for the operating

TABLE 9.2. Operating Ratios, p_2'/p_1', and Acceptance Numbers for Finding a Single Attribute Plan Matching Two Points on the OC Curve, with PR = .05 and CR = .10

p_2'/p_1'	Ac	np_2'	p_2'/p_1'	Ac	np_2'
44.7	0	2.30	2.50	10	15.41
10.9	1	3.89	2.40	11	16.60
6.51	2	5.32	2.31	12	17.78
4.89	3	6.68	2.24	13	18.96
4.06	4	7.99	2.18	14	20.13
3.55	5	9.27	2.12	15	21.29
3.21	6	10.53	2.07	16	22.45
2.96	7	11.77	2.03	17	23.61
2.77	8	12.99	1.99	18	24.76
2.62	9	14.21			

ratio p_2'/p_1' = .04/.01 = 4. (The smaller such a ratio is the more
discriminating power we are asking of our plan.) Then we look for
4.00 in columns 1 and 4. Our ratio 4 lies between 4.06 and 3.55,
so we take the row with 3.55 and find Ac = 5. Now we need a sample
size n. The third entry in this row is np_2' = 9.27. But we started
with p_2' = .04. Thus we have n(.04) = 9.27, giving n = 9.27/.04 = 232.
Therefore, the desired plan is n = 232, Ac = 5, Re = 6. This plan
matches the .04, Pa = .10 point on the OC curve as well as the
Poisson approximation allows. Meanwhile the PR at p_1' = .01 is less
than .05. This is because we did not hit p_2'/p_1' exactly, but instead
took an operating ratio of 3.55. Thus we arrived at a slightly more
discriminating OC curve than we originally asked for. We can find
our actual PR at p_1' = .01 by using Table B. Thus np_1' = 2.32, P(5 or
less) = .968, PR = 1 - .968 = .032. The full OC curve can be drawn
as in Section 9.5.1.

A set of single sampling plans following this approach is given
in Ref. 4.

9.9 SOME PRINCIPLES AND CONCEPTS IN SAMPLING
BY ATTRIBUTES

By way of review and summary, we now give some general material in
acceptance sampling for attributes.

9.9.1 *Prerequisites to Sound Acceptance Sampling
for Attributes*

For sound decision making on a lot or process, we should take care
to provide the following prerequisites:

1. Each type of defect must be clearly defined, so that inso-
    far as possible everyone involved will call any given
    imperfection a "defect," or else everyone will call it "no
    defect." This is especially important in all measurable
    characteristics which can make a piece good or a defective.
    There must be a good way to measure the piece or an accu-
    rate go-not-go gauge. Gauges and measuring instruments
    need frequent checking or calibration.

2. Samples must be taken in an unbiased way, preferably by a
    random procedure. At the very least, the sample or samples
    should be drawn from all portions of the lot. If there are

but few pieces in the lot which can be numbered, then we
may use random numbers of Table D, starting at a random
spot, to select the pieces for our sample. Or if there
are quite a few cartons in the lot, we may use Table D to
select a sampling of cartons to be opened, each of which
is then sampled to obtain the desired total of n pieces.
For example, suppose we have 60 cartons and decide to
select 20, and from each carton four pieces to give n = 80.
Then we number the cartons 1 to 60, bring down a pencil
point in Table D to give a two-digit pair of numbers. Say
it is 37, then carton 37 is taken. Next, 06 is next below
37, so carton 6 is taken. Next, 73, so go to the next.
Suppose it is also 37. Then go to the next pair of digits
until 20 cartons are chosen. *Results are unreliable unless
samples are unbiasedly chosen.*

3. Sound inspection must be somehow *obtained* and *maintained,*
    through careful training and supervision. Also there must
    not be undue pressure for speed in inspecting samples.
    (In one case, it was found that inspectors were making
    bets as to who could find the fewest defects!)

4. The rules for decision making on lots from samples must be
    exactly adhered to. No flinching. Harold Dodge related
    an amusing case to the author. He and a foreman saw an
    inspector take a piece, glance at it, set it back and take
    another. (d was already equal to Ac.) They immediately
    asked the inspector about it. The reply was "Oh, this is
    random sampling so it does not matter which piece I take."

9.9.2 *The Conditional Character of Pa*

Let us emphasize again that Pa is an if-then probability. For example, in Sec. 9.8, if $p' = p_1' = .04$, Pa is .10. That is, there is
a 10% risk of accepting a lot at .04, *if offered.* But if no such
lot (or a worse one) ever comes along, then there would be no risk
at all of accepting a lot with $p' = .04$. Thus we must guard against
statements such as "This plan gives a 10% risk of accepting a lot at
.04." We should add "if offered," or "should one come along." Or
even worse, we must not say "10% of the lots passed by this plan
will be at $p' = .04$."

9.9.3 *What is a Defective?*

As has already been pointed out, defects come in all degrees of
severity and importance. Many are comparatively unimportant, such
as mild blemishes, or a little grease not wiped off from a machine,

whereas others may be very serious indeed. Then there is the question as to whether the defect may be caught in subsequent operations
or in assembly, in which case its seriousness may be lessened. It
may be wise to use the term "nonconforming piece" if it fails to
meet exactly all requirements.

9.9.4 *The Proportion of the Lot Samples,*
n/N is Relatively Unimportant

This concept seems to be difficult for many to grasp. First, let
us suppose that we are talking about Type A operatint characteristic curves, that is, we want the probability of acceptance for a
series of lots, every one of which has exactly the same number of
defectives D in the given constant lot size N. Thus p' = D/N is
constant from lot to lot. Let us now use the same sample size n and
acceptance number Ac for various lot sizes N and see what effect the
lot size has. For the curves in Fig. 9.8, we used n = 10, Ac = 0

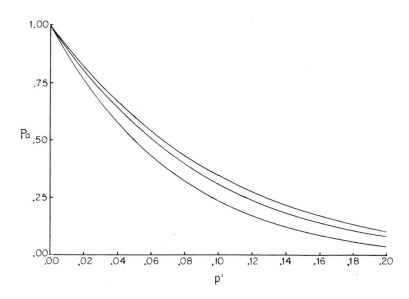

Fig. 9.8. Three operating characteristic curves, all with n = 10,
Ac = 0, Re = 1. They take account of lot size, that is, Type A OC
curves. The bottom curve has N = 20, middle curve N = 50, top curve
N infinite, that is, a Type B OC curve.

for each. But for the bottom curve N = 20, so that this was 50%
sampling. Actually D can only be 0, 1, 2, and so on, so the only
possible p' values are .00, .05, .10, and so on. Thus the "curve"
should be just a series of dots. The same is true for the middle
curve, for which N = 50, giving 20% sampling. (The exact calculation is done by the hypergeometric distribution, Sec. 4.4, but we
used a table [5].) Finally the top curve was for an infinite lot
size N; in practice a very large lot size. Thus for the top curve
we are sampling at 0%!

Now for practical purposes, are not the top two OC curves
practically identical, especially for quite low p' where we are
likely to use such a sampling plan? Even the bottom curve is not
greatly different. Thus the factor giving these OC curves is n and
Ac, far more than n/N.

We also mention that the comparison was for a Type A OC curve.
If our interest lies in Type B OC curves, that is, if we are concerned with Pa for a series of lots, each chosen from a process at
p', then the OC curve for all N's *is the same*. In particular for
n = 10, Ac = 0, the top curve of Fig. 9.8 holds for *all* N's, and
n/N has no influence at all!

9.9.5 *Why Not Sample a Fixed Percentage of Each Lot?*

As we have just seen, the thing which basically determines the shape
and discriminating power of a sampling plan is the sample size and
acceptance-rejection numbers, rather than the proportion n/N of the
lot we sample. Thus, provided we draw our sample randomly, a sample
of 100, say, will do just as well on a lot of 1,000,000 pieces as it
will on a lot of 500 or 1000.

To illustrate this further, let us consider what sampling 10%
of the lot does for us. We will draw the usual type of OC curve
(Type B) for 10% samples from lots of N = 100, 500, 2400, and 10,000.
For each, we will use an acceptance number so that Pa is about .90
when p' = .01. Thus if N = 100, n = .10N = 10, and using Ac = 0
gives Pa = .904. Similarly, if N = 500, n = 50, and letting Ac = 1,
gives Pa = .911 at p' = .01. The other plans are

N = 2400   n = 240   Ac = 4   Pa = .905 at p' = .01

N = 10,000   n = 1000   Ac = 14   Pa = .918 at p' = .01

We cannot hit Pa = .90 exactly because we have to make a choice between whole numbers for our acceptance number.

In Fig. 9.9 we show the four OC curves. Although each plan samples 10% of the lot and each has Pa about .90 when p' is at .01, the curves are markedly dissimilar. The OC curve for n = 10, Ac = 0 is very lazy and undiscriminating, because the sample size is so small. Would this be an acceptable plan? On the other hand, the plan for n = 1000, Ac = 14 is very sharply discriminating. If p' = .008, Pa = .983, whereas if p' = .02, Pa = .103. Is such sharp discrimination with such a steep, square-shouldered OC curve really needed? Very probably not.

The moral is that for small lots we likely need more than 10% of the lot to be in the sample, whereas with large lots, 10% is more

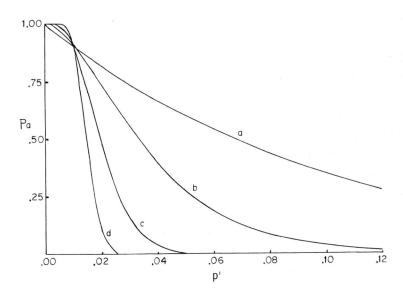

Fig. 9.9. Four operating characteristic curves, based on 10% samples, giving Pa at about .90, when p' = .01. (a) N = 100, n = 10, Ac = 0, (b) N = 500, n = 50, Ac = 1, (c) N = 2400, n = 240, Ac = 4, (d) N = 10,000, n = 1000, Ac = 14.

than enough. Therefore, pay attention to the sample size and acceptance-rejection criteria rather than the lot size. Thus we can gain
by random sampling of relatively large lots, instead of sampling
many small lots.

9.9.6 *There Are Risks Even with 100% Inspection*

A good many people seem to have great confidence in 100% inspection
of entire lots of articles. But risks are still present. Studies
have often been made on the efficiency of 100% sorting. They have
shown that seldom does sorting efficiency reach 80%, that is, will
find 80% of the defectives present in a lot. Efficiency, of course,
depends upon the inspector's training, skill, and motivation. But
it also depends upon the number of characteristics being checked,
the objectivity of the definition of a defect, and the fraction defective present. Efficiency tends to be best when the definitions
are clear, there are few characteristics to be checked, and many defectives are present. But even then fatigue and lack of perfect
attention cause defectives to be missed, and also, on the other hand,
good pieces to be rejected. Even with a gauge and only one dimension being checked, inspection is not perfect. An assistant plant
manager noticed one of his better inspectors sorting piston rings
for thickness by a dial gauge. It had a green sector for good thicknesses. The inspector's eyes were open but seemed a little glassy.
The manager put his hand over the dial while the inspector checked
50 rings! In checking 100% of food jar caps, an inspector is asked
to watch about six lines of caps moving along on parallel conveyors,
for some 15 different defects. The hypnotic effect is such that
100% efficiency is impossible. One more example may be enough. A
book of the Bible was to be reproduced as the "perfect book." After
six expert proofreaders each read the proof 100 times, the proof was
reproduced and sent to college campuses, with a challenge to find
any errors, and a reward as stimulus. Later the book was published
in full confidence. Subsequently six typographical errors came to
light, including one on the first line of the first page!

It has also been shown that sample inspection may possibly give a more accurate figure for the lot fraction defective than 100% sorting, because of the better job of inspection in the sampling work.

The only solution is continually to try to improve all inspection through motivation, objectivity, good tools and not requiring undue speed.

9.9.7 *Why Not Reject All Lots Yielding Any Defectives in a Sample?*

This suggestion is sometimes made, but will it prevent defectives from ever occurring in the lots? Obviously, the answer is no. The OC curve for sampling plans with Ac = 0, are always of the general shape of those in Fig. 9.8, that is, have the concave side of the curve always upward. But always, even with a large sample size n and Ac = 0, there will sometimes be lots passed having p' at some undesirable value, if any such lots reach the sampling inspection station. Meanwhile with such a shape of an OC curve, there may well be lots rejected which have p' at a satisfactorily low amount. Of course, if the defect or defects are so critical that none be tolerated, then the only things to do are (1) to try to perfect the production processes so that no defects are *made* and (2) to do the best you can with 100, 200, 300 or more percent inspection. On the other hand, if a suitably small percentage of defectives can be tolerated, for example, (1) where subsequent operations will find them or (2) where the defects are not all that critical, then sampling inspection can be used. The discriminating power (OC curve) can be set as desired. Of course, the most basic step is the perfecting of the production process. You cannot pass any bad lots, if no such lots are ever offered.

9.9.8 *Will a Sampling Plan Give the Same Decision Every Time?*

Some persons rather thoughtlessly think that if a sampling plan is any good, it will give the same decision every time on a given lot. To "test" the sampling plan, they may try it several times on a lot

and are disillusioned when the decision is not always the same. A
glance at any OC curve will show that repeated use of the *same* sampling plan on any lot can well lead to both kinds of decisions. In
fact for given p', only Pa of the time will we expect the lot to be
passed. For example, for an intermediate p', Pa can be .50, so that
half of the decisions will be to accept and half to reject. But
this is not so bad because such a p' is of "indifference" quality.

9.9.9 Misconception: Sampling Acceptance Plans Can Only Be Used if the Production Process is in Control

This idea is occasionally heard. As a matter of fact almost the
opposite is true. For if we find that the production process is in
control with satisfactory p', then we can even begin to think of
eliminating most of our sampling inspection and using the producer's
records. On the other hand, if the producer is in control but p'
is too high, then *all* lots should be 100% inspected to rectify the
high p', and thus we might well eliminate sampling inspection until
p' is improved.

But if the producer's process is not in control, so that there
are both acceptable and rejectable lots being produced, then sampling inspection is an excellent tool for locating the latter lots
for rejection or rectification, and for passing the former.

9.9.10 Some Notations on OC Curves

Let us now show some old and new notations on a typical OC curve in
Fig. 9.10, namely for the plan: n = 100, Ac = 3. The Poisson distribution of Table B was used for the calculations, the binomial
distribution being the formally correct distribution for this Type
B OC curve.

The vertical dashed lines were drawn to the points on the OC
curve where Pa is .95, .50, .10. The points on the p' axis where
these lines intersect are designated as follows:

$$p'_{95} = \text{p' having Pa} = .95 \qquad\qquad (9.23)$$
$$p'_{50} = \text{indifference quality}$$
$$\quad\ = \text{p' having Pa} = .50 \qquad\qquad (9.24)$$

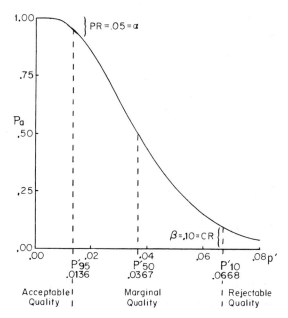

Fig. 9.10. A typical operating characteristic (OC) curve (Type B), calculated from the Poisson distribution, Table B. Notations shown for the sampling plan n = 100, Ac = 3.

$$p'_{10} = p' \text{ having Pa} = .10 \tag{9.25}$$

Usually we think of p' values from zero up to p'_{95} as being regarded by the sampling plan as fully acceptable. Likewise, those p' values above p'_{10} are regarded by the sampling plan as clearly rejectable. Meanwhile, those p' values between represent marginal quality of varying degrees. For these it is not too great an error to accept or to reject.

Note that in Sec. 9.8, p'_1 was the notation for p'_{95} and p'_2 for p'_{10}. Furthermore, in some areas p'_{95} is called the "acceptable quality level" (AQL) and p'_{10} the "lot tolerance proportion defective" (LTPD). But it is more common practice as in the Military Standard 105D (see Chapter 10) to use AQL and LTPD for nominal figures not specifically tied to any Pa's.

In the vertical line above p'_{95}, we find a brace of length .05 which is the probability of *rejection* of a lot at p'_{95}. This is

designated the producer's risk or PR since it is the probability
that his "good" lot will be *rejected* by the sampling plan. The PR
is also often designated by the small Greek alpha, α. Meanwhile on
the vertical line above p'_{10} is a brace for the .10 probability of a
lot of quality p'_{10} being erroneously *accepted* by the plan. This is
called the consumer's risk, CR, since it is the risk the consumer
assumes of accepting a "bad" lot if one should be offered. The CR
is often also designated by the small Greek beta, β.

Now unless some such risks α and β can be assumed, respectively, for p'_1 and p'_2, then *no* sampling plan can be used. But these
four: α, β, $p'_1 < p'_2$ can be set arbitrarily for a sampling plan.
Similarly for defects we could set α, β, $c'_1 < c'_2$.

9.10 SUMMARY

In Sec. 9.9 we have been reviewing principles, concepts, and notations. We complete the review by emphasizing the four curves which
describe and characterize any sampling plan based upon attributes,
either defectives or defects.

1. Protection Curves
    a. OC curve: Pa versus p' or c'
    b. AOQ curve: Average outgoing quality versus p' or c'
2. Cost Related Curves
    a. ASN curve: Average sample number versus p' or c'
    b. ATI curve: Average total inspection versus p' or c'

If willing to pay the price, you can have just about any degree of
protection you desire. Scientific acceptance sampling consists of
obtaining a good balance between protection and cost, while active-
ly ensuring that the prerequisites of Sec. 9.9.1 are carried out.

PROBLEMS

9.1 For the example of single sampling given in Sec. 9.5 check the
results given for Pa, AOQ, and ATI at p' = .02 and at p' = .06.
Find p'_{95} and p'_{10} for the plan by reversing the use of Table B, that
is, finding np' when P(3 or less) is .95 and .10.

9.2 For the example of double sampling given in Sec. 9.6, check the results given for Pa, AOQ, ASN, and ATI at p' = .02 and at p' = .06.

9.3 Sketch the OC, AOQ, ASN, and ATI curves labeling units on the axes, for the following two acceptance sampling plans for lots of 1000: (a) n = 80, Ac = 2, Re = 3 and (b) $n_1 = n_2 = 50$, $Ac_1 = 0$, $Re_1 = 3$, $Ac_2 = 3$, $Re_2 = 4$. Points for p' = .01, .02, ..., .06 should be enough.

9.4 Sketch the OC, AOQ, ASN, and ATI curves, labeling units on the axes for the single sampling plan: N = 2000, n = 100, Ac = 2, Re = 3. Also find the AOQL, and p_{95}' and p_{10}'. Use points for p' = .01, .02, ..., .05 initially.

9.5 Sketch the OC, AOQ, ASN, and ATI curves, labeling units on the axes for the double sampling plan: N = 2000, $n_1 = 50$, $n_2 = 100$, $Ac_1 = 1$, $Re_1 = 4$, $Ac_2 = 3$, $Re_2 = 4$. Use points for p' = .01, .02, ..., .06 which may be enough.

9.6 Find a single sampling plan (using Sec. 9.8) having $p_{95}' = p_1'$ = .01 and $p_{10}' = p_2' = .05$. Similarly, find one for $p_1' = .002$, $p_2' = .01$.

9.7 Find a single sampling plan (using Sec. 9.8) having $p_{95}' = p_1'$ = .01 and $p_{10}' = p_2' = .03$. Also find one for $p_1' = .001$, $p_2' = .003$.

REFERENCES

1. Harvard Univ., Comput. Lab., *Tables of the Cumulative Binomial Probability Distribution*, Harvard Univ. Press, Cambridge, Mass., 1955.

2. I. W. Burr, *Statistical Quality Control Methods*, Dekker, New York, 1976.

3. P. Peach and S. B. Littauer, A note on sampling inspection by attributes, *Ann. Math. Statist.*, *17*, 81-84 (1946).

4. R. L. Kirpatrick, Binomial sampling plans indexed by AQL and LTPD. *Indust. Quality Control*, *22*, 290-292 (1965).

5. G. J. Lieberman and D. B. Owen, *Tables of the Hypergeometric Probability Distribution*, Stanford Univ. Press, Stanford, Calif., 1961.

Chapter Ten

SOME STANDARD SAMPLING PLANS FOR ATTRIBUTES

In this chapter, we shall concentrate upon the ABC standard sampling plan for attributes. We shall also describe briefly some other
sampling plans which are available for their intended purposes.
Interesting as it is, we shall not go into the history of acceptance
sampling plans. An authoritative and readable summary is available
in Ref. 1. The ABC Standard is also called the Military Standard
MIL-STD 105D (there being three earlier versions) and American National Standards Institute ANSI Z1.4. The ABC stands for "American,
British and Canadian," its present form being to some extent the
work of a joint committee [2].

10.1 THE ABC STANDARD OR MILITARY STANDARD 105D

The ABC Standard, *Sampling Procedures and Tables for Inspection by
Attributes*, is widely used throughout the world. It is used for
receiving inspection, within-house inspection between departments,
and final inspection. It is used in attribute inspection, on either
defectives or defects. For the case of acceptance sampling by measurements, there is a companion standard, MIL-STD 414, which will be
briefly described in Chapter 11. The ABC Standard is highly useful,
but its objectives of quality control may only be obtained if *all*
of the instructions are precisely followed.

10.1.1 Aim

The basic aim of the ABC Standard is the maintenance of the outgoing
quality level at a given "acceptable quality level" or better.
There are two criteria of quality level, either of which may be

specified, namely, percent defective, that is *defectives* per hundred
units, or secondly *defects* per hundred units. Then the objective is
to insure that the outgoing quality is at the AQL or better.

Moreover, the standard is designed so that if the producer runs
consistently at precisely the AQL, then the great majority of his
lots can be expected to pass. But if his process is at even a little
worse than the AQL, he can expect to have trouble and must face
"tightened inspection." Thus, the AQL is the *minimum* quality performance at which the producer may safely run, rather than a perfectly safe level at which to run, as some have erroneously thought
of the AQL. The producer is, therefore, advised to run at or better
than the AQL.

Let us here point out that the reader should carefully distinguish between the two quite similar abbreviations: AQL and AOQL.
AQL stands for acceptable quality level, whereas AOQL represents
average outgoing quality limit, that is, the worst long-run average
outgoing quality, no matter what comes in.

10.1.2 *How It Operates to Fulfill the Aim*

Let us look at a typical OC curve, for example, Fig. 9.10. There
we see that for the plan n = 100, Ac = 3, p_{95}' = .0136. This means
that if the producer is running his process at p' = .0136 or 1.36%,
then he can *expect* to have 95% of his lots passed by the plan, that
is, 19 out of 20. Now suppose that the consumer, in using this
plan, has set .0136 as his AQL. Then the "great majority of lots"
from a production process at p' = .0136 will be accepted. So far
so good, but now to obtain such a high Pa when p' = .0136, he must
let Pa be also quite high when p' is a bit above .0136. Specifically, if p' should deteriorate to .020, then Pa is still high,
namely, .857. Therefore, if the consumer continues to use the same
plan n = 100, Ac = 3, he will be passing most of the lots at around
p' = .02, which he does not want to do.

What is the remedy? As we shall see, if a producer runs much
worse than the AQL set, then quite soon the plan will invoke

"tightened inspection." Specifically, it might well tighten to the plan n = 100, Ac = 2. Now, if p' = .02, then Pa = .677 so that only two lots out of three at .02 will be accepted. This is quite a drop from Pa = .857 for the original plan, at p' = .02. But also note that for this tightened plan, n = 100, Ac = 2, the AOQL is at .0136, it happens, so the consumer will have long-run quality at .0136 or better, that is, at his specified AQL. So he will have the quality requested even though the producer was not cooperating. Moreover, according to the standard, if the producer does not im- prove from p' = .02, the plan will soon call for abandonment of sampling altogether, going to 100% inspection.

Thus, in order to give a high probability of acceptance (low producer's risk) when the process is at the AQL or better, and yet to give the consumer adequate protection against worse process aver- ages than the AQL, we must be sure to use tightened inspection when- ever it is called for. The appropriate criteria are discussed fur- ther on.

10.1.3 Saving Inspection by "Reduced Sampling"

There is also provision for using smaller sample sizes than normal when quality is consistently better than the AQL and production is continuous. Under these conditions, bad quality lots are very un- likely and a less steep OC curve, coming from a smaller sample size, can be tolerated. There are, as we shall see, adequate safeguards which can call for reverting to the normal inspection plan. Use of reduced sampling, when available, is still optional and only used to save on inspection, *not* to achieve the basic aim of maintenance of outgoing quality p' = AQL.

10.1.4 Multiple Sampling

The ABC Standard makes available single, double, and multiple samp- ling plans. A typical multiple sampling plan having an AQL of 4.0 percent is the following one

Sample size, n	20	20	20	20	20	20	20
Cumulative sample size	20	40	60	80	100	120	140
Acceptance number, Ac	0	1	3	5	7	10	13
Rejection number, Re	4	6	8	10	11	12	14

This table is almost self explanatory. If in the first sample of 20 pieces, there are no defectives, we can accept the lot at once. Or, if there are four or more defectives we reject the lot at once. But if there are 1, 2, or 3 defectives in the first 20, we take another random sample of 20. Then we compare the cumulative number of defectives $d_1 + d_2$ against $Ac_2 = 1$ and $Re_2 = 6$. If

$$d_1 + d_2 \leq 1 \qquad \text{Accept lot}$$
$$d_1 + d_2 \geq 6 \qquad \text{Reject lot}$$
$$1 < d_1 + d_2 < 6 \qquad \text{Take a third sample of 20}$$

In the last case, we find d_3 defectives and then similarly compare $d_1 + d_2 + d_3$ against $Ac_3 = 3$ and $Re_3 = 8$. This process might possibly continue for seven samples at which point a decision becomes certain because there is no gap between $Ac_7 = 13$ and $Re_7 = 14$. Now, although it is possible to require all seven samples totaling 140, this is quite unlikely. In football language, "It would take a very good broken field runner to reach the end."

The standard provides two companion plans

Single: $n = 80$, $Ac = 7$, $Re = 8$
Double: $n_1 = 50$, $\Sigma\, n = 50$, $Ac_1 = 3$, $Re_1 = 7$
$\quad\quad\quad n_2 = 50$, $\Sigma\, n = 100$, $Ac_2 = 8$, $Re_2 = 9$

These two plans and the multiple plan all have practically the same OC curve and, therefore, AOQ curve. But the multiple plan will save inspection time by having a more favorable (lower) ASN curve than either of the companion plans. Also the double plan has a more favorable ASN curve than does the single plan. Of course, the price you pay for these more favorable ASN curves is greater complexity in the plans, requiring more training and supervision time. But all three plans are practically identical in fulfilling the aims of the standard, and thus which to use is a management decision.

10.1.5   Reproduction of Sample Pages from the
ABC Standard, MIL-STD 105D

SAMPLING PROCEDURES AND TABLES
FOR INSPECTION BY ATTRIBUTES

1. SCOPE

1.1   PURPOSE. This publication establishes sampling plans and procedures for inspection by attributes. When specified by the responsible authority, this publication shall be referenced in the specification, contract, inspection instructions, or other documents and the provisions set forth herein shall govern. The "responsible authority" shall be designated in one of the above documents.

1.2   APPLICATION. Sampling plans designated in this publication are applicable, but not limited, to inspection of the following:

a. End items.

b. Components and raw materials.

c. Operations.

d. Materials in process.

e. Supplies in storage.

f. Maintenance operations.

g. Data or records.

h. Administrative procedures.

These plans are intended primarily to be used for a continuing series of lots or batches.

The plans may also be used for the inspection of isolated lots or batches, but, in this latter case, the user is cautioned to consult the operating characteristic curves to find a plan which will yield the desired protection (see 11.6).

1.3   INSPECTION. Inspection is the process of measuring, examining, testing, or otherwise comparing the unit of product (see 1.5) with the requirements.

1.4   INSPECTION BY ATTRIBUTES. Inspection by attributes is inspection whereby either the unit of product is classified simply as defective or nondefective, or the number of defects in the unit of product is counted, with respect to a given requirement or set of requirements.

1.5   UNIT OF PRODUCT. The unit of product is the thing inspected in order to determine its classification as defective or nondefective or to count the number of defects. It may be a single article, a pair, a set, a length, an area, an operation, a volume, a component of an end product, or the end product itself. The unit of product may or may not be the same as the unit of purchase, supply, production, or shipment.

This page and those through page 288 are reproduced by permission from the Department of Defense from Military Standard MIL-STD 105D, Sampling Procedures and Tables for Inspection by Attributes, 1963.

2. CLASSIFICATION OF DEFECTS AND DEFECTIVES

2.1 METHOD OF CLASSIFYING DEFECTS.
A classification of defects is the enumeration of possible defects of the unit of product classified according to their seriousness. A defect is any nonconformance of the unit of product with specified requirements. Defects will normally be grouped into one or more of the following classes; however, defects may be grouped into other classes, or into subclasses within these classes.

2.1.1 CRITICAL DEFECT. A critical defect is a defect that judgment and experience indicate is likely to result in hazardous or unsafe conditions f o r individuals using, maintaining, or depending upon the product; or a defect that judgment and experience indicate is likely to prevent performance of the tactical function of a major end item such as a ship, aircraft, tank, missile or space vehicle. NOTE: For a special provision relating to critical defects, see 6.3.

2.1.2 MAJOR DEFECT. A major defect is a defect, other than critical, that is likely to result in failure, or to reduce materially the usability of the unit of product for its intended purpose.

2.1.3 MINOR DEFECT. A minor defect is a defect that is not likely to reduce materially the usability of the unit of product for its intended purpose, or is a departure from established standards having little bearing on the effective use or operation of the unit.

2.2 METHOD OF CLASSIFYING DEFECTIVES. A defective is a unit of product which contains one or more defects. Defectives will usually be classified as follows:

2.2.1 CRITICAL DEFECTIVE. A critical defective contains one or more critical defects and may also contain major and or minor defects. NOTE: For a special provision relating to critical defectives, see 6.3.

2.2.2 MAJOR DEFECTIVE. A major defective contains one or more major defects, and may also contain minor defects but contains no critical defect.

2.2.3 MINOR DEFECTIVE. A minor defective contains one or more minor defects but contains no critical or major defect.

3. PERCENT DEFECTIVE AND DEFECTS PER HUNDRED UNITS

3.1 EXPRESSION OF NONCONFORMANCE. The extent of nonconformance of product shall be expressed either in terms of percent defective or in terms of defects per hundred units.

3.2 PERCENT DEFECTIVE. The percent defective of any given quantity of units of product is one hunderd times the number of defective units of product contained therein divided by the total number of units of product, i.e.:

$$\text{Percent defective} = \frac{\text{Number of defectives}}{\text{Number of units inspected}} \times 100$$

3.3 DEFECTS PER HUNDRED UNITS. The number of defects per hundred units of any given quantity of units of product is one hundred times the number of defects contained therein (one or more defects being possible in any unit of product) divided by the total number of units of product, i.e.:

$$\text{Defects per hundred units} = \frac{\text{Number of defects}}{\text{Number of units inspected}} \times 100$$

4.   ACCEPTABLE QUALITY LEVEL (AQL)

4.1   USE. The AQL, together with the Sample Size Code Letter, is used for indexing the sampling plans provided herein.

4.2   DEFINITION. The AQL is the maximum percent defective (or the maximum number of defects per hundred units) that, for purposes of sampling inspection, can be considered satisfactory as a process average (see 11.2).

4.3   NOTE ON THE MEANING OF AQL. When a consumer designates some specific value of AQL for a certain defect or group of defects, he indicates to the supplier that his (the consumer's) acceptance sampling plan will accept the great majority of the lots or batches that the supplier submits, provided the process average level of percent defective (or defects per hundred units) in these lots or batches be no greater than the designated value of AQL. Thus, the AQL is a designated value of percent defective (or defects per hundred units) that the consumer indicates will be accepted most of the time by the acceptance sampling procedure to be used. The sampling plans provided herein are so arranged that the probability of acceptance at the designated AQL value depends upon the sample size, being generally higher for large samples than for small ones, for a given AQL. The AQL alone does not describe the protection to the consumer for individual lots or batches but more directly relates to what might be expected from a series of lots or batches, provided the steps indicated in this publication are taken. It is necessary to refer to the operating characteristic curve of the plan, to determine what protection the consumer will have.

4.4   LIMITATION. The designation of an AQL shall not imply that the supplier has the right to supply knowingly any defective unit of product.

4.5   SPECIFYING AQLs. The AQL to be used will be designated in the contract or by the responsible authority. Different AQLs may be designated for groups of defects considered collectively, or for individual defects. An AQL for a group of defects may be designated in addition to AQLs for individual defects, or subgroups, within that group. AQL values of 10.0 or less may be expressed either in percent defective or in defects per hundred units; those over 10.0 shall be expressed in defects per hundred units only.

4.6   PREFERRED AQLs. The values of AQLs given in these tables are known as preferred AQLs. If, for any product, an AQL be designated other than a preferred AQL, these tables are not applicable.

5.   SUBMISSION OF PRODUCT

5.1   LOT OR BATCH. The term lot or batch shall mean "inspection lot" or "inspection batch," i.e., a collection of units of product from which a sample is to be drawn and inspected to determine conformance with the acceptability criteria, and may differ from a collection of units designated as a lot or batch for other purposes (e.g., production, shipment, etc.).

5.2   FORMATION OF LOTS OR BATCHES. The product shall be assembled into identifiable lots, sublots, batches, or in such other manner as may be prescribed (see 5.4). Each lot or batch shall, as far as is practicable,

5. SUBMISSION OF PRODUCT (Continued)

consist of units of product of a single type, grade, class, size, and composition, manufactured under essentially the same conditions, and at essentially the same time.

5.3 LOT OR BATCH SIZE. The lot or batch size is the number of units of product in a lot or batch.

5.4 PRESENTATION OF LOTS OR BATCHES. The formation of the lots or batches, lot or batch size, and the manner in which each lot or batch is to be presented and identified by the supplier shall be designated or approved by the responsible authority. As necessary, the supplier shall provide adequate and suitable storage space for each lot or batch, equipment needed for proper identification and presentation, and personnel for all handling of product required for drawing of samples.

6. ACCEPTANCE AND REJECTION

6.1 ACCEPTABILITY OF LOTS OR BATCHES. Acceptability of a lot or batch will be determined by the use of a sampling plan or plans associated with the designated AQL or AQLs.

6.2 DEFECTIVE UNITS. The right is reserved to reject any unit of product found defective during inspection whether that unit of product forms part of a sample or not, and whether the lot or batch as a whole is accepted or rejected. Rejected units may be repaired or corrected and resubmitted for inspection with the approval of, and in the manner specified by, the responsible authority.

6.3 SPECIAL RESERVATION FOR CRITICAL DEFECTS. The supplier may be required at the discretion of the responsible authority to inspect every unit of the lot or batch for critical defects. The right is reserved to inspect every unit submitted by the supplier for critical defects, and to reject the lot or batch immediately, when a critical defect is found. The right is reserved also to sample, for critical defects, every lot or batch submitted by the supplier and to reject any lot or batch if a sample drawn therefrom is found to contain one or more critical defects.

6.4 RESUBMITTED LOTS OR BATCHES. Lots or batches found unacceptable shall be resubmitted for reinspection only after all units are re-examined or retested and all defective units are removed or defects corrected. The responsible authority shall determine whether normal or tightened inspection shall be used, and whether reinspection shall include all types or classes of defects or for the particular types or classes of defects which caused initial rejection.

7. DRAWING OF SAMPLES

7.1 SAMPLE. A sample consists of one or more units of product drawn from a lot or batch, the units of the sample being selected at random without regard to their quality. The number of units of product in the sample is the sample size.

7.2 REPRESENTATIVE SAMPLING. When appropriate, the number of units in the sample shall be selected in proportion to the size of sublots or subbatches, or parts of the lot or batch, identified by some rational criterion.

7. DRAWING OF SAMPLES (Continued)

When representative sampling is used, the units from each part of the lot or batch shall be selected at random.

7.3 TIME OF SAMPLING.
Samples may be drawn after all the units comprising the lot or batch have been assembled, or samples may be drawn during assembly of the lot or batch.

7.4 DOUBLE OR MULTIPLE SAMPLING.
When double or multiple sampling is to be used, each sample shall be selected over the entire lot or batch.

8. NORMAL, TIGHTENED AND REDUCED INSPECTION

8.1 INITIATION OF INSPECTION.
Normal inspection will be used at the start of inspection unless otherwise directed by the responsible authority.

8.2 CONTINUATION OF INSPECTION.
Normal, tightened or reduced inspection shall continue unchanged for each class of defects or defectives on successive lots or batchs except where the switching procedures given below require change. The switching procedures given below require a change. The switching procedures shall be applied to each class of defects or defectives independently.

8.3 SWITCHING PROCEDURES.

8.3.1 NORMAL TO TIGHTENED.
When normal inspection is in effect, tightened inspection shall be instituted when 2 out of 5 consecutive lots or batches have been rejected on original inspection (i.e., ignoring resubmitted lots or batches for this procedure).

8.3.2 TIGHTENED TO NORMAL.
When tightened inspection is in effect, normal inspection shall be instituted when 5 consecutive lots or batches have been considered acceptable on original inspection.

8.3.3 NORMAL TO REDUCED.
When normal inspection is in effect, reduced inspection shall be instituted providing that all of the following conditions are satisfied:

a. The preceding 10 lots or batches (or more, as indicated by the note to Table VIII) have been on normal inspection and none has been rejected on original inspection; and

b. The total number of defectives (or defects) in the samples from the preceding 10 lots or batches (or such other number as was used for condition "a" above) is equal to or less than the applicable number given in Table VIII. If double or multiple sampling is in use, all samples inspected should be included, not "first" samples only; and

c. Production is at a steady rate; and

d. Reduced inspection is considered desirable by the responsible authority.

8.3.4 REDUCED TO NORMAL.
When reduced inspection is in effect, normal inspection shall be instituted if any of the following occur on original inspection:

a. A lot or batch is rejected; or

b. A lot or batch is considered acceptable under the procedures of 10.1.4; or

c. Production becomes irregular or delayed; or

d Other conditions warrant that normal inspection shall be instituted.

8.4 DISCONTINUATION OF INSPECTION.
In the event that 10 consecutive lots or batches remain on tightened inspection (or such other number as may be designated by the responsible authority), inspection under the provisions of this document should be discontinued pending action to improve the quality of submitted material.

9. SAMPLING PLANS

9.1 SAMPLING PLAN. A sampling plan indicates the number of units of product from each lot or batch which are to be inspected (sample size or series of sample sizes) and the criteria for determining the acceptability of the lot or batch (acceptance and rejection numbers).

9.2 INSPECTION LEVEL. The inspection level determines the relationship between the lot or batch size and the sample size. The inspection level to be used for any particular requirement will be prescribed by the responsible authority. Three inspection levels: I, II, and III, are given in Table I for general use. Unless otherwise specified, Inspection Level II will be used. However, Inspection Level I may be specified when less discrimination is needed, or Level III may be specified for greater discrimination. Four additional special levels: S–1, S–2, S–3 and S–4, are given in the same table and may be used where relatively small sample sizes are necessary and large sampling risks can or must be tolerated.

NOTE: In the designation of inspection levels S–1 to S–4, care must be exercised to avoid AQLs inconsistent with these inspection levels.

9.3 CODE LETTERS. Sample sizes are designated by code letters. Table I shall be used to find the applicable code letter for the particular lot or batch size and the prescribed inspection level.

9.4 OBTAINING SAMPLING PLAN. The AQL and the code letter shall be used to obtain the sampling plan from Tables II, III or IV. When no sampling plan is available for a given combination of AQL and code letter, the tables direct the user to a different letter. The sample size to be used is given by the new code letter not by the original letter. If this procedure leads to different sample sizes for different classes of defects, the code letter corresponding to the largest sample size derived may be used for all classes of defects when designated or approved by the responsible authority. As an alternative to a single sampling plan with an acceptance number of 0, the plan with an acceptance number of 1 with its correspondingly larger sample size for a designated AQL (where available), may be used when designated or approved by the responsible authority.

9.5 TYPES OF SAMPLING PLANS. Three types of sampling plans: Single, Double and Multiple, are given in Tables II, III and IV, respectively. When several types of plans are available for a given AQL and code letter, any one may be used. A decision as to type of plan, either single, double, or multiple, when available for a given AQL and code letter, will usually be based upon the comparison between the administrative difficulty and the average sample sizes of the available plans. The average sample size of multiple plans is less than for double (except in the case corresponding to single acceptance number 1) and both of these are always less than a single sample size. Usually the administrative difficulty for single sampling and the cost per unit of the sample are less than for double or multiple.

10. DETERMINATION OF ACCEPTABILITY

10.1 PERCENT DEFECTIVE INSPECTION.

To determine acceptability of a lot or batch under percent defective inspection, the applicable sampling plan shall be used in accordance with 10.1.1, 10.1.2, 10.1.3, 10.1.4, and 10.1.5.

10.1.1 SINGLE SAMPLING PLAN.

The number of sample units inspected shall be equal to the sample size given by the plan. If the number of defectives found in the sample is equal to or less than the acceptance number, the lot or batch shall be considered acceptable. If the number of defectives is equal to or greater than the rejection number, the lot or batch shall be rejected.

10.1.2 DOUBLE SAMPLING PLAN.

The number of sample units inspected shall be equal to the first sample size given by the plan. If the number of defectives found in the first sample is equal to or less than the first acceptance number, the lot or batch shall be considered acceptable. If the number of defectives found in the first sample is equal to or greater than the first rejection number, the lot or batch shall be rejected. If the number of defectives found in the first sample is between the first acceptance and rejection numbers, a second sample of the size given by the plan shall be inspected. The

number of defectives found in the first and second samples shall be accumulated. If the cumulative number of defectives is equal to or less than the second acceptance number, the lot or batch shall be considered acceptable. If the cumulative number of defectives is equal to or greater than the second rejection number, the lot or batch shall be rejected.

10.1.3 MULTIPLE SAMPLE PLAN.

Under multiple sampling, the procedure shall be similar to that specified in 10.1.2, except that the number of successive samples required to reach a decision may be more than two.

10.1.4 SPECIAL PROCEDURE FOR REDUCED INSPECTION.

Under reduced inspection, the sampling procedure may terminate without either acceptance or rejection criteria having been met. In these circumstances, the lot or batch will be considered acceptable, but normal inspection will be reinstated starting with the next lot or batch (see 8.3.4 (b)).

10.2 DEFECTS PER HUNDRED UNITS INSPECTION.

To determine the acceptability of a lot or batch under Defects per Hundred Units inspection, the procedure specified for Percent Defective inspection above shall be used, except that the word "defects" shall be substituted for "defectives."

11. SUPPLEMENTARY INFORMATION

11.1 OPERATING CHARACTERISTIC CURVES.

The operating characteristic curves for normal inspection, shown in Table X (pages 30–62), indicate the percentage of lots or batches which may be expected to be accepted under the various sampling plans for a given process quality. The curves shown are for single sampling; curves for double

and multiple sampling are matched as closely as practicable. The O. C. curves shown for AQLs greater than 10.0 are based on the Poisson distribution and are applicable for defects per hundred units inspection; those for AQLs of 10.0 or less and sample sizes of 80 or less are based on the binomial distribution and are applicable for percent defec-

11. SUPPLEMENTARY INFORMATION (Continued)

tive inspection; those for AQLs of 10.0 or less and sample sizes larger then 80 are based on the Poisson distribution and are applicable either for defects per hundred units inspection, or for percent defective inspection (the Poisson distribution being an adequate approximation to the binomial distribution under these conditions). Tabulated values, corresponding to selected values of probabilities of acceptance (P_a, in percent) are given for each of the curves shown, and, in addition, for tightened inspection, and for defects per hundred units for AQLs of 10.0 or less and sample sizes of 80 or less.

11.2 PROCESS AVERAGE.

The process average is the average percent defective or average number of defects per hundred units (whichever is applicable) of product submitted by the supplier for original inspection. Original inspection is the first inspection of a particular quantity of product as distinguished from the inspection of product which has been resubmitted after prior rejection.

11.3 AVERAGE OUTGOING QUALITY (AOQ).

The AOQ is the average quality of outgoing product including all accepted lots or batches, plus all rejected lots or batches after the rejected lots or batches have been effectively 100 percent inspected and all defectives replaced by nondefectives.

11.4 AVERAGE OUTGOING QUALITY LIMIT (AOQL).

The AOQL is the maximum of the AOQs for all possible incoming qualities for a given acceptance sampling plan. AOQL values are given in Table V–A for each of the single sampling plans for normal inspection and in Table V–B for each of the single sampling plans for tightened inspection.

11.5 AVERAGE SAMPLE SIZE CURVES.

Average sample size curves for double and multiple sampling are in Table IX. These show the average sample sizes which may be expected to occur under the various sampling plans for a given process quality. The curves assume no curtailment of inspection and are approximate to the extent that they are based upon the Poisson distribution, and that the sample sizes for double and multiple sampling are assumed to be 0.631n and 0.25n respectively, where n is the equivalent single sample size.

11.6 LIMITING QUALITY PROTECTION.

The sampling plans and associated procedures given in this publication were designed for use where the units of product are produced in a continuing series of lots or batches over a period of time. However, if the lot or batch is of an isolated nature, it is desirable to limit the selection of sampling plans to those, associated with a designated AQL value, that provide not less than a specified limiting quality protection. Sampling plans for this purpose can be selected by choosing a Limiting Quality (LQ) and a consumer's risk to be associated with it. Tables VI and VII give values of LQ for the commonly used consumer's risks of 10 percent and 5 percent respectively. If a different value of consumer's risk is required, the O.C. curves and their tabulated values may be used. The concept of LQ may also be useful in specifying the AQL and Inspection Levels for a series of lots or batches, thus fixing minimum sample size where there is some reason for avoiding (with more than a given consumer's risk) more than a limiting proportion of defectives (or defects) in any single lot or batch.

TABLE I — Sample size code letters

Lot or batch size			Special inspection levels				General inspection levels		
			S-1	S-2	S-3	S-4	I	II	III
2	to	8	A	A	A	A	A	A	B
9	to	15	A	A	A	A	A	B	C
16	to	25	A	A	B	B	B	C	D
26	to	50	A	B	B	C	C	D	E
51	to	90	B	B	C	C	C	E	F
91	to	150	B	B	C	D	D	F	G
151	to	280	B	C	D	E	E	G	H
281	to	500	B	C	D	E	F	H	J
501	to	1200	C	C	E	F	G	J	K
1201	to	3200	C	D	E	G	H	K	L
3201	to	10000	C	D	F	G	J	L	M
10001	to	35000	C	D	F	H	K	M	N
35001	to	150000	D	E	G	J	L	N	P
150001	to	500000	D	E	G	J	M	P	Q
500001	and	over	D	E	H	K	N	Q	R

TABLE II-A—Single sampling plans for normal inspection (Master table)

Sample size code letter	Sample size	0.010		0.015		0.025		0.040		0.065		0.10		0.15		0.25		0.40		0.65		1.0		1.5		2.5		4.0		6.5		10		15		25		40		65		100		150		250		400		650		1000	
		Ac	Re	Ac	Re	Ac	Re	Ac	Re	Ac	Re	Ac	Re	Ac	Re	Ac	Re	Ac	Re	Ac	Re	Ac	Re	Ac	Re	Ac	Re	Ac	Re	Ac	Re	Ac	Re	Ac	Re	Ac	Re	Ac	Re	Ac	Re	Ac	Re	Ac	Re	Ac	Re	Ac	Re	Ac	Re	Ac	Re
A	2	↓		↓		↓		↓		↓		↓		↓		↓		↓		↓		↓		↓		↓		↓		↓		↓		0	1	1	2	2	3	3	4	5	6	7	8	10	11	14	15	21	22	30	31
B	3	↓		↓		↓		↓		↓		↓		↓		↓		↓		↓		↓		↓		↓		↓		↓		0	1	1	2	2	3	3	4	5	6	7	8	10	11	14	15	21	22	30	31	44	45
C	5	↓		↓		↓		↓		↓		↓		↓		↓		↓		↓		↓		↓		↓		↓		0	1	1	2	2	3	3	4	5	6	7	8	10	11	14	15	21	22	30	31	44	45	↑	
D	8	↓		↓		↓		↓		↓		↓		↓		↓		↓		↓		↓		↓		↓		0	1	1	2	2	3	3	4	5	6	7	8	10	11	14	15	21	22	30	31	44	45	↑		↑	
E	13	↓		↓		↓		↓		↓		↓		↓		↓		↓		↓		↓		↓		0	1	1	2	2	3	3	4	5	6	7	8	10	11	14	15	21	22	30	31	44	45	↑		↑		↑	
F	20	↓		↓		↓		↓		↓		↓		↓		↓		↓		↓		↓		0	1	1	2	2	3	3	4	5	6	7	8	10	11	14	15	21	22	↑		↑		↑		↑		↑		↑	
G	32	↓		↓		↓		↓		↓		↓		↓		↓		↓		↓		0	1	1	2	2	3	3	4	5	6	7	8	10	11	14	15	21	22	↑		↑		↑		↑		↑		↑		↑	
H	50	↓		↓		↓		↓		↓		↓		↓		↓		↓		0	1	1	2	2	3	3	4	5	6	7	8	10	11	14	15	21	22	↑		↑		↑		↑		↑		↑		↑		↑	
J	80	↓		↓		↓		↓		↓		↓		↓		↓		0	1	1	2	2	3	3	4	5	6	7	8	10	11	14	15	21	22	↑		↑		↑		↑		↑		↑		↑		↑		↑	
K	125	↓		↓		↓		↓		↓		↓		↓		0	1	1	2	2	3	3	4	5	6	7	8	10	11	14	15	21	22	↑		↑		↑		↑		↑		↑		↑		↑		↑		↑	
L	200	↓		↓		↓		↓		↓		↓		0	1	1	2	2	3	3	4	5	6	7	8	10	11	14	15	21	22	↑		↑		↑		↑		↑		↑		↑		↑		↑		↑		↑	
M	315	↓		↓		↓		↓		↓		0	1	1	2	2	3	3	4	5	6	7	8	10	11	14	15	21	22	↑		↑		↑		↑		↑		↑		↑		↑		↑		↑		↑		↑	
N	500	↓		↓		↓		↓		0	1	1	2	2	3	3	4	5	6	7	8	10	11	14	15	21	22	↑		↑		↑		↑		↑		↑		↑		↑		↑		↑		↑		↑		↑	
P	800	↓		↓		↓		0	1	1	2	2	3	3	4	5	6	7	8	10	11	14	15	21	22	↑		↑		↑		↑		↑		↑		↑		↑		↑		↑		↑		↑		↑		↑	
Q	1250	↓		↓		0	1	1	2	2	3	3	4	5	6	7	8	10	11	14	15	21	22	↑		↑		↑		↑		↑		↑		↑		↑		↑		↑		↑		↑		↑		↑		↑	
R	2000	↓		0	1	1	2	2	3	3	4	5	6	7	8	10	11	14	15	21	22	↑		↑		↑		↑		↑		↑		↑		↑		↑		↑		↑		↑		↑		↑		↑		↑	

↓ = Use first sampling plan below arrow. If sample size equals, or exceeds, lot or batch size, do 100 percent inspection.

↑ = Use first sampling plan above arrow.

Ac = Acceptance number.

Re = Rejection number.

TABLE II-B — Single sampling plans for tightened inspection (Master table)

Acceptable Quality Levels (tightened inspection). Each cell shows **Ac Re** (Acceptance number / Rejection number). ↓ = Use first sampling plan below arrow; ↑ = Use first sampling plan above arrow. If sample size equals or exceeds lot or batch size, do 100 percent inspection.

Code	Sample size	0.010	0.015	0.025	0.040	0.065	0.10	0.15	0.25	0.40	0.65	1.0	1.5	2.5	4.0	6.5	10	15	25	40	65	100	150	250	400	650	1000
A	2	↓	↓	↓	↓	↓	↓	↓	↓	↓	↓	↓	↓	↓	↓	↓	↓	↓	0 1	1 2	2 3	3 4	5 6	8 9	12 13	18 19	27 28
B	3	↓	↓	↓	↓	↓	↓	↓	↓	↓	↓	↓	↓	↓	↓	↓	↓	0 1	1 2	2 3	3 4	5 6	8 9	12 13	18 19	27 28	41 42
C	5	↓	↓	↓	↓	↓	↓	↓	↓	↓	↓	↓	↓	↓	↓	↓	0 1	1 2	2 3	3 4	5 6	8 9	12 13	18 19	27 28	41 42	↑
D	8	↓	↓	↓	↓	↓	↓	↓	↓	↓	↓	↓	↓	↓	↓	0 1	1 2	2 3	3 4	5 6	8 9	12 13	18 19	27 28	41 42	↑	↑
E	13	↓	↓	↓	↓	↓	↓	↓	↓	↓	↓	↓	↓	↓	0 1	1 2	2 3	3 4	5 6	8 9	12 13	18 19	27 28	41 42	↑	↑	↑
F	20	↓	↓	↓	↓	↓	↓	↓	↓	↓	↓	↓	↓	0 1	1 2	2 3	3 4	5 6	8 9	12 13	18 19	27 28	41 42	↑	↑	↑	↑
G	32	↓	↓	↓	↓	↓	↓	↓	↓	↓	↓	↓	0 1	1 2	2 3	3 4	5 6	8 9	12 13	18 19	27 28	41 42	↑	↑	↑	↑	↑
H	50	↓	↓	↓	↓	↓	↓	↓	↓	↓	↓	0 1	1 2	2 3	3 4	5 6	8 9	12 13	18 19	27 28	41 42	↑	↑	↑	↑	↑	↑
J	80	↓	↓	↓	↓	↓	↓	↓	↓	↓	0 1	1 2	2 3	3 4	5 6	8 9	12 13	18 19	27 28	41 42	↑	↑	↑	↑	↑	↑	↑
K	125	↓	↓	↓	↓	↓	↓	↓	↓	0 1	1 2	2 3	3 4	5 6	8 9	12 13	18 19	27 28	41 42	↑	↑	↑	↑	↑	↑	↑	↑
L	200	↓	↓	↓	↓	↓	↓	↓	0 1	1 2	2 3	3 4	5 6	8 9	12 13	18 19	27 28	41 42	↑	↑	↑	↑	↑	↑	↑	↑	↑
M	315	↓	↓	↓	↓	↓	↓	0 1	1 2	2 3	3 4	5 6	8 9	12 13	18 19	27 28	41 42	↑	↑	↑	↑	↑	↑	↑	↑	↑	↑
N	500	↓	↓	↓	↓	↓	0 1	1 2	2 3	3 4	5 6	8 9	12 13	18 19	27 28	41 42	↑	↑	↑	↑	↑	↑	↑	↑	↑	↑	↑
P	800	↓	↓	↓	↓	0 1	1 2	2 3	3 4	5 6	8 9	12 13	18 19	27 28	41 42	↑	↑	↑	↑	↑	↑	↑	↑	↑	↑	↑	↑
Q	1250	↓	↓	↓	0 1	1 2	2 3	3 4	5 6	8 9	12 13	18 19	27 28	41 42	↑	↑	↑	↑	↑	↑	↑	↑	↑	↑	↑	↑	↑
R	2000	↓	↓	0 1	1 2	2 3	3 4	5 6	8 9	12 13	18 19	27 28	41 42	↑	↑	↑	↑	↑	↑	↑	↑	↑	↑	↑	↑	↑	↑
S	3150	↓	0 1	1 2	2 3	3 4	5 6	8 9	12 13	18 19	27 28	41 42	↑	↑	↑	↑	↑	↑	↑	↑	↑	↑	↑	↑	↑	↑	↑

↓ = Use first sampling plan below arrow. If sample size equals or exceeds lot or batch size, do 100 percent inspection.
↑ = Use first sampling plan above arrow.
Ac = Acceptance number.
Re = Rejection number.

TABLE II-C—Single sampling plans for reduced inspection (Master table)

Acceptable Quality Levels (reduced inspection)†

(Each cell below gives the Acceptance number (Ac) and Rejection number (Re). ↓ = use first sampling plan below arrow; ↑ = use first sampling plan above arrow.)

Code	Sample size	0.010	0.015	0.025	0.040	0.065	0.10	0.15	0.25	0.40	0.65	1.0	1.5	2.5	4.0	6.5	10	15	25	40	65	100	150	250	400	650	1000
A	2	↓	↓	↓	↓	↓	↓	↓	↓	↓	↓	↓	↓	↓	↓	↓	↓	0 2	1 2	2 3	3 4	5 6	7 8	10 11	14 15	21 22	30 31
B	2	↓	↓	↓	↓	↓	↓	↓	↓	↓	↓	↓	↓	↓	↓	↓	↓	1 3	1 3	2 4	3 4	5 6	7 8	10 11	14 15	21 22	30 31
C	2	↓	↓	↓	↓	↓	↓	↓	↓	↓	↓	↓	↓	↓	↓	1 3	1 4	1 4	1 4	2 5	3 6	5 8	7 10	10 13	14 17	21 24	↑
D	3	↓	↓	↓	↓	↓	↓	↓	↓	↓	↓	↓	↓	↓	0 1	0 2	1 3	1 4	2 5	3 6	5 8	7 10	10 13	14 17	21 24	↑	↑
E	5	↓	↓	↓	↓	↓	↓	↓	↓	↓	↓	↓	↓	0 1	0 2	1 3	1 4	2 5	3 6	5 8	7 10	10 13	14 17	21 24	↑	↑	↑
F	8	↓	↓	↓	↓	↓	↓	↓	↓	↓	↓	↓	0 1	0 2	1 3	1 4	2 5	3 6	5 8	7 10	10 13	14 17	21 24	↑	↑	↑	↑
G	13	↓	↓	↓	↓	↓	↓	↓	↓	↓	↓	0 1	0 2	1 3	1 4	2 5	3 6	5 8	7 10	10 13	14 17	21 24	↑	↑	↑	↑	↑
H	20	↓	↓	↓	↓	↓	↓	↓	↓	↓	0 1	0 2	1 3	1 4	2 5	3 6	5 8	7 10	10 13	14 17	21 24	↑	↑	↑	↑	↑	↑
J	32	↓	↓	↓	↓	↓	↓	↓	↓	0 1	0 2	1 3	1 4	2 5	3 6	5 8	7 10	10 13	14 17	21 24	↑	↑	↑	↑	↑	↑	↑
K	50	↓	↓	↓	↓	↓	↓	↓	0 1	0 2	1 3	1 4	2 5	3 6	5 8	7 10	10 13	14 17	21 24	↑	↑	↑	↑	↑	↑	↑	↑
L	80	↓	↓	↓	↓	↓	↓	0 1	0 2	1 3	1 4	2 5	3 6	5 8	7 10	10 13	14 17	21 24	↑	↑	↑	↑	↑	↑	↑	↑	↑
M	125	↓	↓	↓	↓	↓	0 1	0 2	1 3	1 4	2 5	3 6	5 8	7 10	10 13	14 17	21 24	↑	↑	↑	↑	↑	↑	↑	↑	↑	↑
N	200	↓	↓	↓	↓	0 1	0 2	1 3	1 4	2 5	3 6	5 8	7 10	10 13	14 17	21 24	↑	↑	↑	↑	↑	↑	↑	↑	↑	↑	↑
P	315	↓	↓	↓	0 1	0 2	1 3	1 4	2 5	3 6	5 8	7 10	10 13	14 17	21 24	↑	↑	↑	↑	↑	↑	↑	↑	↑	↑	↑	↑
Q	500	↓	↓	0 1	0 2	1 3	1 4	2 5	3 6	5 8	7 10	10 13	14 17	21 24	↑	↑	↑	↑	↑	↑	↑	↑	↑	↑	↑	↑	↑
R	800	0 1	0 2	1 3	1 4	2 5	3 6	5 8	7 10	10 13	14 17	21 24	↑	↑	↑	↑	↑	↑	↑	↑	↑	↑	↑	↑	↑	↑	↑

⇩ = Use first sampling plan below arrow. If sample size equals or exceeds lot or batch size, do 100 percent inspection.

⇧ = Use first sampling plan above arrow.

Ac = Acceptance number.

Re = Rejection number.

† = If the acceptance number has been exceeded, but the rejection number has not been reached, accept the lot, but reinstate normal inspection (see 10.1.4).

TABLE III-A — Double sampling plans for normal inspection (Master table)

Each Acceptable Quality Level (AQL) cell shows the pair "Ac Re" for that sampling stage (normal inspection).

Sample size code letter	Sample	Sample size	Cumulative sample size	0.010	0.015	0.025	0.040	0.065	0.10	0.15	0.25	0.40	0.65	1.0	1.5	2.5	4.0	6.5	10	15	25	40	65	100	150	250	400	650	1000
A	First			↓	↓	↓	↓	↓	↓	↓	↓	↓	↓	↓	↓	↓	↓	↓	↓	↓	↓	↓	↓	↓	↓	↓	↓	↓	↓
	Second																												
B	First	2	2	↓	↓	↓	↓	↓	↓	↓	↓	↓	↓	↓	↓	↓	↓	↓	•	0 2	0 3	1 4	2 5	3 7	5 9	7 11	11 16	17 22	25 31
	Second	2	4																	1 2	3 4	4 5	6 7	8 9	12 13	18 19	26 27	37 38	56 57
C	First	3	3	↓	↓	↓	↓	↓	↓	↓	↓	↓	↓	↓	↓	↓	↓	•	0 2	0 3	1 4	2 5	3 7	5 9	7 11	11 16	17 22	25 31	↑
	Second	3	6																1 2	3 4	4 5	6 7	8 9	12 13	18 19	26 27	37 38	56 57	
D	First	5	5	↓	↓	↓	↓	↓	↓	↓	↓	↓	↓	↓	↓	↓	•	0 2	0 3	1 4	2 5	3 7	5 9	7 11	11 16	17 22	25 31	↑	↑
	Second	5	10															1 2	3 4	4 5	6 7	8 9	12 13	18 19	26 27	37 38	56 57		
E	First	8	8	↓	↓	↓	↓	↓	↓	↓	↓	↓	↓	↓	↓	•	0 2	0 3	1 4	2 5	3 7	5 9	7 11	11 16	17 22	25 31	↑	↑	↑
	Second	8	16														1 2	3 4	4 5	6 7	8 9	12 13	18 19	26 27	37 38	56 57			
F	First	13	13	↓	↓	↓	↓	↓	↓	↓	↓	↓	↓	↓	•	0 2	0 3	1 4	2 5	3 7	5 9	7 11	11 16	17 22	25 31	↑	↑	↑	↑
	Second	13	26													1 2	3 4	4 5	6 7	8 9	12 13	18 19	26 27	37 38	56 57				
G	First	20	20	↓	↓	↓	↓	↓	↓	↓	↓	↓	↓	•	0 2	0 3	1 4	2 5	3 7	5 9	7 11	11 16	17 22	25 31	↑	↑	↑	↑	↑
	Second	20	40												1 2	3 4	4 5	6 7	8 9	12 13	18 19	26 27	37 38	56 57					
H	First	32	32	↓	↓	↓	↓	↓	↓	↓	↓	↓	•	0 2	0 3	1 4	2 5	3 7	5 9	7 11	11 16	17 22	25 31	↑	↑	↑	↑	↑	↑
	Second	32	64											1 2	3 4	4 5	6 7	8 9	12 13	18 19	26 27	37 38	56 57						
J	First	50	50	↓	↓	↓	↓	↓	↓	↓	↓	•	0 2	0 3	1 4	2 5	3 7	5 9	7 11	11 16	17 22	25 31	↑	↑	↑	↑	↑	↑	↑
	Second	50	100										1 2	3 4	4 5	6 7	8 9	12 13	18 19	26 27	37 38	56 57							
K	First	80	80	↓	↓	↓	↓	↓	↓	↓	•	0 2	0 3	1 4	2 5	3 7	5 9	7 11	11 16	17 22	25 31	↑	↑	↑	↑	↑	↑	↑	↑
	Second	80	160									1 2	3 4	4 5	6 7	8 9	12 13	18 19	26 27	37 38	56 57								
L	First	125	125	↓	↓	↓	↓	↓	↓	•	0 2	0 3	1 4	2 5	3 7	5 9	7 11	11 16	17 22	25 31	↑	↑	↑	↑	↑	↑	↑	↑	↑
	Second	125	250								1 2	3 4	4 5	6 7	8 9	12 13	18 19	26 27	37 38	56 57									
M	First	200	200	↓	↓	↓	↓	↓	•	0 2	0 3	1 4	2 5	3 7	5 9	7 11	11 16	17 22	25 31	↑	↑	↑	↑	↑	↑	↑	↑	↑	↑
	Second	200	400							1 2	3 4	4 5	6 7	8 9	12 13	18 19	26 27	37 38	56 57										
N	First	315	315	↓	↓	↓	↓	•	0 2	0 3	1 4	2 5	3 7	5 9	7 11	11 16	17 22	25 31	↑	↑	↑	↑	↑	↑	↑	↑	↑	↑	↑
	Second	315	630						1 2	3 4	4 5	6 7	8 9	12 13	18 19	26 27	37 38	56 57											
P	First	500	500	↓	↓	↓	•	0 2	0 3	1 4	2 5	3 7	5 9	7 11	11 16	17 22	25 31	↑	↑	↑	↑	↑	↑	↑	↑	↑	↑	↑	↑
	Second	500	1000					1 2	3 4	4 5	6 7	8 9	12 13	18 19	26 27	37 38	56 57												
Q	First	800	800	↓	↓	•	0 2	0 3	1 4	2 5	3 7	5 9	7 11	11 16	17 22	25 31	↑	↑	↑	↑	↑	↑	↑	↑	↑	↑	↑	↑	↑
	Second	800	1600				1 2	3 4	4 5	6 7	8 9	12 13	18 19	26 27	37 38	56 57													
R	First	1250	1250	↓	•	0 2	0 3	1 4	2 5	3 7	5 9	7 11	11 16	17 22	25 31	↑	↑	↑	↑	↑	↑	↑	↑	↑	↑	↑	↑	↑	↑
	Second	1250	2500			1 2	3 4	4 5	6 7	8 9	12 13	18 19	26 27	37 38	56 57														

↓ = Use first sampling plan below arrow. If sample size equals or exceeds lot or batch size, do 100 percent inspection.
↑ = Use first sampling plan above arrow
Ac = Acceptance number
Re = Rejection number
• = Use corresponding single sampling plan (or alternatively, use double sampling plan below, where available).

TABLE III-B — Double sampling plans for tightened inspection (Master table)

Acceptable Quality Levels (tightened inspection)

Sample size code letter	Sample	Sample size	Cumulative sample size
A	—	—	—
B	First	2	2
	Second	2	4
C	First	3	3
	Second	3	6
D	First	5	5
	Second	5	10
E	First	8	8
	Second	8	16
F	First	13	13
	Second	13	26
G	First	20	20
	Second	20	40
H	First	32	32
	Second	32	64
J	First	50	50
	Second	50	100
K	First	80	80
	Second	80	160
L	First	125	125
	Second	125	250
M	First	200	200
	Second	200	400
N	First	315	315
	Second	315	630
P	First	500	500
	Second	500	1000
Q	First	800	800
	Second	800	1600
R	First	1250	1250
	Second	1250	2500
S	First	2000	2000
	Second	2000	4000

AQL columns (Ac / Re): 0.010, 0.015, 0.025, 0.040, 0.065, 0.10, 0.15, 0.25, 0.40, 0.65, 1.0, 1.5, 2.5, 4.0, 6.5, 10, 15, 25, 40, 65, 100, 150, 250, 400, 650, 1000

The body of the table consists of paired acceptance (Ac) and rejection (Re) numbers arranged diagonally, with the recurring double-sampling block (first sample / cumulative second sample):

First (Ac Re)	Second (Ac Re)
0 2	1 2
0 3	3 4
1 4	4 5
2 5	6 7
3 7	11 12
6 10	15 16
9 14	23 24
15 20	20 21
20 23	35 52
29	53

Downward / upward arrows direct the user to adjacent sampling plans; a dot (•) directs the user to the corresponding single sampling plan.

↓ = Use first sampling plan below arrow. If sample size equals or exceeds lot or batch size, do 100 percent inspection.
↑ = Use first sampling plan above arrow.
Ac = Acceptance number Re = Rejection number
• = Use corresponding single sampling plan (or, alternatively, use double sampling plan below, where available).

TABLE III-C — Double sampling plans for reduced inspection (Master table)

Acceptable Quality Levels (reduced inspection)†

Legend:
- ↓ = Use first sampling plan below arrow. If sample size equals or exceeds lot or batch size, do 100 percent inspection.
- ↑ = Use first sampling plan above arrow.
- Ac = Acceptance number.
- Re = Rejection number.
- • = Use corresponding single sampling plan (or alternatively, use double sampling plan below, when available.)
- † = If, after the second sample, the acceptance number has been exceeded, but the rejection number has not been reached, accept the lot, but reinstate normal inspection (see 10.14).

Code	Sample	Sample size	Cum. sample size	0.010 Ac	Re	0.015 Ac	Re	0.025 Ac	Re	0.040 Ac	Re	0.065 Ac	Re	0.10 Ac	Re	0.15 Ac	Re	0.25 Ac	Re	0.40 Ac	Re	0.65 Ac	Re	1.0 Ac	Re	1.5 Ac	Re	2.5 Ac	Re	4.0 Ac	Re	6.5 Ac	Re	10 Ac	Re	15 Ac	Re	25 Ac	Re	40 Ac	Re	65 Ac	Re	100 Ac	Re	150 Ac	Re	250 Ac	Re	400 Ac	Re	650 Ac	Re	1000 Ac	Re
A				↓		↓		↓		↓		↓		↓		↓		↓		↓		↓		↓		↓		↓		↓		↓		↓		↓		↓		↓		↓		↓		↓		↓		↓		•		•	
B				↓		↓		↓		↓		↓		↓		↓		↓		↓		↓		↓		↓		↓		↓		↓		↓		↓		↓		↓		↓		↓		↓		↓		↓		•		•	
C				↓		↓		↓		↓		↓		↓		↓		↓		↓		↓		↓		↓		↓		↓		↓		↓		↓		↓		↓		↓		↓		↓		↓		↓		•		•	
D	First	2	2	↓		↓		↓		↓		↓		↓		↓		↓		↓		↓		↓		↓		↓		↓		0	2	0	3	0	4	0	4	1	5	2	7	3	8	5	10	7	12	11	17	•		•	
	Second	2	4	↓		↓		↓		↓		↓		↓		↓		↓		↓		↓		↓		↓		↓		↓		0	2	0	4	1	5	3	6	4	7	6	9	8	12	12	16	18	22	26	30	•		•	
E	First	3	3	↓		↓		↓		↓		↓		↓		↓		↓		↓		↓		↓		↓		↓		0	2	0	3	0	4	0	4	1	5	2	7	3	8	5	10	7	12	11	17	↑		•		•	
	Second	3	6	↓		↓		↓		↓		↓		↓		↓		↓		↓		↓		↓		↓		↓		0	2	0	4	1	5	3	6	4	7	6	9	8	12	12	16	18	22	26	30	↑		•		•	
F	First	5	5	↓		↓		↓		↓		↓		↓		↓		↓		↓		↓		↓		↓		0	2	0	3	0	4	0	4	1	5	2	7	3	8	5	10	7	12	11	17	↑		↑		•		•	
	Second	5	10	↓		↓		↓		↓		↓		↓		↓		↓		↓		↓		↓		↓		0	2	0	4	1	5	3	6	4	7	6	9	8	12	12	16	18	22	26	30	↑		↑		•		•	
G	First	8	8	↓		↓		↓		↓		↓		↓		↓		↓		↓		↓		↓		0	2	0	3	0	4	0	4	1	5	2	7	3	8	5	10	7	12	11	17	↑		↑		↑		•		•	
	Second	8	16	↓		↓		↓		↓		↓		↓		↓		↓		↓		↓		↓		0	2	0	4	1	5	3	6	4	7	6	9	8	12	12	16	18	22	26	30	↑		↑		↑		•		•	
H	First	13	13	↓		↓		↓		↓		↓		↓		↓		↓		↓		↓		0	2	0	3	0	4	0	4	1	5	2	7	3	8	5	10	7	12	11	17	↑		↑		↑		↑		•		•	
	Second	13	26	↓		↓		↓		↓		↓		↓		↓		↓		↓		↓		0	2	0	4	1	5	3	6	4	7	6	9	8	12	12	16	18	22	26	30	↑		↑		↑		↑		•		•	
J	First	20	20	↓		↓		↓		↓		↓		↓		↓		↓		↓		0	2	0	3	0	4	0	4	1	5	2	7	3	8	5	10	7	12	11	17	↑		↑		↑		↑		↑		•		•	
	Second	20	40	↓		↓		↓		↓		↓		↓		↓		↓		↓		0	2	0	4	1	5	3	6	4	7	6	9	8	12	12	16	18	22	26	30	↑		↑		↑		↑		↑		•		•	
K	First	32	32	↓		↓		↓		↓		↓		↓		↓		↓		0	2	0	3	0	4	0	4	1	5	2	7	3	8	5	10	7	12	11	17	↑		↑		↑		↑		↑		↑		•		•	
	Second	32	64	↓		↓		↓		↓		↓		↓		↓		↓		0	2	0	4	1	5	3	6	4	7	6	9	8	12	12	16	18	22	26	30	↑		↑		↑		↑		↑		↑		•		•	
L	First	50	50	↓		↓		↓		↓		↓		↓		↓		0	2	0	3	0	4	0	4	1	5	2	7	3	8	5	10	7	12	11	17	↑		↑		↑		↑		↑		↑		↑		•		•	
	Second	50	100	↓		↓		↓		↓		↓		↓		↓		0	2	0	4	1	5	3	6	4	7	6	9	8	12	12	16	18	22	26	30	↑		↑		↑		↑		↑		↑		↑		•		•	
M	First	80	80	↓		↓		↓		↓		↓		↓		0	2	0	3	0	4	0	4	1	5	2	7	3	8	5	10	7	12	11	17	↑		↑		↑		↑		↑		↑		↑		↑		•		•	
	Second	80	160	↓		↓		↓		↓		↓		↓		0	2	0	4	1	5	3	6	4	7	6	9	8	12	12	16	18	22	26	30	↑		↑		↑		↑		↑		↑		↑		↑		•		•	
N	First	125	125	↓		↓		↓		↓		↓		0	2	0	3	0	4	0	4	1	5	2	7	3	8	5	10	7	12	11	17	↑		↑		↑		↑		↑		↑		↑		↑		↑		•		•	
	Second	125	250	↓		↓		↓		↓		↓		0	2	0	4	1	5	3	6	4	7	6	9	8	12	12	16	18	22	26	30	↑		↑		↑		↑		↑		↑		↑		↑		↑		•		•	
P	First	200	200	↓		↓		↓		↓		0	2	0	3	0	4	0	4	1	5	2	7	3	8	5	10	7	12	11	17	↑		↑		↑		↑		↑		↑		↑		↑		↑		↑		•		•	
	Second	200	400	↓		↓		↓		↓		0	2	0	4	1	5	3	6	4	7	6	9	8	12	12	16	18	22	26	30	↑		↑		↑		↑		↑		↑		↑		↑		↑		↑		•		•	
Q	First	315	315	↓		↓		↓		0	2	0	3	0	4	0	4	1	5	2	7	3	8	5	10	7	12	11	17	↑		↑		↑		↑		↑		↑		↑		↑		↑		↑		↑		•		•	
	Second	315	630	↓		↓		↓		0	2	0	4	1	5	3	6	4	7	6	9	8	12	12	16	18	22	26	30	↑		↑		↑		↑		↑		↑		↑		↑		↑		↑		↑		•		•	
R	First	500	500	↓		↓		0	2	0	3	0	4	0	4	1	5	2	7	3	8	5	10	7	12	11	17	↑		↑		↑		↑		↑		↑		↑		↑		↑		↑		↑		↑		•		•	
	Second	500	1000	↓		↓		0	2	0	4	1	5	3	6	4	7	6	9	8	12	12	16	18	22	26	30	↑		↑		↑		↑		↑		↑		↑		↑		↑		↑		↑		↑		•		•	

TABLE IV-A—*Multiple sampling plans for normal inspection (Master table)*

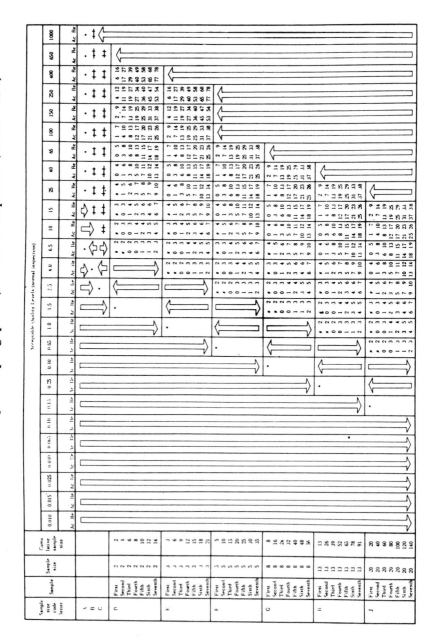

Acceptable Quality Levels (normal inspection)

Table of multiple sampling plans by sample size code letter (K, L, M, N, P, Q, R), showing Sample (First–Seventh), Sample size, Cumulative sample size, and Ac (Acceptance number) / Re (Rejection number) for Acceptable Quality Levels: 0.010, 0.015, 0.025, 0.040, 0.065, 0.10, 0.15, 0.25, 0.40, 0.65, 1.0, 1.5, 2.5, 4.0, 6.5, 10, 15, 25, 40, 65, 100, 150, 250, 400, 650, 1000.

Sample size code letter	Sample	Sample size	Cumulative sample size
K	First	32	32
	Second	32	64
	Third	32	96
	Fourth	32	128
	Fifth	32	160
	Sixth	32	192
	Seventh	32	224
L	First	50	50
	Second	50	100
	Third	50	150
	Fourth	50	200
	Fifth	50	250
	Sixth	50	300
	Seventh	50	350
M	First	80	80
	Second	80	160
	Third	80	240
	Fourth	80	320
	Fifth	80	400
	Sixth	80	480
	Seventh	80	560
N	First	125	125
	Second	125	250
	Third	125	375
	Fourth	125	500
	Fifth	125	625
	Sixth	125	750
	Seventh	125	875
P	First	200	200
	Second	200	400
	Third	200	600
	Fourth	200	800
	Fifth	200	1000
	Sixth	200	1200
	Seventh	200	1400
Q	First	315	315
	Second	315	630
	Third	315	945
	Fourth	315	1260
	Fifth	315	1575
	Sixth	315	1890
	Seventh	315	2205
R	First	500	500
	Second	500	1000
	Third	500	1500
	Fourth	500	2000
	Fifth	500	2500
	Sixth	500	3000
	Seventh	500	3500

= Use first sampling plan below arrow. If sample size equals or exceeds lot or batch size, do 100 percent inspection.

= Use first sampling plan above arrow (refer to preceding page, when necessary).

Ac = Acceptance number.

Re = Rejection number.

= Use corresponding single sampling plan (or alternatively, use multiple plan below, where available).

Ac, Re, . . = Acceptance not permitted at this sample size.

TABLE IV-B—Multiple sampling plans for tightened inspection (Master table)

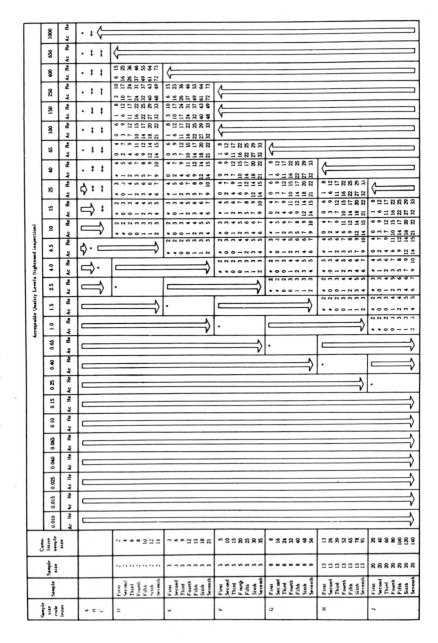

Acceptable Quality Levels (tightened inspection)

Table columns (each with Ac = Acceptance number and Re = Rejection number sub‑columns):

0.010 · 0.015 · 0.025 · 0.040 · 0.065 · 0.10 · 0.15 · 0.25 · 0.40 · 0.65 · 1.0 · 1.5 · 2.5 · 4.0 · 6.5 · 10 · 15 · 25 · 40 · 65 · 100 · 150 · 250 · 400 · 650 · 1000

Sample size code letter	Sample	Sample size	Cumulative sample size
K	First	32	32
	Second	32	64
	Third	32	96
	Fourth	32	128
	Fifth	32	160
	Sixth	32	192
	Seventh	32	224
L	First	50	50
	Second	50	100
	Third	50	150
	Fourth	50	200
	Fifth	50	250
	Sixth	50	300
	Seventh	50	350
M	First	80	80
	Second	80	160
	Third	80	240
	Fourth	80	320
	Fifth	80	400
	Sixth	80	480
	Seventh	80	560
N	First	125	125
	Second	125	250
	Third	125	375
	Fourth	125	500
	Fifth	125	625
	Sixth	125	750
	Seventh	125	875
P	First	200	200
	Second	200	400
	Third	200	600
	Fourth	200	800
	Fifth	200	1000
	Sixth	200	1200
	Seventh	200	1400
Q	First	315	315
	Second	315	630
	Third	315	945
	Fourth	315	1260
	Fifth	315	1575
	Sixth	315	1890
	Seventh	315	2205
R	First	500	500
	Second	500	1000
	Third	500	1500
	Fourth	500	2000
	Fifth	500	2500
	Sixth	500	3000
	Seventh	500	3500
S	First	800	800
	Second	800	1600
	Third	800	2400
	Fourth	800	3200
	Fifth	800	4000
	Sixth	800	4800
	Seventh	800	5600

The body of the table gives the cumulative Ac / Re values for each code letter across the AQL columns. Numerical plans appear along a diagonal band; the remaining cells contain directional arrows. Representative readings of the numerical band (cumulative Ac, Re for the seven stages) include, for code letter K at AQL 10: First ▪,8 — Second 6,12 — Third 11,17 — Fourth 16,22 — Fifth 22,25 — Sixth 27,29 — Seventh 32,33.

Legend / symbols:

↓ (arrow) = Use first sampling plan below arrow. If sample size equals or exceeds lot or batch size, do 100 percent inspection.

↑ (arrow) = Use first sampling plan above arrow (refer to preceding page, when necessary).

Ac = Acceptance number

Re = Rejection number

= Use corresponding single sampling plan (or alternatively, use multiple sampling plan below, where available)

▪ = Acceptance not permitted at this sample size.

TABLE V-B — Average Outgoing Quality Limit Factors for Tightened Inspection (Single sampling)

Acceptable Quality Level

Code letter	Sample size	0.010	0.015	0.025	0.040	0.065	0.10	0.15	0.25	0.40	0.65	1.0	1.5	2.5	4.0	6.5	10	15	25	40	65	100	150	250	400	650	1000
A	2																			42	69	97	160	260	400	620	970
B	3																		28	46	65	110	170	270	410	650	1100
C	5																	17	27	39	63	100	160	250	390	610	
D	8																11	17	24	40	64	99	160	240	380		
E	13																11	15	24	40	61	95	150	240			
F	20														7.4	6.5	9.7	16	26	40	62						
G	32													4.6	4.2	6.9	9.9	16	25	39							
H	50												2.8	2.6	4.3	6.1	10	16	25								
J	80											1.8	1.7	2.7	3.9	6.3	9.9	16									
K	125										1.2	1.1	1.7	2.4	4.0	6.4	9.9										
L	200									0.74	0.67	1.1	1.6	2.5	4.1	6.4											
M	315								0.46	0.42	0.69	0.97	1.6	2.6	4.0	6.2											
N	500							0.29	0.27	0.44	0.62	1.0	1.6	2.5	3.9												
P	800						0.18	0.17	0.27	0.39	0.63	1.0	1.6	2.5													
Q	1250					0.12	0.11	0.17	0.24	0.40	0.64	0.99	1.6														
R	2000			0.046	0.074	0.067	0.11	0.16	0.25	0.41	0.64	0.99															
S	3150	0.018	0.029	0.027	0.042	0.069	0.097	0.16	0.26	0.40	0.62																

Note: For the exact AOQL, the above values must be multiplied by (1 - $\dfrac{\text{Sample size}}{\text{Lot or Batch size}}$)

(see 11.4)

TABLE VI-A—Limiting Quality (in percent defective) for which $P_a = 10$ Percent (for Normal Inspection, Single sampling)

Code letter	Sample size	Acceptable Quality Level															
		0.010	0.015	0.025	0.040	0.065	0.10	0.15	0.25	0.40	0.65	1.0	1.5	2.5	4.0	6.5	10
A	2															68	
B	3														54	41	58
C	5													37	27	36	54
D	8												25	18	25	30	44
E	13											16	12	16	20	27	42
F	20										11	7.6	10	13	18	22	34
G	32									6.9	4.8	6.5	8.2	11	14	19	29
H	50								4.5	3.1	4.3	5.4	7.4	9.4	12	16	24
J	80							2.8	2.0	2.7	3.3	4.6	5.9	7.7	10	14	23
K	125						1.8	1.2	1.7	2.1	2.9	3.7	4.9	6.4	9.0		
L	200					1.2	0.78	1.1	1.3	1.9	2.4	3.1	4.0	5.6			
M	315				0.73	0.49	0.67	0.84	1.2	1.5	1.9	2.5	3.5				
N	500			0.46	0.31	0.43	0.53	0.74	0.94	1.2	1.6	2.3					
P	800		0.29	0.20	0.27	0.33	0.46	0.59	0.77	1.0	1.4						
Q	1250	0.18															
R	2000																

TABLE VI-B—Limiting Quality (in defects per hundred units) for which P_a = 10 Percent (for Normal Inspection, Single sampling)

Code letter	Sample size	0.010	0.015	0.025	0.040	0.065	0.10	0.15	0.25	0.40	0.65	1.0	1.5	2.5	4.0	6.5	10	15	25	40	65	100	150	250	400	650	1000
A	2															120			200	270	330	460	590	770	1000	1400	1000
B	3														77			130	180	220	310	390	510	670	940	1300	1900
C	5													46			78	110	130	190	240	310	400	560	770	1100	1800
D	8												29			49	67	84	120	150	190	250	350	480	670		
E	13											18			30	41	51	71	91	120	160	220	300	410			
F	20										12			20	27	33	46	59	77	100	140						
G	32									7.2			12	17	21	29	37	48	63	88							
H	50								4.6			7.8	11	13	19	24	31	40	56								
J	80							2.9			4.9	6.7	8.4	12	15	19	25	35									
K	125						1.8		4.6	3.1	4.3	5.4	7.4	9.4	12	16	23										
L	200					1.2			2.0	2.7	3.3	4.6	5.9	7.7	10	14											
M	315				0.73			1.2	1.7	2.1	2.9	3.7	4.9	6.4	9.0												
N	500		0.29	0.46			0.78	1.1	1.3	1.9	2.4	3.1	4.0	5.6													
P	800					0.49	0.67	0.84	1.2	1.5	1.9	2.5	3.5														
Q	1250				0.31	0.43	0.53	0.74	0.94	1.2	1.6	2.3															
R	2000	0.18		0.20	0.27	0.33	0.46	0.59	0.77	1.0	1.4																

Acceptable Quality Level

TABLE VIII — Limit Numbers for Reduced Inspection

Acceptable Quality Level

Number of sample units from last 10 lots or batches	0.010	0.015	0.025	0.040	0.065	0.10	0.15	0.25	0.40	0.65	1.0	1.5	2.5	4.0	6.5	10	15	25	40	65	100	150	250	400	650	1000
20 - 29	•	•	•	•	•	•	•	•	•	•	•	•	•	•	•	0	0	2	4	8	14	22	40	68	115	
30 - 49	•	•	•	•	•	•	•	•	•	•	•	•	•	•	0	0	1	3	7	13	22	36	63	105	178	181
50 - 79	•	•	•	•	•	•	•	•	•	•	•	•	•	0	0	2	3	7	14	25	40	63	110	181	301	277
80 - 129	•	•	•	•	•	•	•	•	•	•	•	•	0	0	2	4	7	14	24	42	68	105	181	297		
130 - 199	•	•	•	•	•	•	•	•	•	•	•	0	0	2	4	7	13	25	42	72	115	177	301	490		
200 - 319	•	•	•	•	•	•	•	•	•	•	0	0	2	4	8	14	22	40	68	115	181	277	471			
320 - 499	•	•	•	•	•	•	•	•	•	0	0	1	4	8	14	24	39	68	113	189						
500 - 799	•	•	•	•	•	•	•	•	0	0	2	3	7	14	25	40	63	110	181							
800 - 1249	•	•	•	•	•	•	•	0	0	2	4	7	14	24	42	68	105	181								
1250 - 1999	•	•	•	•	•	•	0	0	2	4	7	13	24	40	69	110	169									
2000 - 3149	•	•	•	•	•	0	0	2	4	8	14	22	40	68	115	181										
3150 - 4999	•	•	•	•	0	0	1	4	8	14	24	38	67	111	186											
5000 - 7999	•	•	•	0	0	2	3	7	14	25	40	63	110	181												
8000 - 12499	•	•	0	0	2	4	7	14	24	42	68	105	181													
12500 - 19999	•	0	0	2	4	7	13	24	40	69	110	169														
20000 - 31499	0	0	2	4	8	14	22	40	68	115	181															
31500 - 49999	0	1	4	8	14	24	38	67	111	186																
50000 & Over	2	3	7	14	25	40	63	110	181	301																

• Denotes that the number of sample units from the last ten lots or batches is not sufficient for reduced inspection for this AQL. In this instance more than ten lots or batches may be used for the calculation, provided that the lots or batches used are the most recent ones in sequence, that they have all been on normal inspection, and that none has been rejected while on original inspection.

TABLE IX — Average sample size curves for double and multiple sampling (normal and tightened inspection)

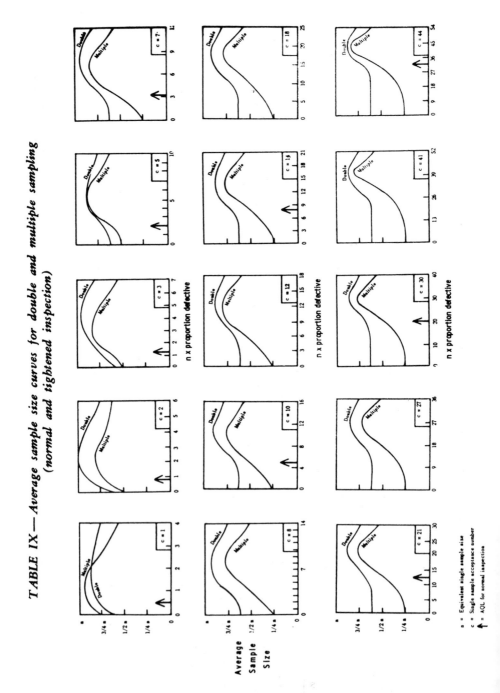

TABLE X-J — Tables for sample size code letter: J

CHART J - OPERATING CHARACTERISTIC CURVES FOR SINGLE SAMPLING PLANS

(Curves for double and multiple sampling are matched as closely as practicable)

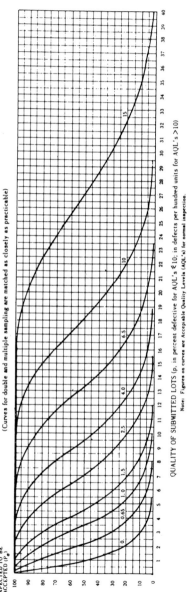

PERCENT OF LOTS EXPECTED TO BE ACCEPTED (P$_a$)

QUALITY OF SUBMITTED LOTS (p, in percent defective for AQL's ≤10; in defects per hundred units for AQL's >10)

Note: Figures on curves are Acceptable Quality Levels (AQL's) for normal inspection.

TABLE X-J-1 - TABULATED VALUES FOR OPERATING CHARACTERISTIC CURVES FOR SINGLE SAMPLING PLANS

Acceptable Quality Levels (normal inspection)

p (in defects per hundred units)

P$_a$	0.15	0.65	1.0	1.5	2.5	4.0	6.5	10	15
99.0	0.013	0.186	0.545	1.03	2.23	3.63	5.96	9.35	15.7
95.0	0.064	0.444	1.02	1.71	3.27	4.98	7.71	11.6	18.6
90.0	0.131	0.665	1.38	2.18	3.94	5.82	8.78	12.9	20.3
75.0	0.360	1.20	2.16	3.17	5.27	7.45	10.8	15.3	23.4
50.0	0.866	2.10	3.34	4.59	7.09	9.59	13.3	18.3	27.1
25.0	1.73	3.37	4.90	6.39	9.28	12.1	16.3	21.8	31.2
10.0	2.88	4.86	6.65	8.35	11.6	14.7	19.3	25.2	35.2
5.0	3.75	5.93	7.87	9.69	13.1	16.4	21.2	27.4	37.8
1.0	5.76	8.30	10.5	12.6	16.4	20.0	25.2	31.8	42.9
	0.25	1.0	1.5	2.5	4.0	6.5	10	15	X

Acceptable Quality Levels (tightened inspection)

p (in percent defective)

P$_a$	0.15	0.65	1.0	1.5	2.5	4.0	6.5	10
99.0	0.013	0.188	0.550	1.05	2.30	3.72	6.13	9.75
95.0	0.064	0.444	1.03	1.73	3.32	5.06	7.91	11.9
90.0	0.132	0.666	1.38	2.20	3.98	5.91	8.95	13.2
75.0	0.359	1.202	2.16	3.18	5.30	7.50	10.9	15.5
50.0	0.863	2.09	3.33	4.57	7.06	9.55	13.3	18.3
25.0	1.72	3.33	4.84	6.31	9.14	11.9	16.0	21.3
10.0	2.84	4.78	6.52	8.16	11.3	14.2	18.6	24.2
5.0	3.68	5.80	7.66	9.39	12.7	15.8	20.3	26.0
1.0	5.59	8.00	10.1	12.0	15.6	18.9	23.6	29.5
	0.25	1.0	1.5	2.5	4.0	6.5	10	X

Note: All values given in above table based on Poisson distribution as approximation to the Binomial.

TABLE X-J-2 — SAMPLING PLANS FOR SAMPLE SIZE CODE LETTER: J

Acceptable Quality Levels (normal inspection) — each numeric cell is shown as "Ac Re"

Type of sampling plan	Cumulative sample size	Less than 0.15	0.15	0.25	0.40	0.65	1.0	1.5	2.5	4.0	6.5	10	15	Higher than 15
Single	80	▽	0 1	Use Letter H	Use Letter H	1 2	2 3	3 4	5 6	7 8	10 11	14 15	21 22	△
Double	50	▽	•	Use Letter H	Use Letter L	0 2	0 3	1 4	2 5	3 7	5 9	7 11	11 16	△
Double	100					1 2	3 4	4 5	6 7	8 9	12 13	18 19	26 27	
Multiple	20	▽	•	Use Letter K	Use Letter K	* 2	* 2	* 3	* 4	0 4	0 5	1 7	2 9	△
Multiple	40					* 2	0 3	0 3	1 5	1 6	3 8	4 10	7 14	
Multiple	60					0 2	0 3	1 4	2 6	3 8	6 10	8 13	13 19	
Multiple	80					0 3	1 4	2 5	3 7	5 10	8 13	12 17	19 25	
Multiple	100					1 3	2 4	3 6	5 8	7 11	11 15	17 20	25 29	
Multiple	120					1 3	3 5	4 6	7 9	10 12	14 17	21 23	31 33	
Multiple	140					2 3	4 5	6 7	9 10	13 14	18 19	25 26	37 38	

Acceptable Quality Levels (tightened inspection):

Less than 0.25	0.25	0.40	0.65	1.0	1.5	2.5	4.0	6.5	X	X	X	Higher than 15

Legend:

△ = Use next preceding sample size code letter for which acceptance and rejection numbers are available.

▽ = Use next subsequent sample size code letter for which acceptance and rejection numbers are available.

Ac = Acceptance number

Re = Rejection number

‖ = Use single sampling plan above (or alternatively use letter M)

• = Acceptance not permitted at this sample size.

10.1.6 General Description of the ABC Standard

There are nine basic tables of specific sampling plans. Tables
II-A, II-B, and II-C are all single sampling plans, respectively,
for normal, tightened, and reduced sampling. Likewise Tables III-A,
III-B, and III-C are for double sampling, and IV-A, IV-B, and IV-C
for multiple sampling. To find a specific plan in any one of these
nine tables, one needs a "sample size code letter," A, B, ..., R,
and an AQL, which may be in terms of *defectives* per 100 units (that
is, percent defective) or *defects* per 100 units. Either may be used
for AQL's of .010 up to 10, but beyond 10, the basis will be defects
per 100 units. The sample size code letter is found from Table I,
using the lot size N to determine the row and the "inspection level"
to give the column, from which the letter is found. Inspection
levels I, II, and III are the commonly used ones, the other four
S-1, ..., S-4 being for special purposes. Level II is used in general unless for some reason smaller samples than normal are desired
and level I specified, or larger samples and level III specified.

Let us now summarize the steps for finding a specific sampling
acceptance plan in the Standard.

1. Decide on the size of lot N, which is to be sample
    inspected. This need not be a production lot size.

2. Decide upon an inspection level, in general II.

3. Using 1 and 2, enter Table I to find the corresponding
    sample size code letter A, B, ..., R, the last calling
    for largest sample sizes.

4. Decide upon single, double, or multiple sampling.

5. Decide whether to start with normal (almost always),
    tightened, or reduced sampling.

6. Table II-A, III-A, or IV-A will thus be determined by
    4 and 5 (if normal inspection is to be used).

7. Decide upon the inspection basis of defectives or defects.

8. Decide upon the desired AQL: for percent defectives
    (10.0 or less) or defects per 100 units; using only what
    is available in tables.

9. Enter the table determined in 6, using the AQL for the
    column from 8 and the row from the sample size code letter
    in 3. This commonly gives the acceptance-rejection numbers

in the block, the sample size(s) being given to the
left of this block.

10. Following the above, we may reach an asterisk or an arrow.
     An asterisk means to use the single sampling plan for the
     desired AQL and code letter, instead of double or multiple
     sampling. If an arrow is encountered, follow it to the
     first block with acceptance-rejection numbers, using sam-
     ple sizes to the left of *this block*, *not* to the left of
     the *original block*.

For example, suppose that we have lot sizes of 500 units and
will use inspection level II, for normal sampling with an AQL of
0.65% defectives. The first two items specified, on entering Table
I tell us to use sample size code letter H. For normal double sampling, we use Table III-A. In this table in the 0.65 column and the
H row, we find in the block a downward arrow. Following it down to
the first block containing numbers (that is, the next lower block
here), we find $Ac_1 = 0$, $Re_1 = 2$, $Ac_2 = 1$, $Re_2 = 2$. Now going over
to the left of *this block* we find $n_1 = n_2 = 50$. Note very particularly that we do *not* use the sample sizes $n_1 = n_2 = 32$, which are
normally used for code letter H. Thus our plan is to take a random
sample of 50 from the 500 in the lot, inspect, and finding d_1 defectives (1) accept if $d_1 = 0$, (2) reject if $d_1 \geq 2$, or (3) take another sample if $d_1 = 1$. Now this second sample of 50 from the 450
remaining in the lot yields d_2 defectives. Then (1) accept the lot
if $d_1 + d_2 = 1$ or (2) reject if $d_1 + d_2 \geq 2$.

Under the same conditions but for *tightened* inspection, we use
Table III-B, and find $n_1 = n_2 = 80$, $Ac_1 = 0$, $Re_1 = 2$, $Ac_2 = 1$,
$Re_2 = 2$. This is indeed a tighter plan making for a lowered Pa at
any p'. Further, under the same conditions, except using *reduced*
inspection, we go to Table III-C. Code letter H calls for $n_1 = n_2$
$= 13$, but in the column for AQL = 0.65 we again find an arrow pointing downward. It yields a block with $Ac_1 = 0$, $Re_1 = 2$, $Ac_2 = 0$,
$Re_2 = 2$, but these are for $n_1 = n_2 = 20$ (the reduced double sample
sizes for code letter J, not H). The author hopes that you have
wondered at the gap between Ac_2 and Re_2. This seems to mean that
no decision is reached if $d_1 + d_2$ happens to be 1. This is covered
in Sec. 10.1.4 of the *Standard*. There it says that if such an event

occurs, we still accept this lot (because the previous quality has
been excellent relative to the AQL), but we are now alerted to the
possibility that quality has slipped from its previous excellence.
Therefore, we abandon reduced sampling with its quite lenient OC
curve and go back to normal sampling.

Now let us have a look at Table X-J, which we have included as
a sample. There is in the Standard such a pair of pages for each
code letter A to R. They give a wealth of information on the respective sampling plans for each code letter. All sampling plans--
single, double, and multiple--are listed on the second pages as on
the page, X-J-2. At the top of the first page X-J-1 are given the
OC curves for all AQL's for single normal inspection (most being
calculated using the Poisson distribution as in our Table B, but
the binomial is used if the single plan has $n \leq 80$ *and also* the
AQL ≤ 10.0). As usual, Pa is the vertical scale, the bottom scale
being for percent defective or defects per 100 units for the process. The OC curve for the normal single sampling plan for code
letter J and AQL = 0.65%, namely, n = 80, Ac = 1, Re = 2 is shown
in the graph next to the number "0.65." Thus, for example, if p'
= 3% defective Pa = .30, or if p' = 2% defective Pa = .55.

The table in the lower half of the first page gives certain
convenient points on the OC curve, namely, in percents p'_{99}, p'_{95}
p'_{90}, and so on to p'_{01}. These figures are for single normal sampling plans, as are the OC curves. The OC curves for double and
multiple normal sampling "are matched as closely as practicable."
Our examples of Secs. 9.5 and 9.6 were such a pair of matched plans,
and appear on Table X-J-2, for an AQL of 1.5%. Note that the table
at the bottom of the first page of Table X-J is entered from the
top for *normal* sampling AQL's and from the *bottom* for *tightened*
sampling AQL's. Some of the plans duplicate; others do not. For
example, the normal plan with an AQL of 1.5% is the same plan as
that for the tightened plan with an AQL of 2.5%. OC curves for
some tightened plans are not shown. There are two sides to this
table, one for defectives (on the left), the other for defects.

With this description of the tables in mind, carefully read
the first eight pages of the Standard. Section 7.1 specifies random sampling. Section 7.2 provides for proportional sampling from
boxes on containers within the lot. Section 8 is of much importance
in using the Standard, providing as it does the "switching rules."
The rule 8.3.1 for going from normal to tightened inspection is
clear and easily followed as long as a continuing record of action
on consecutive lots is preserved in order. Note Sec. 8.3.2 in con-
junction with 8.4. When on tightened inspection, the quality must
be quite quickly improved so as to return to normal inspection by
having five consecutive lots passing; or else inspection will be
discontinued until action is taken to improve the quality submitted.

Bear in mind that invoking tightened inspection, when called
for, *must* be carried out. This is *not* an optional part of the Standard. It is essential in order to provide the consumer with adequate protection, and to ensure that the outgoing quality is at the
AQL or better, in the long run. If not followed, then you are not
using the ABC Standard!

On the other hand, use of reduced sampling when available *is
optional*. It is not necessary to use it in order to ensure quality
at the AQL or better. Its only function is to safely save inspection with adequate safeguards. *All four* requirements of Sec. 8.3.3
must be met in order to go onto reduced inspection. They ensure
consistently excellent quality better than the AQL. Use is made of
Table VIII, which gives limit numbers of defectives or defects found
in cumulations of the single samples or in first samples for double
or multiple sampling, for at least 10 lots. But to go from reduced
sampling back to normal, *only one* of the four conditions of Sec.
8.3.4 need occur. One of the provisions involves an acceptance of
a lot under conditions which raise a question as to whether quality
may have deteriorated, as noted before and covered in Sec. 10.1.4.

Let us now describe the remaining tables which we have reproduced. Only the normal and tightened *multiple* sampling plans IV-A,
IV-B, have been included, omitting the corresponding reduced plans.

Table V-B provides the average outgoing quality limits, AOQL's, for all tightened, single-sampling plans. This table thus gives a measure of the consumer protection provided when on tightened inspection. Except for those single tightened plans for which the acceptance number, Ac = 0, all of these AOQL's are at about the corresponding AQL or better. This means that when we go onto tightened inspection we are, in general, using a plan providing outgoing quality averaging no worse than the AQL.

Tables VI-A and VI-B give quality levels having a .10 probability of acceptance if offered, respectively, in terms of percent defective and defects per 100 units. These are particularly useful in case we are sample inspecting isolated or infrequent lots, since these figures tell what quality levels, of which a given plan provides a 90% assurance against acceptance, if offered.

Table IX contains very useful information on the average sample number (ASN) curves, in a convenient compact form, by means of which we may compare single, double, and multiple sampling. They are for both normal and tightened inspection, but do not include cases for which the single plan has Ac = 0. All other acceptance numbers for a single plan are included.

For example, if we wish the ASN curves for normal sampling, code letter J, AQL = 1.5% defective, we will use the third square which says c = 3, the letter "c" being an alternative to "Ac." Now pay close attention to the way the scales are made up, as we now describe. The three plans call for the following

$$n = 80, Ac = c = 3, Re = 4$$
$$n_1 = 50, \Sigma n = 50, Ac_1 = 1, Re_1 = 4$$
$$n_2 = 50, \Sigma n = 100, Ac_2 = 4, Re_2 = 5$$

and the multiple plan with n's of 20. Now consider the vertical scale: n, 3/4 n, and so on. Here n = 80, so that in sampled pieces these five points are 80, 60, 40, 20, and 0 for our example. Now for the horizontal scale: it is np', that is, the expected number of defectives in a sample. Since the n we are talking about is 80, these numbers represent 80p'. When 80p' = 4, say, then dividing by

80 gives p' = 4/80 = .05. Therefore, when in the horizontal scale
we see 4, we could put a p' value of .05 for a p' scale. Similarly,
when we see 2 in the scale shown, this is equivalent to p' = .025,
6 to p' = .075, and so on. So we now can compare ASN's for the
three plans. The single plan is uniformly at 80. The double plan
starts off at n = 50 (just above 1/2 n) and reaches a maximum of
about 75 when np' = 4.5 or so, which corresponds to p' = 4.5/80
= .056, then it starts to decrease back toward 50. Meanwhile the
multiple plan starts off at 40, because it takes two samples, each
of 20, before there is a zero acceptance number. Thereafter, the
ASN increases, reaching a maximum of about 65 when np' = 3 or p'
= 3/80 = .038.

Now what is the meaning of the arrow? It shows the relative
position of the AQL in normal sampling. For n = 80, this plan with
Ac = 3 has the arrow at about 1.3. Equating this to 80AQL = 1.3,
we obtain in this example AQL = 1.3/80 = .016. It is listed for
our example as 1.5%, which is close enough for the scale reading.

For the single normal plan for n = 500, Ac = 3, and AQL = 0.25,
the vertical scale would read 500, 375, 250, 125, 0. Now when we
have 4 on the horizontal scale this is 500p' = 4 or p' = 4/500
= .008, and so on. Meanwhile, the AQL arrow is at 1.3/500 = .0026
= .26%, close to AQL = .25%.

Thus each of the various blocks show the *relative* ASN's for a
whole collection of plans. Is not this a rather neat, compact summary form?

This completes our discussion of the ABC Standard with its aim
of maintenance of outgoing quality at the AQL or better, but meanwhile giving a small risk of a lot being rejected when the producer
is running at the AQL or better.

10.2 THE DODGE-ROMIG SAMPLING TABLES

The earliest program of sampling plans was developed in the Bell
System in the 1920s and was published by Harold F. Dodge and Harry
G. Romig. A revised edition was published in 1959 [3]. These plans
have by no means outlived their usefulness.

10.2.1 Aim

The general aim of the Dodge-Romig Tables is, first, to provide the consumer with desired protection, either a specified AOQL or a specified lot tolerance percent defective with 10% risk, that is, a specified p'_{10}. Then secondly, the specified consumer protection is provided on a minimum overall inspection cost, that is, on a minimum average total inspection (ATI). This minimum ATI is attainable as a function of the incoming fraction defective $\bar{p}$.

Let us explain this approach further. Suppose that we decide that we want an AOQL of 1.0% as protection. For any given lot size, there is a large number of single and double sampling plans, each one of which has an AOQL of 1.0% approximately. Which of these should we take? We first must decide whether to use single or double sampling. Let us say it is to be double sampling. Now then if the average fraction defective has been running at $\bar{p} = .005 = .5\%$, we must sort through all double sampling plans with AOQL's of 1% to find which one has the minimum ATI when the incoming quality level is $\bar{p} = .005$ and for a given lot size N. Using this plan on product at $\bar{p}$, we will (1) have AOQL protection at 1.0 percent *and* (2) obtain this protection through minimum ATI as long as the incoming quality is at $\bar{p}$. But if $\bar{p}$ changes, then probably some other plan will provide the specified AOQL protection and minimum ATI. A change in lot size may also require a different plan to minimize the ATI while supplying the required protection.

This aim is useful in many cases. But note that it is quite distinct from the ABC Standard.

10.2.2 Description

There are four different sets of tables, being the combinations of Single (S) versus Double (D), and Lot Tolerance Percent Defective (L) versus Average Outgoing Quality Limit (A). Thus, the four sets are SL, DL, SA, DA. The available LTPD's are (in percents) .5, 1.0, 2.0, 3.0, 4.0, 5.0, 7.0, and 10.0. The AOQL's available are (in percents) .1, .25, .50, .75, 1.0, 1.5, 2.0, 2.5, 3.0, 4.0, 5.0,

7.0, 10.0. Therefore a DA-2 table means double sampling plans with
an AOQL = 2.0%. In each such table there are *rows* of classes of
lot size up to 100,000, some 20 classes. Then there are six columns of classes of $\bar{p}$ in each table.

To enter the Dodge-Romig tables to find a plan, one uses the
following:

1. Decide on single or double sampling.

2. Decide on LTPD or AOQL type of protection.

3. Specify the amount of the type of protection chosen in 2.

4. The information in 1, 2, 3 locates the table to use.

5. In this table, locate the row class containing the given
   lot size.

6. Find the column class of $\bar{p}$'s which contains our observed
   $\bar{p}$. If $\bar{p}$ is unknown, use the largest class of $\bar{p}$'s.

7. From 5 and 6 we find a cell in the table which lists for
   single sampling: the sample size n and acceptance number
   called c. Or for double sampling, samples n_1 and n_2
   (which differ) and the acceptance number c_1 for the first
   sample and acceptance number c_2 for the total defectives
   in the two samples.

8. Also in each cell is given the amount of the opposite kind
   of consumer protection. Thus in the AOQL tables the cells
   list the LTPD, called p_t, for the plan.

9. The plans are listed for inspection for defectives. If,
   however, inspection is for defects, use average defects
   per 100 units, instead of $\bar{p}$ in percent. Then we count
   defects in samples, rather than defectives.

10.3 OTHER SAMPLING INSPECTION PLANS

There were a number of good sampling plans as forerunners of the
ABC Standard. Many of their best features were carried into the
ABC Standard. (See Ref. 1.) We mention again the collection of
sampling plans indexed on p'_{95} and p'_{10}, whose usefulness lies in
isolated lots or infrequently submitted lots. (See Ref. 4 in Chapter 9.)

Another plan which has its place in some applications was published by H. C. Hamaker and associates in the Phillips Review (Ein-
hoven, The Netherlands) in 1949 and 1950 (three papers). The plans

were indexed on the "point of indifference," namely p'_{50}, and the slope of the OC curve at that point. Steep slopes of course call for sharp discrimination and larger sample sizes. The general objective is to equalize producer and consumer risks.

Still another type of sampling inspection by attributes is pure "sequential sampling," wherein, after each piece is inspected for defects, we can (1) accept the lot or process, (2) reject it, or (3) ask for another piece to inspect. No limit is set on the sample size which varies. The objective is to achieve a given OC curve for protection, on a minimum ASN.

10.4 CONTINUOUS SAMPLING PLANS

A rather different type of sampling plan from those we have been studying will now be discussed. In many processes, units of product come along slowly enough so that each unit may be inspected for defects. That is, it is feasible to inspect 100%, and it is being done in this manner to control quality. In such a situation, it may be feasible to *sample* inspect at least some of the time in order to save inspection time. The continuous sampling plan which is now to be discussed does this, with adequate safeguards, in order to deliver an AOQL type of protection.

Let us first give the conditions for which the plan was designed:

1. There are no natural lots submitted for inspection and a decision. Instead, there is a continuous flow of product such as on a conveyor belt.

2. There are discrete units of product, each of which is either good or nonconforming, that is, has none or has one or more defects.

3. Production is not so rapid as to make 100% inspection infeasible.

4. The quality of production has an acceptably low fraction defective for at least much of the time.

Continuous sampling plans for a particular defect, or for a class of defects, call for alternating periods of 100% sorting and of sampling inspection. The relative amount of time spent on each

depends upon the quality level of the defect or defects in question.
If quality is excellent, most of the time is spent on sampling, or
if relatively poor most of time is at 100% sorting.

10.4.1 The Original CSP-1 Plan

The first continuous sampling plan was developed by Dodge [4,5].
The procedure is as follows:

1. At the beginning, inspect for the defect(s) in question,
    100% of the units consecutively, as produced, and continue
    such inspection until i units in succession are found to
    be free from the defect(s).

2. When i consecutive good units are found, discontinue 100%
    inspection, and inspect only a fraction f of the units,
    selecting the sample units one at a time from the flow of
    units, in such a way as to insure an unbiased sample.

3. Whenever under sampling a defective is found, revert imme-
    diately to 100% inspection of succeeding units, continuing
    until again i consecutive units are found free from defects,
    as in step 1, when sampling inspection is resumed.

4. Correct or replace with good units all defective units
    found.

It is thus to be seen that there are two constants governing
the plan, namely i and f. The larger i is the more difficult it is
to qualify for sampling and thus the greater the protection supplied
against relatively poor quality. Similarly, the larger f is the
higher the proportion of units sampled, and thus the more difficult
it is to stay on sampling, for any given incoming p'. How do we
find i and f? First, we must specify our desired protection. This
is in terms of the AOQL, that is, the worst long-run average quality
outgoing, no matter what quality level p' comes in. Therefore, we
first decide on a desired AOQL amount of protection. We might then
specify f and find i or specify i and find f, from the charts in
Refs. 4 and 5. The former is the usual approach. Accordingly, we
present in Table 10.1 some combinations. If others are desired,
see the original sources or a reproduction in Ref. 2 of Chapter 9.

From the table, we see that in order to obtain AOQL consumer
protection of 1.00% we will use i = 110 if we specify an f of .100,

TABLE 10.1. Combinations of f and i for Given AOQL Protection for Continuous Sampling Plan CSP-1. Values of i to Qualify are in Body of Table.

Sample 1 in:	f	.05	.10	.20	.30	.50	1.00	2.00	3.00	5.00
					AOQL (%)					
5	.200	1430	720	360	240	142	72	35	24	14
8	.125	1950	960	480	320	193	96	48	32	19
10	.100	2210	1100	550	370	220	110	55	37	21
15	.0667	2700	1320	660	440	270	132	67	44	26
20	.050	3080	1500	750	510	310	150	75	50	29

or one in each 10 units. So we will need 110 in succession all good, in order to qualify for sampling. Once on sampling we will choose one at random out of each 10. This is best done by use of a table of random numbers, such as our Table D, or an abstract of this table. The main thing to avoid is choosing every 10th unit, because of the possibility of a cycle in the occurrence of defects. Or if we choose a larger sampling proportion f = .200, or one in each five, then it only takes 72 good units in succession to qualify. It is easily seen that f and i are inversely related for any given AOQL.

10.4.2 *Variations from CSP-1*

If quality is more or less intermediate, then there may be considerable alternation between 100% and sampling inspection periods. This can be a troublesome factor in adjustment of personnel loads. One way to help is by using *inspection* personnel for the sampling periods and *production* personnel for the 100% inspection. One advantage of this is to encourage production to improve its quality performance.

Another way to avoid excessive changing from sorting to sampling and vice versa is to use the CSP-2 plan [6]. In this plan, we again have the two constants i and f for given AOQL, and the initial qualification for sampling is the same, that is, i consecutive good

units. But when sampling and a defective is found, we do not immediately return to 100% inspection of units. Instead, we are alerted
and watch the sampled units. If among the next i *sampled* units a
second defect is found, we immediately revert to 100% inspection.
But if no further defective is found among the next i sampled units,
we are requalified fully, so to speak. And a single defective does
not force us off from sampling, but we would as before watch the
next i sampled units for a second defective which would terminate
the sampling period.

Now it is obvious that for a given AOQL and f, CSP-2 is more
lenient for a fixed qualification number i, because we are less
likely to be thrown off of sampling. Therefore, for given AOQL and
f, we must use a larger i under CSP-2 than when under CSP-1. In
fact, i for CSP-2 is roughly a third more than for CSP-1. This rule
can be used on entries of Table 10.1 if you wish to use CSP-2.

A variation of CSP-2 is called CSP-3. It is basically like
CSP-2, but when the first defective is encountered, we look at all
of the next four units, instead of sampling. If these are all good
then we continue on sampling, counting these four among the next i,
all of which must be good in order to be fully requalified. Thus
CSP-3 is essentially a CSP-2 plan, but with added protection against
a short run of spotty quality, causing repeated defects.

There is also a military standard incorporating features of
CSP-1 through CSP-3, and some other features [7]. Currently, it is
scheduled for revision.

10.5* CHAIN SAMPLING PLAN, ChSP-1

Whenever a quality characteristic involves destructive or costly
tests so that only a small number of tests n can be justified per
lot, an acceptance number of Ac = 0 is naturally used. But for all
sample sizes n, when Ac = 0, the shape of the OC curve is like those
of Fig. 9.8, that is, with the concave side everywhere up. This
makes it relatively difficult to pass even quite good material; Pa
falls down rapidly from 1.00, as p' increases. Moreover with an n
of 5 or 10, the OC curve is rather undiscriminating. For these

reasons, the chain sampling plan ChSP-1 was developed to improve
the shape of the OC curve, making it more discriminating with an
elongated S shape more like that in Fig. 9.10 [8]. Further conditions for using ChSP-1, as indicated in the reference, are (1) the
product to be inspected consists of a series of lots produced by an
essentially continuing process, which under normal conditions can
be expected to be of the same quality p', and (2) the product comes
from a source in which the consumer has confidence.

The procedure is as follows: (1) Take a random sample of n
units or specimens and inspect or test each against the requirement
specified. (2) Use the acceptance number Ac = 0 for defectives;
except use Ac = 1 if no defective was found in the immediately preceding i samples of n. (i is a constant number specified in the
plan and may be 1, 2, 3, ...) Thus we specify n and i, and rou-
tinely use Ac = 0, except that when *one* defective shows up in the
sample of n, we must check back over the preceding i lots for the
presence of any defectives. If there was one, we reject the present lot (but do not reject any of the previous lots, which are probably out of our hands anyway). There could not be two, for then we
would have had an earlier rejection. Thus whenever a defective is
found, the next i lots must show perfect samples (d = 0) for all of
them to be accepted. But if they are, then we can once more toler-
ate a single defective, but would then be alerted and could not have
a second defective in the next i samples without a rejection. Of
course, d = 2 in any n gives immediate rejection.

The OC curve has a slightly different meaning than those we
have been presenting in this and the preceding chapters. Now the
vertical scale is the percentage of *lots* expected to be accepted
for a given process p'. Table 10.2 gives p' values for which Pa is
.95 and .10 for several ordinary Ac = 0 plans and for ChSP-1 plans
with the same n. Note how in each case p'_{95} increases as the number
i of previous lots decreases. Thus these ChSP-1 plans give a better chance of acceptance at relatively very good quality levels,
while retaining the same consumer protection, p'_{10}. This is from

TABLE 10.2   Points p' on the OC Curve for Which Pa = .95 and .10
for Ordinary Ac = 0 Plans and for ChSP Plans with Same Sample Size n

	n	Ac	i	Proportion of lots expected to be accepted	
				.95	.10
Ordinary	4	0	--	.013	.44
ChSP-1	4		4	.031	.44
	4		2	.042	.44
Ordinary	5	0	--	.010	.37
ChSP-1	5		4	.024	.37
	5		2	.030	.37
Ordinary	6	0	--	.008	.32
ChSP-1	6		4	.020	.32
	6		2	.027	.32
Ordinary	10	0	--	.005	.21
ChSP-1	10		4	.012	.21
	10		2	.017	.21

the shape of the OC curves being an elongated S rather than being
everywhere concave upward.

10.6* SKIP-LOT SAMPLING PLAN, SkSP-1

The skip-lot sampling plan developed by Dodge [9] is a plan for
omitting the testing or inspection of a certain proportion of a
series of lots, while still providing specified protection. It is
basically a continuous sampling plan like CSP-1, but is applied to
lots instead of *units* or *pieces*. The lots in the stream are each
tested or inspected. This may be on the basis of some measurable
characteristic such as chemical content, or it may be by a sample
inspection decision to accept or reject, or even 100% inspection.
But after a sufficiently long series of consecutive lots i have
been found to meet requirements, then the inspection or testing can
go to doing this check on only a proportion f of the lots. But as

soon as a nonconforming lot is found among this proportion f of lots, we go back to inspecting or testing every lot, and must requalify. Thus, this SkSP-1 is in reality a CSP-1 plan applied to *lots*.

Now we are concerned with the proportion of nonconforming lots being put out by the production process. The plan seeks to provide AOQL protection of the following type: For each *incoming* fraction of nonconforming lots, the plan will provide a lower fraction defective *outgoing*, that is, an AOQ fraction of nonconforming lots. Considering all the AOQ's, there will be a maximum, the AOQL. The plan is indexed on this AOQL (for lots nonconforming).

There are two procedures: A1, where each nonconforming lot is to be either corrected or replaced by a conforming one, and A2, where each nonconforming lot is to be rejected and is not to be replaced by a conforming lot. The required number i to qualify differs by one in the two cases. The following plans are given in Ref. 9:

	AOQL (%)	f	i	
			Proc. A1	Proc. A2
Standard plan	2	1/2	14	15
Other plans	3	1/2	9	10
	5	1/2	5	6
	5	1/3	9	10
	5	1/4	12	13

The skip-lot plan seems a reasonable approach to inspection of lots for noncritical defects. But if the defect in question is really critical, then it would seem best to at least have a substantial sample checked from every lot. In fact, commonly lots are at least 100% inspected for critical defects.

10.7 SUMMARY

We have been discussing a variety of sampling plans in this chapter,
each of which is ready to be used, without making any calculations.
The objective of the ABC Standard or MIL-STD 105D is to obtain a
quality level outgoing which is at least as good as the specified
AQL and to accomplish this while using sampling decisions on lots.
The other plans discussed were designed to provide some sort of
consumer protection while saving on inspection costs. The Dodge-
Romig tables enable the consumer to obtain either a specified AOQL
or LTPD, and further, to achieve this on a minimum ATI for given
$\bar{p}$ incoming.

The chain sampling, ChSP-1, plan operates on lots being given
expensive or destructive tests, so that small sample sizes are im-
perative. But use of an Ac= 0 gives an undesirable concave-up shape
to the OC curve. By using information on previous lots, the shape
of the OC curve can be improved, while retaining the small sample
size. On the other hand, when the objective is to control the percentage of unsatisfactory *lots* to a previously set AOQL, then under
appropriate conditions we can use the skip-lot plan, SkSP-1, to provide this protection while requiring only a proportion f of *lots* to
be tested or inspected.

If concern is with the fraction defective or nonconforming of
units of product in continuous production we can use the continuous
plan, CSP-1. This plan provides for maintenance of AOQL protection
for units. It saves on inspection by providing for sampling after
i consecutive units are all good. One hundred percent inspection
is reinstated whenever a defective unit is found in the sampled
pieces. Plans for CSP-2 and CSP-3 can be used to decrease the frequency of alternation of sampling and sorting 100%.

PROBLEMS

10.1 Find the following plans from the ABC Standard:

	Lot size	Inspection level	S, D, or M	N, T, or R	Defects or defectives	AQL (%)
(a)	500	II	Single	Normal	Defectives	0.40
(b)	500	II	Single	Tightened	Defectives	0.40
(c)	500	II	Single	Reduced	Defectives	0.40
(d)	500	I	Single	Normal	Defectives	0.40
(e)	2000	II	Double	Normal	Defectives	0.65
(f)	2000	II	Double	Normal	Defectives	0.065
(g)	2000	II	Double	Normal	Defects	15
(h)	5000	III	Double	Tightened	Defectives	0.40
(i)	5000	III	Multiple	Tightened	Defectives	0.40
(j)	5000	II	Multiple	Normal	Defectives	0.15
(k)	1000	II	Double	Reduced	Defectives	2.5
(l)	1000	II	Single	Reduced	Defectives	2.5

10.2 Find single sampling plans in the ABC Standard for lot size N = 500, inspection level II, with AQL = 2.5 (defectives) for (a) normal, (b) tightened, and (c) reduced. Draw the three OC curves on the same axes, labeling the scales. Comment on the curves.

10.3 A Dodge-Romig sampling plan for an AOQL of 1% and lots of 3000 calls for n_1 = 80, n_2 = 170, Ac_1 = 0, Re_1 = 5, Ac_2 = 4, Re_2 = 5. Find Pa when p' = .04. Is this compatible with the listed p'_{10} = .036?

10.4 A Dodge-Romig single sampling plan for lots of N = 3000 and $\bar{p}$ = .21 to .40% is n = 80, Ac = 1. It is for an AOQL of 1.0%. Draw the OC and AOQ curves and check this AOQL.

10.5 Find the CSP-1 continuous sampling plan for an AOQL of 1.00% with f = .125. Explain the meaning of i and f. What is the corresponding CSP-2 plan? Again explain i and f.

10.6 In what sense is the OC curve for the ChSP-1 chain sampling plan n = 10, Ac = 0, i = 2 more favorable than that for the ordinary sampling plan n = 10, Ac = 0? Explain in simple terms.

10.7 Explain in simple terms the operation of a SkSP-1 skip sampling plan having AOQL = 2% and f = 1/2. What advantages and disadvantages does it have?

REFERENCES

1. H. F. Dodge, Notes on the evolution of acceptance sampling plans, Parts I to IV. *J. Quality Techn.*, *1*, 77-88, 155-162, 225-232, and *2*, 1-8, 19 (1969, 1970).

2. Military Standard MIL-STD-105D (ABC), *Sampling Procedures and Tables for Inspection by Attributes*, Department of Defense, 1963. Also American National Standards Institute (Z1.4), and available also from American Society for Quality Control, Milwaukee, Wisc.

3. H. F. Dodge and H. G. Romig, *Sampling Inspection Tables*, Wiley, New York, 1959.

4. H. F. Dodge, A sampling inspection plan for continuous production. *Ann. Math. Statist.*, *14*, 264-279 (1943).

5. H. F. Dodge, Sampling plans for continuous production. *Indust. Quality Control*, *4* (No. 3), 5-9 (1947).

6. H. F. Dodge and M. N. Torrey, Additional continuous sampling inspection plans. *Indust. Quality Control*, *7* (No. 5), 7-12 (1951).

7. Military Standard MIL-STD-1235(ORD), *Single and Multi-level Continuous Sampling Procedures and Tables for Inspection by Attributes*, Department of the Army, 1962.

8. H. F. Dodge, Chain sampling inspection plan. *Indust. Quality Control*, *11* (No. 4), 10-13 (1955).

9. H. F. Dodge, Skip-lot sampling plan. *Indust. Quality Control*, *11* (No. 5), 3-5 (1955).

Chapter Eleven

SAMPLING BY VARIABLES

In the preceding two chapters, the decision for acceptance or rejection of a lot or process was based upon attributes, that is, counts of the number of defective units or of defects on the units in samples. Such an approach is always possible. For, if the quality characteristic is a measurement, such as a dimension, a weight of contents or a tensile strength, this can always be converted to an attribute by comparison to a maximum or a minimum limit, or to two limits. Thus the measurement either lies inside limits or it does not. On the other hand, many attribute characteristics are not capable of being converted to a measurement.

But when the quality characteristic is a measurement, there is the possibility of basing the decision on the measurements on units within the sample. This has the possibility of making sound decisions on *smaller samples*, because the measurements provide more information than just whether the measurement lies within or outside of limits. For example, it will tell how "defective" the unit was. But there are three "prices" to pay for the smaller sample size for a comparable power of discrimination: (1) It usually takes more time to make and record measurements than to merely find out whether the unit is in or outside of limits, (2) knowledge of the shape or type of measurement distribution is essential, and (3) decisions by measurements are somewhat more complicated and commonly require at least some simple calculations. The second of these is unfortunately often overlooked and does require a substantial amount of previous data.

But properly supported, decisions based upon samples of measurements can be very effective. Basically the decision is whether
the lot or process seems to have a satisfactory *distribution*.

11.1 KNOWLEDGE OF DISTRIBUTION TYPE

As we have just been pointing out, if we are to take full advantage
of our observed sample measurements, we need to know what *shape* or
type the distribution takes. This is quite distinct from the param-
eters: the population average μ and standard deviation σ.

If we are making a decision on a *process*, we may well know
from past experience that the distribution is approximately normal.
This is often the case, especially where conditions have been controlled, and where the process has not been reset while the product
in question was produced. In fact, the sampling decision may be
whether the process *level* is in need of resetting or adjusting.

On the other hand, when we are examining a lot a vendor has
sent us, we must be more careful, and a normal distribution is not
so readily assumed. For one thing, the producer may have run the
process at several distinct levels. Thus, if we were to measure up
the entire lot and tabulate a frequency table, such as Table 2.2,
we might find several humps of high frequencies, or possibly a lop-
sided unsymmetrical distribution. Such is often the case when conditions have not been well controlled. Another type of abnormality
is that in which the producer has been having trouble meeting our
specification limits and thus has had the output sorted, more or
less perfectly, to specification limits. Such a situation can cause
a frequency distribution that stops abruptly at the ends, with one
or both tails chopped off by inspection, human or mechanical. A
third cause of nonnormality of distribution is where there is a
physical limitation in one direction. Out-of-round, out-of-square
dimensions cannot be below zero, nor can eccentricity (distance between center lines). These give rise to distributions tailing out
on the high side. Similar tendencies to unsymmetrical distributions
may occur in percent impurity, content weights in containers,

strengths, and weights of castings or gloss readings. For such measurements, we cannot soundly use methods which are designed under
the assumption that the distribution is normal. Unless otherwise
stated, the methods to be discussed all assume a normal distribution
of measurements. Such an assumption is likely to be safe if the
process is running under homogeneous or controlled conditions. But
it is desirable to tabulate at least 100 measurements, or better
200, to check the shape.

11.2 GENERAL AIM: TO JUDGE WHETHER DISTRIBUTION IS SATISFACTORY

The general aim in acceptance sampling by variables is to use measurements on a sample of units from a lot or process in order to
judge whether the population distribution seems to be satisfactory.
By a satisfactory distribution, we usually mean that the percentage
of units having measurements outside the specification limit or
limits is acceptably small, perhaps 1%, for example. The two cases
give rise to "one-tail" and "two-tail" tests, that is, where concern is with only one tail of the distribution or with both tails.
In the former case are strengths subject to a minimum, or blowing
times of fuses under marginal overload, subject to a maximum limit.
In the latter are most dimensions, that is, those which have two
specification limits (± limits). Also, for example, hardness limits
may be in both directions.

If in addition to knowing that the distribution is approximately normal, we also know the population or process standard deviation from past data, we are indeed in a strong position. And we
can set up simple acceptance criteria for the process or lot, based
on the mean $\bar{x}$ only.

11.3 DECISIONS ON LOT MEAN, KNOWN σ, NORMAL DISTRIBUTION

It may seem strange that we might know the lot or process standard
deviation σ and yet fail to know the mean μ. But this can well happen. The process level is subject to many factors in practice which
may not affect the variability. Or, to put it in control chart

terms, the R chart may well show good control and have a quite constant $\bar{R}$, while the $\bar{x}$ chart shows lack of control. In such a case,
we may estimate σ by $\bar{R}/d_2$. This is σ for individual x's, that is,
σ_x. It tells about how far away from μ we must expect an individual
x to lie. And in fact an x may rarely be as much as $3\sigma_x$ away from μ.

11.3.1 *Review of Distributions of x and of $\bar{x}$*

Let us suppose that we have a normal distribution of individual x's,
with average μ and standard deviation σ_x. Then we recall from Chapter 7 that the distribution of sample *averages* will also be normal,
and the average of $\bar{x}$'s is the same as that for the x's, namely

$$\mu_{\bar{x}} = E(\bar{x}) = \mu \qquad (11.1)$$

and the standard deviation of the $\bar{x}$'s is

$$\sigma_{\bar{x}} = \frac{\sigma_x}{\sqrt{n}} \qquad (11.2)$$

that is, a typical $\bar{x}$ may be expected to lie closer to μ than we expect an x, by a factor of $\sqrt{n}$.

To explain further, let us say that for a dimension, $\mu = .1720$
in. and $\sigma_x = .0002$ in. Then in Figure 11.1a we show the distribution
of x. Now for sample means from samples of n = 4 dimensions, we have
by (11.2)

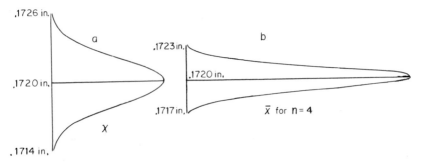

Fig. 11.1. Comparison of distributions of x and of $\bar{x}$ for n = 4.
when $\mu = .1720$ in. and $\sigma_x = .0002$ in. Normal distribution of x is
assumed.

$$\sigma_{\bar{x}} = \frac{.0002 \text{ in.}}{\sqrt{4}} = .0001 \text{ in.}$$

while the average for the $\bar{x}$'s is the same as for the x's, namely, .1720 in. Thus in Figure 11.1b, the $\bar{x}$'s are more closely clustered around μ than are the x's. If n were larger, $\sigma_{\bar{x}}$ would be even smaller. (The relative frequencies for $\bar{x}$'s are twice as great as those for x's so that the total areas enclosed by the curves are the same.)

We shall make constant use of these facts in the following examples of sampling plans. Be careful to make sure whether you are concerned with the distribution of individual x's or with that for averages, $\bar{x}$'s.

11.3.2 A Specific Example

Suppose that we are to sample test a lot of parts for tensile strength. This being an expensive and destructive test, we are interested in a small sample size. Past results indicate the strengths to be normally distributed, with a standard deviation running at about 3000 psi. The specified minimum tensile strength is 70,000 psi. The main question with each lot is whether the lot mean is safely above 70,000 psi. Accordingly, we plan to make tensile tests on a sample of n parts and to make our decision from the sample mean strength $\bar{x}$.

Now what do we mean by "safely above 70,000 psi?" A reasonable possibility is to take as an acceptable lot mean strength the value μ = lower specification + $3\sigma_x$ = 70,000 + 3(3000) = 79,000. If μ = 79,000, what percentage of parts will have strengths below 70,000? To answer this we make use of (7.12):

$$z = \frac{x - \mu}{\sigma_x} = \frac{70,000 - 79,000}{3000} = -3.00$$

Then we use Table A to find the probability of a strength below 70,000. It is .0013 = .13%, or about one out of 770 parts. This would seem to be small enough to regard μ = 79,000 psi as clearly safe, depending, of course, upon the criticality of the part.

Next suppose we set μ = lower specification + $2\sigma_x$ = 70,000 + 2(3000) = 76,000 as an undesirable lot mean strength. What percentage would be below 70,000 if μ = 76,000 psi? Again use (7.12):

$$z = \frac{70,000 - 76,000}{3000} = -2.00$$

which by Table A gives .0228 = 2.28% or one in 44, which might well be undesirable.

Thus we wish our sampling plan to distinguish between lots having μ_1 = 76,000 psi and lots having μ_2 = 79,000 psi. As we have already mentioned, we will use $\bar{x}$ to decide about the lot. Now if $\bar{x}$ were to be above 79,000, we might well feel safe and accept the lot, whereas if $\bar{x}$ is below 76,000, we would undoubtedly wish to reject the lot. But if $\bar{x}$ lies between 76,000 and 79,000, what would we wish to do? Somewhere between them there must be a cut-off point, say K, so that

$\bar{x}$ > K   Accept lot

$\bar{x}$ < K   Reject lot

We need to find a reasonable K value, and not only that, but also the sample size n.

In order to find K and n for the test, we must make our decision making more precise by deciding upon risks as follows: Let the small Greek alpha, α, stand for the probability of erroneous *rejection* if μ is actually at μ_2 the *acceptable level*. Likewise let the small Greek beta, β, stand for the probability of erroneous *acceptance* if μ is actually at μ_1 the *rejectable level*. α and β are called "risks of wrong decisions" and can be set at will. Summarizing

α = P(rejection, given μ is acceptable at μ_2)

β = P(acceptance, given μ is rejectable at μ_1)

Usually risks are set as desired, and then the sample size is determined. If n seems too large to be economic, then one or both risks must be increased and/or the difference between μ_1 and μ_2 increased.

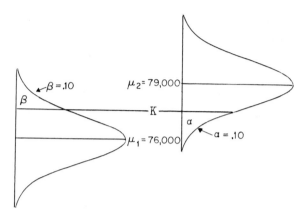

Fig. 11.2. Two distributions of averages $\bar{x}$, one acceptable at μ_2 = 79,000 psi, the other rejectable at μ_1 = 76,000 psi. Also shown are the respective risks, α of rejection at μ_2, β of acceptance at μ_1. Each set to be .10.

Let us take risks of .10, that is, for example, if μ = 79,000 psi there will be 1 chance in 10 of erroneous rejection, and a 90% chance of acceptance. We may now proceed to find n and K. To do this we note first that n must be large enough to make the $\bar{x}$ distributions sufficiently narrow, as shown in Fig. 11.2. (The respective distributions of the x's are far wider, of course.)

In order to find n and K, consider first the distribution of the $\bar{x}$'s, centered at $\mu = \mu_2$ = 79,000 psi. We shall need to find the standard normal curve value z for which the *lower tail* is .10. For this we look in the first part of Table A (where the z's are negative) for .10. It is not there exactly, but we do find

$$z = -1.280 \qquad p = .1003$$
$$z = \qquad p = .1000$$
$$z = -1.290 \qquad p = .0985$$

Interpolating for the required z, we must go 3/18 of the way from -1.280 toward -1.290, or -1.280 - (3/18)(.010) = -1.282. We are now in a position to use (7.12) for the $\bar{x}$ distribution:

$$z = \frac{x - \mu}{\sigma} \quad \text{becomes} \quad z = \frac{K - \mu_2}{\sigma_{\bar{x}}} \quad \text{or} \quad -1.282 = \frac{K - 79,000}{3000/\sqrt{n}}$$

Thus we have arrived at an equation in two unknowns n and K. We need another, which comes from the distribution at μ_1.

This time we have an *upper* tail with a positive z value having .10 probability above it or .90 below. We could look for .9000 in Table A and interpolate to find z = +1.282. Or we could use the symmetry of the normal curve and merely change the sign of our other z. Then

$$+1.282 = \frac{K - 76,000}{3000/\sqrt{n}} \quad \text{or} \quad +1.282 = (K - 76,000)\frac{\sqrt{n}}{3000}$$

$$-1.282 = \frac{K - 79,000}{3000/\sqrt{n}} \quad \text{or} \quad -1.282 = (K - 79,000)\frac{\sqrt{n}}{3000}$$

Subtracting yields $2.564 = \sqrt{n}$, and squaring both sides gives

$$n = 6.57$$

But since we can only make six or seven tests we choose n = 7.

Now to find K, we could substitute into either equation. But since we rounded n up to the whole number 7, the other equation (not substituted into) will not be exactly satisfied. Or since the two risks were taken equal, we can merely take K half-way between μ_1 and μ_2, that is,

$$K = 77,500$$

Our sampling plan is then to take n = 7 tensile tests and find $\bar{x}$:

$$\bar{x} \geq 77,500 \qquad \text{Accept lot}$$
$$\bar{x} < 77,500 \qquad \text{Reject lot}$$

The actual risks will be just a bit less than .10. For we have

$$z = \frac{77,500 - 79,000}{3000/\sqrt{7}} = -1.32$$

giving $\alpha = .0934 = \beta$.

Now if seven tests are too expensive, then we may lower μ_1 below 76,000, or raise μ_2 above 79,000, or increase one or both of the risks α and β. That is, we must sacrifice some of the sharpness of discrimination in order to make n smaller. Or if we can afford a larger n we can make a sharper discrimination, subject to smaller risks.

We may sketch the OC curve Pa versus μ by drawing a smooth curve through the points (.09, 76,000), (.50, 77,500), (.91, 79,000) such that Pa approaches zero below 76,000, and one above 79,000.

11.3.3 One-Specification Limit Test

Now let us generalize the specific example just discussed. For this we will use the notation

z_p = standard normal curve z with probability p
above it   (11.3)

Thus $z_{.10}$ has an upper tail of .10, and is +1.282, $-z_{.10}$ by symmetry has a lower tail of .10, and is -1.282, z_α has an upper tail of α, and so on.

Case 1. Acceptable mean μ_2 > rejectable mean μ_1, risk of rejection if $\mu = \mu_2$ is α, risk of acceptance if $\mu = \mu_1$ is β. The equations then are

$$-z_\alpha = \frac{K - \mu_2}{\sigma/\sqrt{n}} \qquad z_\beta = \frac{K - \mu_1}{\sigma/\sqrt{n}} \qquad (11.4)$$

in which α, β, μ_1, μ_2, and σ are known, and n and K are to be found. z_α and z_β, of course, are found in Table A. For n we have then

$$n = \frac{(z_\alpha + z_\beta)^2 \sigma^2}{(\mu_2 - \mu_1)^2} \qquad (11.5)$$

which we round up to the first whole number above. Then K may be found from either equation of (11.4).

The OC curve goes through the three points (β, μ_1), (.5, K), $(1 - \alpha, \mu_2)$.

Case 2. Rejectable mean μ_2 > acceptable mean μ_1, risk of rejection if $\mu = \mu_1$ is α, risk of acceptance if $\mu = \mu_2$ is β. The equations then are

$$z_\alpha = \frac{K - \mu_1}{\sigma/\sqrt{n}} \qquad -z_\beta = \frac{K - \mu_2}{\sigma/\sqrt{n}} \tag{11.6}$$

The solution for n is still (11.5), and then we may substitute into either of (11.6) to find K.

An OC curve may be sketched through the points $(1 - \alpha, \mu_1)$, $(.5, K)$, (β, μ_2).

11.3.4 *A Two-Way Example*

When there are two specification limits for a measurement, we have a problem of protecting against the lot or process mean μ being too high or too low. For example, a lot of parts is to be tested for hardness on the Rockwell C scale, to specifications of 60 and 70. Past experience has shown that lot distributions are approximately normal with σ = 1.5 points. The main thing to test for is the lot mean. We pick as the acceptable mean μ_0 = 65. For the two reject-able lot means, we may well come in from the specification limits 2σ = 3, giving μ = 60 + 3 = 63 and μ = 70 - 3 = 67. If μ = 67, what proportion of parts will be outside specifications? None will be below 60 which is 4.67σ below 67, but some will be above 70. Use

$$z = \frac{70 - 67}{1.5} = +2$$

The probability below z = +2 is .9772, leaving .0228 above. Like-wise if μ = 63, there will be .0228 below 60, that is, about 2% too soft.

Next we need to set risks of wrong decisions. As usual, we let α be the *risk of rejecting an acceptable lot*, namely, one with $\mu = \mu_0$ = 65. Suppose we set this at α = .01. Next, β is the *risk of accepting a rejectable lot*, that is, one with μ = 63 or 67, say, μ_1 and μ_2. The value for β will depend upon how critical the hardness is for the part in question. Suppose we set β = .10.

The form of the test is to set two limits around μ_0, between which the average sample hardness, $\bar{x}$, is expected to lie, say,

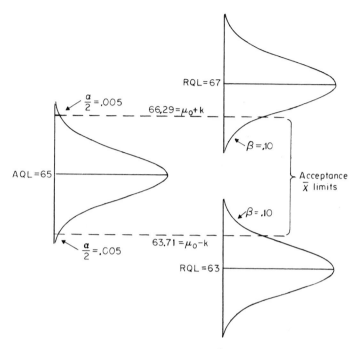

Fig. 11.3. Two-way test for the mean $\bar{x}$ for the hardness example.
AQL = μ_0 = 65, RQL = $\mu_0 \pm 2$ = 63, 67. Also shown are the areas for
α and β risks and the acceptance limits for $\bar{x}$ at $\mu_0 \pm k$ = 65 $\pm$ 1.29.

$\mu_0 \pm k$. If $\bar{x}$ lies within we accept the lot, if outside we reject
the lot. Our job is to find n and k, subject to the risks α and β,
and for the critical means μ_1, μ_0, and μ_2.

See Fig. 11.3, which pictures the distributions of $\bar{x}$ at the
three critical means, to provide the risks α = .01 and β = .10.
Notice that α = .01 is split into two parts .005 above μ_0 + k and
.005 below μ_0 - k. On the other hand, if $\mu = \mu_2$ = 67, then all of
β = .10 lies below μ_0 + k, and similarly if $\mu = \mu_1$ = 63, all of
β = .10 will lie above μ_0 - k. Now let us set up appropriate equations to find n and k. Remembering that we have distributions for
$\bar{x}$'s, we, therefore, have as standard deviation $\sigma_{\bar{x}} = \sigma_x/\sqrt{n}$. The two
distributions, which are cut by the line μ_0 + k = 65 + k, have means
65 and 67, and thus

$$z_{\alpha/2} = \frac{(65 + k) - 65}{1.5/\sqrt{n}} = z_{.005}$$

$$-z_{\beta} = \frac{(65 + k) - 67}{1.5/\sqrt{n}} = -z_{.10}$$

Using Table A and interpolating we find $z_{.005} = 2.575$, $z_{.10} = 1.282$. Therefore, we have

$$\frac{k}{1.5/\sqrt{n}} = 2.575$$

$$\frac{k - 2}{1.5/\sqrt{n}} = -1.282$$

$$\sqrt{n}\,\frac{k}{1.5} = 2.575$$

$$\sqrt{n}\,\frac{k - 2}{1.5} = -1.282$$

By subtracting

$$\sqrt{n}\,\frac{2}{1.5} = 3.857$$

$$\sqrt{n} = 2.89$$

$$n = 8.37 \text{ or } 9$$

We may now substitute n = 9 into either equation to find k. (They will give slightly differing k's because of using n = 9 instead of 8.37.) Suppose we take the first equation

$$\sqrt{9}\,\frac{k}{1.5} = 2.575 \qquad \text{or} \qquad k = 1.29$$

There are also two equations available from the cut-off line at $\mu_0 - k$, but by the symmetry they prove to be identical to the two we just solved. Thus our plan is to take n = 9 hardness readings and find $\bar{x}$ from them, then

$\bar{x}$ between 65 ± 1.29   Accept lot or process
$\bar{x}$ outside 65 ± 1.29   Reject lot or process

An OC curve, Pa versus μ, might be sketched.

11.3.5 Two-Specification Limit Test

Knowing σ, we can compare it with the specified tolerance, U - L. If U - L is, say, 8σ or more we may best use two one-way tests, one near L and one near U. But if U - L is less than 8σ or so, then we probably will do best with a two-way test such as just given. There U - L = 6.67σ.

The general set-up then is that we have a desired nominal or acceptable mean μ_0, and equally spaced on opposite sides of μ_0, two rejectable means μ_1 and μ_2, that is, $\mu_0 - \mu_1 = \mu_2 - \mu_0$. Also we have to set risks: α of rejection if $\mu = \mu_0$ and β of acceptance if $\mu = \mu_1$ or μ_2. Then the equations at $\mu_0 + k$ are

$$z_{\alpha/2} = \frac{\mu_0 + k - \mu_0}{\sigma/\sqrt{n}} \qquad -z_\beta = \frac{\mu_0 + k - \mu_2}{\sigma/\sqrt{n}} \qquad (11.7)$$

The solution for n is

$$n = (z_{\alpha/2} + z_\beta)^2 \left(\frac{\sigma}{\mu_2 - \mu_0}\right)^2 \qquad (11.8)$$

Then rounding (11.8) up to the next whole number, we can substitute n into either of (11.7) to find k. Then for sample means $\bar{x}$:

$\bar{x}$ between $\mu_0 \pm k$ Accept lot or process
$\bar{x}$ outside $\mu_0 \pm k$ Reject lot or process

As in the example, we could give two more equations at $\mu_0 - k$, but they prove to be identical to (11.7).

An OC curve may be sketched through the five points (β, μ_1), $(.5, \mu_0 - k)$, $(1 - \alpha, \mu_0)$, $(.5, \mu_0 + k)$, (β, μ_2).

11.3.6* Controlling the Percentage Nonconforming

We could take the following one-way approach to control the percentage outside of limits, say, at an upper specification limit U: (1) Set μ_1 at such a level that there is a tolerably small proportion p_1' above U, that is, call $\mu_1 = U - z_{p_1'}\sigma$. (2) Set μ_2 at such a higher level that there is a rejectable proportion p_2' above U, that is, call $\mu_2 = U - z_{p_2'}\sigma$. (3) Using these μ_1 and μ_2 in Sec.

11.3.3, case 2, we proceed as there described. Substituting the
foregoing values of μ_1 and μ_2 into (11.5), we obtain

$$n = \left(\frac{z_\alpha + z_\beta}{z_{p_1'} - z_{p_2'}}\right)^2 \tag{11.9}$$

Then find the acceptance criterion K from either of (11.6), namely

$$z_\alpha = \frac{K - U + z_{p_1'}\sigma}{\sigma/\sqrt{n}} \quad \text{or} \quad -z_\beta = \frac{K - U + z_{p_2'}\sigma}{\sigma/\sqrt{n}}$$

$$K = U - z_{p_1'}\sigma + \frac{z_\alpha\sigma}{\sqrt{n}} \quad \text{or} \quad K = U - z_{p_2'}\sigma - \frac{z_\beta\sigma}{\sqrt{n}} \tag{11.10}$$

In the case of a lower specification L, we use the following
equations

$$K = L + z_{p_1'}\sigma - \frac{z_\alpha\sigma}{\sqrt{n}} \quad \text{or} \quad K = L + z_{p_2'}\sigma + \frac{z_\beta\sigma}{\sqrt{n}} \tag{11.10a}$$

11.4 DECISIONS ON LOT BY MEASUREMENTS, σ UNKNOWN, NORMAL DISTRIBUTION

We may make the following one-way test under these conditions. The
objective is to make a sound decision on the *percentage* of parts
lying beyond the one specification limit, say, an upper limit U.
With σ for the lot *unknown*, we must take account of the wide variety
of combinations of μ and σ which would be satisfactory. Thus for
example, suppose U = 1000, and that 1% above 1000 is permissible.
We first seek $z_{.01}$ in Table A, finding

Cumulative probability	z
.9898	2.32
.9901	2.33

giving $z_{.01}$ = 2.32 + (2/3).01 = 2.327. Then if σ = 1, μ of 1000
- 2.327(1) = 997.673 or lower would be satisfactory for then p'
would be .01 or less. But if σ = 10, then μ must be 1000 - 2.327(10)

= 976.73 or less for p' to be .01 or less. The larger σ is, the farther below U we must have μ.

In order to use the technique being discussed, we now set up the problem much as in Sec. 9.8, which was using a single sampling test by attributes, that is, n and Ac. Now we will be working with the same sort of requirements, but making the test using $\bar{x}$ and s from a sample of measurements. (The assumption of normality is vital here.) Specifically we have

Acceptable quality level = AQL = p_1'
Rejectable quality level = RQL = p_2'
P(reject if p_1') = α
P(accept if p_2') = β

We thus have two critical fractions defective for parts beyond U and corresponding risks of wrong decisions, α and β, all four of which may be set at will.

The test takes the form:

1. Take n observations and find $\bar{x}$ and s.
2. Then $\bar{x} + ks \leq U$ Accept
             $\bar{x} + ks > U$ Reject

The multiple k of the sample standard deviation s, provides a "buffer" so that $\bar{x}$ does not come too close to U. How close is safe depends upon both k and s. Our task, therefore, is to find n and k for the desired control of fraction defective p'. Specifically, we have the following formulas from Ref. 1:

$$k = \frac{z_\alpha z_{p_2'} + z_\beta z_{p_1'}}{z_\alpha + z_\beta} \tag{11.11}$$

$$n = \frac{k^2 + 2}{2}\left(\frac{z_\alpha + z_\beta}{z_{p_1'} - z_{p_2'}}\right)^2 \tag{11.12}$$

In general, using (11.12) does not give a whole number for n, so we round upward to the next nearest integer for the sample size.

If our concern is with a lower specification limit L, we still use (11.11) and (11.12) then the decisions

$$\bar{x} - ks \geq L \qquad \text{Accept}$$
$$\bar{x} - ks < L \qquad \text{Reject}$$

Let us take as an example

$$AQL = p_1' = .01 \qquad \alpha = .05$$
$$RQL = p_2' = .05 \qquad \beta = .10$$

From Table A, we have $z_{.01} = 2.327$, $z_{.05} = 1.645$, $z_{.10} = 1.282$, and then (11.11) gives

$$k = \frac{1.645(1.645) + 1.282(2.327)}{1.645 + 1.282} = 1.944$$

$$n = \frac{1.944^2}{2} + 2\left(\frac{1.645 + 1.282}{2.327 - 1.645}\right)^2 = 53.2 \text{ or } 54$$

So we take 54 measurements and find $\bar{x}$ and s. Then

$$\bar{x} + 1.944s \leq U \qquad \text{Accept}$$
$$\bar{x} + 1.944s > U \qquad \text{Reject}$$

We might compare this with the pure attribute sample size, using Sec. 9.8: $p_2'/p_1' = 5 = R_0$, then Ac = 3, $np_2' = 6.68$ and n = 134. Or if we knew σ and used it, Sec. 11.3.6 gives by (11.9) n = 18.4 or 19. Note the large gain achieved by knowing σ. (Also there is no need to calculate s in order to make the test!)

11.5 SINGLE-SAMPLE TEST ON VARIABILITY

This test is a convenient one for deciding whether a lot or process standard deviation σ is acceptable. It is easily applied. It can be used for making a decision on the variability of a lot. An important application is that of testing whether a production process is capable of meeting some tolerance. Say the tolerance T = U - L is given. Then if σ for the process is about one-eighth of T, we would say that the process is fully capable of meeting T, even with set-up error and some drift of the mean. But if σ is as much as one-fifth or even one-sixth of T, then the process will have to be closely set up and the average tightly controlled to meet specifications. So we would have to distinguish between σ_1 = T/8 and σ_2 = T/5.

Another useful application is for an acceptance test on a gauge or measurement technique. We might be able to set an *acceptable* standard deviation of measurement error σ_1 and a *rejectable* standard deviation of measurement error σ_2. The objective is to *accept* most of the time (1 - α) if σ_e = σ_1, and to *reject* most of the time (1 - β) if σ_e = σ_2. This is where σ_e is the true standard deviation for repeated measurements or analyses on the same or on homogeneous material. Then we have the following four test quantities to be set for the test:

$$\text{AQL} = \sigma_1 \qquad \alpha = P(\text{reject if } \sigma = \sigma_1)$$
$$\text{RQL} = \sigma_2 \qquad \beta = P(\text{accept if } \sigma = \sigma_2)$$

Table F can be easily used if we agree to use equal risks $\alpha = \beta$, and to let them take one of the four values .10, .05, .02, or .01.

There are two steps to take once the four test quantities are set. First we find the quotient σ_2/σ_1, which describes the discrimination ratio. Then we choose whichever of columns (2) to (5) corresponds to our chosen $\alpha = \beta$. We now seek in that column the desired σ_2/σ_1, taking the nearest entry less than or equal to our σ_2/σ_1. Then for this entry, we find in column (1) the sample size n. The second step is to find in this row and the appropriate column (6) to (9) the multiplier for σ_1^2 to obtain K. Then the plan is to find s^2 for n observations and

$$s^2 \leq K \qquad \text{Accept}$$
$$s^2 > K \qquad \text{Reject}$$

Risk α is preserved, but β may be slightly decreased by this test.

As an example, suppose a lathe is to be used under certain conditions to meet a tolerance of .0004 in. We would like to approve the process as fully capable if $\sigma = \sigma_1$ = .00004 in., would like to reject if $\sigma = \sigma_2$ = .00007 in. Suppose further that we set $\alpha = \beta$ = .10. Then in column (2) we look for σ_2/σ_1 = 1.75. The sample size is n = 13. To find the test criterion K look in column (6) opposite n = 13, finding 1.55.

$$K = 1.55(.00004 \text{ in.})^2 = .248(.0001 \text{ in.})^2$$

Now run off 13 parts under homogeneous conditions, measure them, find s^2. If $s^2 \leq K$ approve the lathe for the job, but if $s^2 > K$, regard the lathe as not capable, under present conditions.

A final word might be said about the OC curve. We have, of course, Pa = 1 - α when $\sigma = \sigma_1$ and Pa = β at $\sigma = \sigma_2$. Also as σ decreases below σ_1, Pa approaches one, and as σ increases above σ_2, Pa approaches zero. Also as σ goes from σ_1 to σ_2 Pa smoothly decreases from 1 - α down to β. These facts are enough for a sketch of the OC curve. Further information on the OC curve and the derivation of the method are available in Ref. 2.

11.6 DESCRIPTION OF MILITARY STANDARD MIL-STD 414

The Military Standard 414 [3] is an outgrowth of research beginning in 1939 [4] and continuing through World War II and afterward. As its title implies, the standard is for controlling the percentage beyond one or two specification limits, through the use of measurement statistics $\bar{x}$, s, and R. Since concern is with the tail or tails of the distribution of the individual x's, the assumption of normality is quite crucial, and is probably not sufficiently emphasized in the Standard and other references. This assumption may also be given insufficient attention in practice.

Now let us proceed with a description of the 414 plans.

1. All are single sampling plans, making decisions on the basis of n measurements, rather than upon counts of the number outside of specifications. All assume that the distribution of individual measurements x in the lot is *normal*.

2. Nevertheless, the plans are concerned with the percentage beyond a single specification limit L or U, or outside of two limits L and U. These are one-way and two-way plans, respectively.

3. Plans are available for two general cases: σ known-- Section D in the Standard using σ for variability; σ unknown--Section B using sample s or Sec. C using sample ranges R's.

4. Plans are to control the lot fraction defective. They include normal, tightened, and reduced plans.

5. There are two forms: 1 and 2. (This is unfortunate because it makes the standard more bulky and complicated than necessary.) Decisions on a lot would be the same under either form. Since Form 2 is used whenever a two-way decision (both L and U given) is made, we recommend using Form 2 throughout, that is, for both one- and two-way decisions.

6. OC curves for Pa versus lot fraction defective are provided for normal and tightened inspection plans. They are in the Standards' Table A-3, a page for each sample size code letter, listed for normal inspection; but also correct for tightened inspection by finding what normal inspection plan corresponds to the desired tightened plan.

7. To start to find a plan, specify an acceptable quality level AQL in percent defective. Table A-1 converts this to one of the index AQL's, for example, .65%. Then for inspection levels I to V (smaller to larger n's) and lot size N, Table A-2 gives a sample size code letter B, C, ..., Q. Use level IV if none is specified. Thus we have a code letter and an AQL in *percent*.

8. Now choose Sec. B, C, or D according to knowledge of σ and preference for R or s, as in our step 3.

9. Form 2--Using sample standard deviation s: Sec. B in Standard.

    a. One-way, say, an upper specification U.

        i. For the AQL and code letter from 7, find in Table B-3 the sample size n (first column) and the maximum allowed estimated percent defective M in the column for our AQL. (This M will be larger than the AQL, so that lots with p' = AQL will have a high Pa.) For a random sample of n measurements x, find $\bar{x}$ and s.

        ii. Form the "quality index"

$$Q_U = (U - \bar{x})/s \qquad\qquad (11.13)$$

(This is much like (7.12) z = $(x - \mu)/\sigma$ for obtaining the percentage beyond some limit x, but instead of μ and σ it uses the sample $\bar{x}$ and s.) From Q_U and n, enter Table B-5 to find the estimated percentage above U, say p_U.

        iii. Decision:

$$p_U \leq M \qquad \text{Accept lot}$$
$$\qquad\qquad\qquad\qquad\qquad\qquad (11.14)$$
$$p_U > M \qquad \text{Reject lot}$$

iv. If protection is relative to a lower specification L, use step i as it is, but in step ii use

$$Q_L = (\bar{x} - L)/s \qquad (11.15)$$

Then use Table B-5 to find p_L and make the decision by

$$\begin{aligned} p_L &\leq M \qquad \text{Accept lot} \\ p_L &> M \qquad \text{Reject lot} \end{aligned} \qquad (11.16)$$

v. Tightened plans are found in Table B-3 by entering from the AQL's at the bottom, rather than those at the top as in normal plans. Reduced plans are found in Table B-4.

b. Two-way plans, L and U both given

i. Same as step a.i.

ii. Form Q_L by (11.15) and Q_U by (11.13) and use Table B-5 to find estimated percentages p_L and p_U.

iii. Decision

$$\begin{aligned} p_L + p_U &\leq M \qquad \text{Accept lot} \\ p_L + p_U &> M \qquad \text{Reject lot} \end{aligned} \qquad (11.17)$$

iv. Same as step a.v.

c. Switching rules are based in considerable part upon the estimated fraction defective for a series of lots, that is, upon $\bar{p}_U + \bar{p}_L$, usually for 10 lots. The rules are similar to those in MIL-STD 105D.

10. Form 2--Using sample ranges R: Sec. C in Standard.

a. If n = 3, 4, 5, or 7, use R. Or if n = 10, 15, ..., break up the sample into subsamples of five each (first five measurements made, second five, and so on) finding range for each and use $\bar{R}$.

b. Proceed as in 9a and 9b, except using R or $\bar{R}$ and Tables C-3 to C-5 and

$$Q_U = c(U - \bar{x})/\bar{R} \qquad Q_L = c(\bar{x} - L)/\bar{R} \qquad (11.18)$$

which provide estimated percentages p_U and p_L from Table C-5. Tables C-3 and C-4 give the c values for each n; they are like d_2 values for control charts.

11. Form 2--Using known population σ: Sec. D in Standard.

a. Procedure very similar to step 9, but use Tables D-3 to D-5, the first two providing n, M, and a v-quantity. Then use

$$Q_U = v(U - \bar{x})/\sigma \qquad Q_L = v(\bar{x} - L)/\sigma \qquad (11.19)$$

and find estimates p_U and p_L from Table D-5.

12. If there is real doubt as to the normality of the dis-
    tribution of the x's, then one possibility is to *accept*
    only by MIL-STD 414. But if the variables approach using
    414 would call for *rejection*, then do not reject yet, but
    continue onward, counting the number of pieces outside of
    specifications, until the sample size n for an appropriate
    attribute plan is completed. Such an attribute plan might
    be from MIL-STD 105D. This approach is called "variables-
    attributes sampling." In particular, it protects a pro-
    ducer who may have a process running outside of L to U,
    but who has sorted his product carefully to these limits.

13. If the shape of the distribution of x's is not normal,
    but is known and is unsymmetrical, then one might use
    Ref. 5. If the longer tail in a skewed curve is toward
    U, one can use a smaller n than for the normal, but if
    the short tail is toward U a larger n is needed.

11.7 CHECKING A PROCESS SETTING

Our two suggested plans for checking the level μ of a process follow
plans given in Ref. 2. The typical way to keep a production process
"on the beam" is to use a measurement control chart. However, it
may occur that from past experience it is found that the process
maintains reasonably good control and that σ_x is known. It may then
be desirable to merely check the process level from time to time,
and in particular to check the initial setting of the process.
There are two cases to consider. There is first the case of two
specification limits L and U for which the tolerance T = U - L is
only six or seven σ_x's. In this case then, we must maintain μ quite
close to the nominal (middle of specification range (U + L)/2) in
order to avoid pieces out of specifications. Then secondly there
are cases where the tolerance T is eight or more times σ_x. We may
then use a one-way check at each specification. Or, of course,
there may be just one limit, a minimum L *or* a maximum U, not both.

11.7.1 *Two-Way Check of Level*

We set the safe or desired level at the nominal (L + U)/2 = μ_0, and
then two unsafe or undesirable levels equally spaced on opposite
sides of μ_0, as follows:

Safe process level μ_0:

$P(\text{approval if } \mu = \mu_0) = .942 \doteq .95$

Unsafe process levels $\mu_0 \pm 1\sigma_x$:

$P(\text{reject if } \mu = \mu_0 \pm \sigma_x) = .897 \doteq .90$

Plan to give such control of risks of wrong decisions is:

Take n = 10 measurements and find $\bar{x}$

Then $\bar{x}$ between $\mu_0 - .6\sigma_x$ and $\mu_0 + .6\sigma_x$, approve setting

$\bar{x}$ outside, reject and reset process

11.7.2 One-Way Check of Level

For a check as to whether μ is too close to an upper specification U, we regard $\mu = U - 3\sigma_x$ as a safe level, for then (from Table A) there will only be .0013 of the pieces above U. Likewise, we regard $\mu = U - 2\sigma_x$ as an unsafe level for then (by Table A) there will be .0228 of the pieces above U. A practical one-way check with approximate .10 risks is the following:

Take n = 7 measurements and find $\bar{x}$. Then

$\bar{x} \leq U - 2.5\sigma_x$ Approve setting

$\bar{x} > U - 2.5\sigma_x$ Reject setting

$P(\text{accept if } \mu = U - 3\sigma_x) = .907$

$P(\text{accept if } \mu = U - 2.5\sigma_x) = .500$

$P(\text{reject if } \mu = U - 2\sigma_x) = .907$

If μ is to be checked near lower specification L, use

Take n = 7 measurements and find $\bar{x}$. Then

$\bar{x} \geq L + 2.5\sigma_x$ Approve setting

$\bar{x} < L + 2.5\sigma_x$ Reject setting

The risks on this plan are symmetrical to those on the check at U.

11.8 SUMMARY

The sampling plans in this chapter have been using measurements to determine whether the lot or process distribution is acceptable. The basic assumption has been that the distribution was normal.

Thus we were specifically concerned, in the decision, with μ and/or σ. In some cases, we assumed that σ was known from previous records, whereas in other cases we assumed that σ was unknown. Tests can be made directly on μ and/or σ as in Secs. 11.3 and 11.5. Or the test can ascertain whether the combination of μ and σ is such as to give an acceptably small percentage of pieces outside of specifications, as in Secs. 11.3.6 and 11.4.

It is recommended that control chart records be maintained in a series of lots, even though the decision on each individual lot is made using the present chapter. This will determine the control or lack of it, and in particular may permit accurate estimation of σ, so that known-σ plans may be used instead of unknown-σ plans, thus saving on the sample size and calculations.

Military Standard 414 is an effective system of integrated plans, providing normal, tightened, and reduced inspection for protection against one or two specification limits. It is for control of the fraction or percent defective. The plans assume normality, but one hedge, if this is not a safe assumption, is to use variables-attributes, permitting quick acceptance via measurements, but only rejecting via attributes.

Simple checks on process average level were also given.

PROBLEMS

11.1 The time of blow for fuses under a marginal circuit condition is subject to a *maximum* specification limit of 150 sec = U. Experience has shown that σ_x = 25 sec. Regarding μ_1 = 100 sec as the AQL and μ_2 = 125 sec as the RQL, and setting respective risks of α = .10 and β = .10, find an appropriate sampling plan. Sketch the OC curve for Pa versus μ, labeling axes.

11.2 The "Scott value" for material for a battery is subject to a *maximum* specification of 26, and σ_x is known from experience to be .56. Assume normality. Set up a sampling plan so that if $\mu = \mu_1$ = 24.5 the probability of rejection is .05, whereas if $\mu = \mu_2$ = 25.0 the probability of acceptance is .05. Sketch the OC curve Pa versus μ, labeling axes.

11.3   For the weight of contents of a package of a food product, the *minimum* specification for weights is 500 g. From past experience σ_x is known to be .6 g. Taking $\mu = \mu_1 = 501$ g as the RQL, and $\mu = \mu_2 = 502$ g as the AQL, and using respective risks β and α both at .05, find a sampling plan. Assume normal distribution of the x's. Sketch the OC curve, Pa versus μ, labeling axes.

11.4   For the length of a spring under a compressive force of 50 kg, there is set a *minimum* limit of 10.0 cm. Assume that $\sigma_x = .2$ cm, and that the distribution of the x's is normal. Set $\mu = \mu_1 = 10.4$ cm as the RQL and $\mu = \mu_2 = 10.6$ cm as the AQL, and take risks of $\beta = \alpha = .05$. Find a sampling plan for the test and sketch the OC curve Pa versus μ, labeling axes.

11.5   For doubled thickness of rubber gaskets for metal tops for food jars σ_x was known to be .002 in., and the distribution of x's to be normal. Consider $\mu = .106$ in. = AQL and $\mu = .103$ in. or .109 in. as RQL's. Set $\alpha = \beta = .01$. (Actually, if μ is at either RQL, the rubber is used in another application.) Set up an appropriate sampling plan. Sketch the OC curve, Pa versus μ, labeling axes.

11.6   A small part has diameter specifications of .1100 and .1102 in., each lot being produced under a single set-up. Experience has shown a normal distribution of diameters with $\sigma_x = .00003$ in. Taking $\mu = \mu_0 = .1101$ in. as AQL and $\mu = \mu_0 \pm .00003$ in. as RQL's and setting α and β risks both at .05, find a sampling plan. Assume normality. Sketch the OC curve, Pa versus μ, labeling axes.

11.7   Devise a test of variability for lots having $\sigma_1 = 1.715$, $\sigma_2 = 3.43$, $\alpha = \beta = .10$, assuming normality. Sketch the OC curve Pa versus σ, labeling axes. Test once each on distributions A and C of Table 7.3, by experimentation.

11.8   Muzzle velocity of target ammunition is to be tested for variability. Take $\sigma_1 = 5$ ft/sec and $\sigma_2 = 12$ ft/sec, and use risks $\alpha = \beta = .05$. Set up a sampling plan to make the test. Sketch the OC curve, Pa versus σ, labeling axes.

11.9 A new gauge is being considered. If the standard deviation, σ_1, on homogeneous material is .00002 in. the gauge should be approved, whereas if at σ_2 = .00003 in. it should be rejected. Use risks of α = β = .10, and find a sampling plan.

11.10 An analytical technique is supposed to have a standard deviation error of measurement (on homogeneous material) of 2 ppm. Set σ_1 = 1.5 ppm and σ_2 = 2.5 ppm, and both risks at .05. Determine an appropriate test for the proposed technique.

11.11 For fuses as in Prob. 11.1, considering σ_x as unknown, devise a sampling plan for measurements, so that if p' = p_1' = .005, Pa = .95, while if p' = p_2' = .02, Pa = .05.

11.12 For the contents weights in Prob. 11.3, considering σ_x as unknown, devise a sampling plan for measurements, so that if p' = p_1' = .005, Pa = .95, while if p' = p_2' = .05, Pa = .05.

11.13 Verify the risks given in the one-way check of a process setting in Sec. 11.7.2.

11.14 Verify the risks given in the two-way check of a process setting in Sec. 11.7.1.

Assuming MIL-STD 414 is available, find a sampling plan for the following requirements, assuming normal distribution of the x's:

11.15 Lot size N = 1000, inspection level IV, one specification U, σ unknown, normal inspection, using sample s, AQL = .65%.

11.16 Lot size N = 3000, inspection level IV, two specifications, σ unknown, normal inspection, using sample R's, AQL = .40%.

REFERENCES

1. Statistical Research Group, Columbia Univ., *Techniques of Statistical Analysis*, McGraw-Hill, New York, 1947.

2. I. W. Burr, *Statistical Quality Control Methods*, Dekker, New York, 1976.

3. Military Standard MIL-STD 414, *Sampling Procedures and Tables for Inspection by Variables for Percent Defective*, U. S.

Government Printing Office, Washington, D. C., 1957. Also American National Standards Institute (Z1.8).

4. H. G. Romig, Allowable Average in Sampling Inspection, Ph.D. Thesis, Columbia University, New York, 1939.

5. W. J. Zimmer and I. W. Burr, Variable sampling plans based on non-normal populations, *Indust. Quality Control*, *20* (No. 1), 18-26 (1963).

Chapter Twelve

TOLERANCES FOR MATING PARTS AND ASSEMBLIES

Nearly all individual piece parts are manufactured to be assembled
with other piece parts built to match them. For example, a shaft
is made for assembly within a bearing. It must be possible for the
shaft to go into the bearing. That is, assuming perfectly round
pieces, x_1 the inside diameter of the bearing must be larger than
x_2 the outside diameter of the shaft in order to permit assembly.
On the other hand, the clearance in diameters, $x_1 - x_2$, should not
be too large or the fit will be too loose. Therefore, we have a
problem as to what tolerances to set for the bearing and the shaft
to permit assembly in a very high percentage of cases, but to avoid
fits which are too loose. The objective is to accomplish this with
random assembly, which means picking at random the bearing and shaft
to be assembled. One other way sometimes used is to measure all
bearings and all shafts and *selectively assemble* the large diameter
bearings and shafts together, and the small diameter shafts and
bearings together. This has disadvantages: (1) it is expensive
and time consuming, and (2) when we need to replace a bearing we
may have trouble.

In general terms then, our problem is, given the distributions
of component dimensions or characteristics, what is the distribution
of the assembly characteristic? And how may we put statistical laws
to work in helping us?

12.1 AN EXAMPLE OF BEARING AND SHAFT

Let us suppose that we have a process that will produce shafts about
2 cm in diameter with a standard deviation σ_x = .0003 cm, and another process for bearings with σ_y = .0004 cm. Moreover, these are

333

achievable without sorting and with good control of averages. Then
we might well set 3σ specification limits for shafts of 2.0009
$\pm$ .0009 = 2.0000, 2.0018 cm. Next, so as not to overlap causing
interference, we might set limits for the bearings of 2.0020,
2.0044 = 2.0032 $\pm$ .0012 cm. Thus our two nominal diameters are
2.0032 and 2.0009 cm.

Since these σ's are achieved without sorting and the averages
are in control, we will also assume the distributions to be normal.
Let us set

    y = inside diameter of a bearing

    x = outside diameter of a shaft

    $w = y - x$ = diametral clearance of a random pair

Then we have

    σ_y = .0004 cm = standard deviation of bearing diameters

    σ_x = .0003 cm = standard deviation of shaft diameters

and we also set the processes to run at

    μ_y = 2.0032 cm = mean diameter of bearings

    μ_x = 2.0009 cm = mean diameter of shafts

Now then we can be assured that the clearance w will virtually always lie between the following two extremes:

    $\min_w = \min_y - \max_x = 2.0020 - 2.0018 = .0002$ cm

    $\max_w = \max_y - \min_x = 2.0044 - 2.0000 = .0044$ cm

This is a common way of thinking among engineers and has been for a
very long time. There is nothing really wrong about it, but there
is a way to do better by using the laws of statistics and taking
advantage of *compensating errors*. For one thing, what is the chance
of a clearance of only .0002 cm? This could occur with a 1-in-1000
minimum bearing and 1-in-1000 *maximum shaft* (for 3σ extremes). This
gives a 1-in-1,000,000 chance for such a small clearance. (This is
a rough analogy of what is operating, not an accurate approach.)

    Now given

    $w = y - x$ (12.1) and μ_y, μ_x, σ_y, σ_x and that x and y are independent (as in random assembly), it can then be proved that

$$\mu_w = \mu_y - \mu_x \tag{12.2}$$

$$\sigma_w = \sqrt{\sigma_y^2 + \sigma_x^2} \tag{12.3}$$

Take a good look at (12.3). σ_w is *not* the sum $\sigma_y + \sigma_x$ as was assumed implicitly in the foregoing way of setting tolerances for y and x. σ_w is in fact much *less* than $\sigma_y + \sigma_x$ since the square of $\sigma_y + \sigma_x$ is $\sigma_y^2 + 2\sigma_y\sigma_x + \sigma_x^2$, not $\sigma_y^2 + \sigma_x^2$. Or to be numerical, σ_w is *not* $\sigma_y + \sigma_x$ which is .0004 + .0003 = .0007 cm. Instead

$$\sigma_w = \sqrt{.0004^2 + .0003^2} = .0005 \text{ cm}$$

Therefore, using (12.2) and (12.3), we have for diametral clearance w in (12.1) the following for our two processes:

$$\mu_w = \mu_y - \mu_x = 2.0032 - 2.0009 = .0023 \text{ cm}$$

$$\sigma_w = \sqrt{\sigma_y^2 + \sigma_x^2} = \sqrt{.0004^2 + .0003^2} = .0005 \text{ cm}$$

Therefore, with the production processes set as assumed, we are *actually* meeting limits for w in a $\pm 3\sigma$ sense of

$$\mu_w \pm 3\sigma_w = .0023 \pm 3(.0005)$$
$$= .0023 \pm .0015$$
$$= .0008, .0038$$

These limits are considerably closer than .0002, .0044 cm as found from the purely additive basis on σ's, or tolerances.

Just how rare would a diametral clearance w of .0002 cm or less be. Using our μ_w, σ_w in (7.12)

$$z = \frac{.0002 - \mu_w}{\sigma_w} = \frac{.0002 - .0023}{.0005} = -4.2$$

Using a larger table than our Table A, we find a probability for z of -4.2 or less to be .000013.

Therefore, if we wish to have an actual minimum diametral

clearance w of .0002 cm, while retaining x limits at 2.0009 ± .0009
cm, we can make use of σ_w = .0005 cm as follows:

$$\mu_w = .0002 + 3(.0005) = .0017$$

But $\mu_w = \mu_y - \mu_x$, and so substituting what we know, .0017 = μ_y -
2.0009 giving μ_y = 2.0026. Therefore, we can set limits for the
bearings at

$$\mu_y \pm 3\sigma_y = 2.0026 \pm 3(.0004)$$
$$= 2.0026 \pm .0012$$
$$= 2.0014, 2.0038$$

This will give 3σ limits for clearance of .0017 ± .0015 = .0002,
.0032, thus avoiding the looser fits up to .0044 cm, originally in
mind.

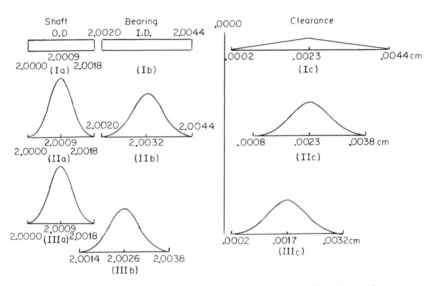

Fig. 12.1. Three tolerance-meeting sets of distributions for a
shaft and bearing. Section I is for the traditional purely additive tolerances. The two component distributions Ia and Ib are
uniform between the limits and give approximately the clearance
distribution shown. II and III are for ±3σ normal-curve meeting of
tolerances. IIa and IIb give IIc for clearances. The overlapping
tolerance distributions of III still give IIIc for clearances, with
extremely small chance for a bearing smaller than a shaft in random
assembly.

See Fig. 12.1, which pictures the two approaches: traditional
and statistical. The distribution Ic shown for clearances is only
approximately correct for the two uniform or rectangular distributions for shaft and bearing, which fill the tolerance interval. If
the processes give ±3σ normal curve meeting of tolerances as shown,
then the clearances behave as pictured in IIc and IIIc. Distribution IIc is entirely safely away from zero. But we can use *overlapping* tolerances for shaft and bearing as shown, *provided* we meet
them with ±3σ normal curve distributions, and will then have IIIc,
which is still safely above zero, but has a more favorable maximum
clearance of .0032 cm.

It is worth mentioning with regard to this example that the
assumption we used of independence of bearing and shaft may not
fully apply if there is drift due to tool wear. Thus shaft outside
diameters may tend to gradually increase, and bearing inside diameters may tend to gradually decrease. And so, initially the clearance may be relatively large and later on relatively small. We
have assumed that such tool-wear drifting has been held under control by resetting as needed.

12.2 AN EXAMPLE OF AN ADDITIVE COMBINATION

Assemblies where component characteristics add are very common in
industry, perhaps more so than subtractive combinations such as discussed in Sec. 12.1 on clearances. Examples are assemblies on a
shaft, resistances in a series circuit, thickness of two pieces
bolted together, thickness after two coatings, and contents or impurity in a mixture of two or more liquids.

Let us consider a hypothetical example on resistances in a circuit. Suppose we connect three resistors in a series. Then the
component resistances may be called w, x, and y, and the resistances
of the wire and connections z. Then for the total resistance we
have

$$r = w + x + y + z \qquad (12.4)$$

Let us illustrate with

$$\mu_w = 800 \qquad \mu_x = 300 \qquad \mu_y = 195 \qquad \mu_z = 4$$
$$\sigma_w = 6 \qquad \sigma_x = 3 \qquad \sigma_y = 3 \qquad \sigma_z = .5 \text{ ohms}$$

We have for (12.4)

$$\mu_r = \mu_w + \mu_x + \mu_y + \mu_z \tag{12.5}$$

which in the example gives

$$\mu_r = 1299 \text{ ohms}$$

Next, if these resistances are independent, which seems to be quite a safe assumption here, then we also have

$$\delta_r = \sqrt{\sigma_w^2 + \sigma_x^2 + \sigma_y^2 + \sigma_z^2} \tag{12.6}$$

This gives here

$$\sigma_r = \sqrt{6^2 + 3^2 + 3^2 + .5^2} = 7.37$$

That is, just a little more than σ_w alone. Therefore, as long as the averages are maintained at the levels shown, we can expect that nearly all circuit resistances will lie between

$$\mu_r \pm 3\sigma_r = 1299 \pm 3(7.37)$$
$$= 1299 \pm 22$$
$$= 1277, 1321$$

Now let us suppose that instead of working with the normal distributions with averages and standard deviations, which we considered known, we only used specification limits at $\mu \pm 3\sigma$, that is

$$800 \pm 3(6) = 782, 818$$
$$300 \pm 3(3) = 291, 309$$
$$195 \pm 3(3) = 186, 204$$
$$4 \pm 3(.5) = 2.5, 5.5$$

Many would now add the four low specification limits and the four high ones. This gives 1261.5, 1336.5. Now are these reasonable limits? Of course, if the four component resistances all lie inside their respective limits, then the total resistance for the circuit will indeed lie within 1261.5 to 1336.5. But even when we

have uniform distributions of resistances for the components, between limits as in Fig. 12.1 Ia and Ib, the distribution of circuit
resistances will not be uniform between 1261.5 and 1336.5. Instead,
there will be considerable tapering out to these limits, making extreme values rarer. But if these limits are met with normal curve
distributions, then as we have seen, practically all total resistances will lie between 1277 and 1321. The width of this band is 44
as against 75 for the purely additive approach. Thus if we can be
assured of a normal curve ($\pm 3\sigma$) meeting of specifications, then we
could set specification limits of, say, 1275 and 1325 rather than
1261 and 1337.

It may also be stated that even when the distributions of component characteristics are not normal, the distribution of the combined or assembly characteristic tends toward a normal distribution
with μ and σ like (12.5) and (12.6). (This fact is from the Central
Limit Theorem.)

Thus, we have another example in which the tolerance limits for
the characteristic of the assembly are narrower than might have been
expected from the limits for the components as a result of compensating errors or variations.

12.3 GENERAL FORMULAS

Let us set down the general formulas for the kinds of relationships
we have been studying. Let us define

$$y = x_1 \pm x_2 \pm \cdots \pm x_k \qquad (12.7)$$

which we call an additive-subtractive relation. The signs could be
positive or negative throughout or any combination. There are k
characteristics of component parts affecting the assembly characteristic y. Then we have

$$\mu_y = \mu_1 \pm \mu_2 \pm \cdots \pm \mu_k \qquad (12.8)$$

For the next formula, we need to assume that the various component
characteristics x_1, x_2, ..., x_k act independently, that is, that

whatever value one x takes has no influence upon or relation to the
other x's in the assembly or combination. If the production processes are in control or if assembly is done randomly, this is likely to be a safe assumption. Then

$$\sigma_y = \sqrt{\sigma_1^2 + \sigma_2^2 + \cdots + \sigma_k^2} \qquad\qquad (12.9)$$

Note that although there are plus and minus signs in (12.7) and
(12.8), *all* signs under the radical in (12.9) are plus. Also we
mention that σ_y is much less than the simple sum $\sigma_1 + \sigma_2 + \cdots + \sigma_k$
(as assumed in purely additive tolerances).

Finally, we may say that if the component characteristics, x_1,
x_2, ..., x_k have normal distributions, then y will also. Moreover,
even if the x's are not normally distributed, y still tends to be,
especially as k increases. (This could be made more exact by quoting
the Central Limit Theorem in one of its several forms.)

The author consulted on one job where 10 different dimensions
had a formula for clearance like (12.7). A wider tolerance was being requested on one dimension, so as to eliminate one refined
threading operation. Process capabilities were well known. Analysis showed that, given the wider tolerance, there would still be
only one chance in 10,000 of inability to assemble the parts. The
saving on this one order was $300,000.

In another case, analysis showed that there were 38 dimensions
on the shaft of an auto generator which could affect end-play via a
relation like (12.7). Thus there was at least some possibility,
through using statistical tolerancing, of having limits $\sqrt{38} = 6.2$
times as wide as in additive tolerancing, *provided* averages are well
controlled. Many tolerances were accordingly widened.

12.4 SETTING REALISTIC TOLERANCES

Basic to the setting of realistic tolerances is knowledge of the
capabilities of the processes which will produce the components.
To find the process capability, we need to have achieved reasonable
process control, for otherwise we simply do not know what the

process can do. Then we can use (7.10) or (7.23), that is, $\bar{R}/d_2$ or
$\bar{s}/c_4$ to estimate σ for each component. Now such a σ is for a short
time interval. It is reasonable to add a bit to this short-term σ,
in order to estimate the long-term σ, which will contain some set-
up error and some drifting of the average. We might use

$$\sigma(\text{long term}) = \frac{\sigma(\text{short term})4}{3} \qquad (12.10)$$

It is the $\sigma(\text{long term})$'s, which we should substitute into (12.9) to
find σ_y. Now setting μ_y at the desired nominal for the assembly,
we can set the component nominals μ_i's in (12.8) to yield the desired μ_y. Then we use the $\sigma(\text{long term})$'s of the components to substitute into (12.9) to find σ_y. Then the processes will nearly
always (99.7% or so) meet specification limits on the assembly of

$$\mu_y \pm 3\sigma_y \qquad (\text{limits for assembly}) \qquad (12.11)$$

If these limits are satisfactory, then we have realistic specification limits, and need only exercise the controls over components
that we have been using. But if the limits in (12.11) are not narrow enough, then we will need to make at least one of the components
more narrow. The place to start looking is the component with the
largest $\sigma(\text{long term})$. There are two ways to proceed: (1) *Exercise
tight control over the process average* and thus bring down $\sigma(\text{long}$
term) to close to $\sigma(\text{short term})$ and, therefore, cut (12.9) narrow-
ing (12.11). (2) To arbitrarily set specification limits on a component narrower than $\mu \pm 3\sigma(\text{short term})$, which is to be achieved
through 100% sorting and possibly rework. Which component or components to restrict thus is an engineering and economic decision.
But since such sorting does *not* leave a normal distribution, we can
conservatively use $\sigma = (U - L)/4$, instead of the normal curve
$\sigma = (U - L)/6$ for $\sigma(\text{short term})$. In this latter case, we will have
the expensive inspection job to do, with possible reworking as well.

Of course, there is also a possibility: (3) To decide to
widen the specification limits to be at those in (12.11), which the
processes can economically supply to the assembly.

These general considerations can also be used in telling
whether two particular lots of product will satisfactorily assemble.
Material review boards can use this technique for analysis and a
decision.

12.5 RELATIONS OTHER THAN ADDITIVE-SUBTRACTIVE

The majority of functional relationships between assembly and components would seem to take the form of (12.7). But there are a
great many other functions that might possibly occur, such as
$y = x_1 \cdot x_2$ or $y = x_1/x_2$. Such can be handled. See Ref. 1. μ_y
and σ_y are found by algebra and calculus.

12.6 SUMMARY

This chapter has provided an introduction to the way in which dimensions and other component characteristics combine into the characteristic of an assembly. Specifically discussed was the case where
the relation is additive-subtractive (12.7) and in which there is
independence or random assembly. Then the tolerances which we can
set for the components are usually considerably greater than we can
with purely additive tolerancing, sometimes called "worst case"
tolerancing. But to reap the benefits, we must exercise reasonable
control over the various process *averages*. The distribution of the
assembly characteristic tends to be normal. An approach to practical tolerancing was given in Sec. 12.4, where several alternatives
were suggested. These can often eliminate much inspection, rework,
and scrapping, if implemented. The methods can also be used for
decisions on lots of mating parts for objective decisions by material review boards.

If the function for the assembly characteristic is not additive-subtractive there are other methods which can be used. See
Ref. 1.

PROBLEMS

12.1 To illustrate the approach to normality as the number of components increase, perform this experiment on dice totals. For a single die, the faces may show 1, 2, 3, 4, 5, 6 with equal probability, that is, 1/6 each. This is a flat-topped uniform distribution. Now throw three dice and count the total of the faces showing. The total can be anywhere from 3 to 18 inclusive, but the extremes are much less likely than, say, 10 or 11. Throw the three dice 108 times, tabulating directly into a frequency table. Compare with the theoretical frequency distribution given in the answers.

12.2 Consider shafts with specification limits 2.0002 and 2.0019 in. The matching bearings also carry specification limits of 2.0009 and 2.0033 in. Note the sizable overlapping of the two sets of limits. Suppose that these limits are met in a $\pm 3\sigma$ sense with normal distributions. Make a guess as to the proportion of random choices of bearing and shaft, which will have *negative* clearance, that is, will not go together. To illustrate, throw three dice for the number of .0001 in. above 2.0000 in. for a shaft, for example, 12 total is 2.0012 in. Likewise, throw six dice (or three, twice) for the inside diameter of a bearing. The difference then is the diametral clearance. You can tabulate the differences, for example, 20 - 12 = 8. Obtain 50 clearances by dice throws. For three-dice totals $\mu = 10.5$, $\sigma = 2.96$, and for six-dice totals $\mu = 21$, $\sigma = 4.18$. Use (12.7) through (12.9) to find μ and σ for the clearance, then use (7.12) and Table A to approximate the percentage of clearances below zero. Is this less than you expected?

12.3 A pin shows good control on O.D. = outside diameter with $\bar{\bar{y}}$ = .21440, $\bar{R}$ = .00032 in. for samples of n = 5. A mating collar also shows good control on I.D., with $\bar{\bar{x}}$ = .21503, $\bar{R}$ = .00040 in. and n = 5. Estimate the process standard deviations for each. Let the diametral clearance be w = x - y, and estimate μ_w and σ_w. Assuming normal distributions, what percentage of pairs of pin and collar will have clearance less than .0001 in.?

12.4 In manufacture of lamps, the clearance w = x - y is of interest, where x = distance from rim of base to the glass, and y = distance from rim of base to top of innershell. For x's, large samples
gave estimated μ = 1.0499, σ = .0428 in., whereas for y's μ = .9079,
σ = .0120 in. Good normality was present. Describe the distribution of w = x - y. What percentage of w's are negative in random
assembly?

12.5 A brass washer and a mica washer are to be assembled one on
top of the other. For the former μ = .1155, σ = .00045 in., whereas for the latter μ = .0832, σ = .00180 in. Find μ and σ for the
combined thickness. What assumptions were needed? Assuming normality, set ±3σ limits for the combined thickness.

12.6 A general concept in interpreting measurements of material is
the following: We let w = the observed measurement, x = the "true"
measurement, and y = the error of measurement (which may be + or -).
Then w = x + y. Since x and y are independent we may use (12.9)
with k = 2. Also if the measurement is "unbiased" then μ_y = 0.
Suppose μ_x = 115.0, σ_x = 2.0, σ_y = .5 mm. Assuming that x and y
are normally distributed, describe the distribution of the observed
measurement w. How influential was σ_y? (Take note that although
x = w - y, we could not use $\sigma_x = \sqrt{\sigma_w^2 + \sigma_y^2}$ because w and y are *not*
independent. In fact, this equation would give more variation in
true dimension than in the measured dimension!)

12.7 At one time 200-ohm resistors, made to = .75%, that is, to
200 ± 1.5 ohm were quite costly, due in part to the necessity of
sorting out those outside the limits. But production was able to
manufacture 100-ohm resistors to ±1% without sorting and with good
normality of distribution. Take μ = 100, σ = .33 ohm. Now assem-
bling at random two 100-ohm resistors in series gives what distribution for total resistance? Will such a distribution meet limits
of 200 ± 1.5 ohms? Take note that we should check on the independence of resistances in 100-ohm resistors, because wire diameter
might cause nonindependence. (A company made quite large savings
using this approach.)

12.8 Five pieces are assembled on a shaft, two of the first kind with μ = 2.0140, σ = .0014, two of a second kind with μ = 3.2061, σ = .0012, and one of a third kind with μ = 3.8402, σ = .0022 (all in inches). For total length w, use $w = x_1 + x_2 + \cdots + x_5$, since random assembly is to be used. What are μ and σ for length w, and what $\pm 3\sigma$ limits can it meet? Are these limits narrower than we would find using the five maximum limits 2.0140 + 3(.0014), etc., and the five minimum limits?

12.9 In packaged weight control, it is not very convenient to obtain directly the weight of the net contents. Instead, we may proceed indirectly by weighing empty containers x, covers y, and total filled packages z. Letting the net weight of contents be w, we thus have $w = z - x - y$. Now we *cannot* find σ_w by use of (12.9), because z, x, and y are not independent. Instead, we can let $z = x + y + w$, in which x, y, and w are likely to be quite independent. Then we use (12.9) for the relation between σ^2's. Knowing three of these, we can find σ_w^2 and σ_w. However, we *can* use (12.8) on either form of (12.7) to find μ_w. Now, given the actual observed data z: μ = 189.5, σ = 7.1; x: μ = 10.7, σ = 3.5; y: μ = 3.1, σ = 1.1 (in pounds), find μ and σ for w. Assuming normality estimate the percentage of contents weights below the minimum specification of 165 lb.

12.10 The following is an example of aid to a materials review board. In a plant manufacturing soap and cosmetics, a question arose as to whether to use a lot of caps for cologne bottles. Accordingly tests were made on the distribution of cap strength (the torque which would break the cap), and the distribution of the torque which the capping machine would apply to a cap. They found in inch-pounds, respectively,

$$\bar{x} = 11.1 \qquad s_x = 2.80 \qquad \bar{y} = 8.0 \qquad s_y = 1.51$$

Assuming independence (very safe here) and normality, estimate the average and standard deviation of the margin of safety $w = x - y$.

Estimate the percentage of caps that can be expected to break by a negative w value.

REFERENCE

1.   I. W. Burr, *Statistical Quality Control Methods*, Dekker, New York, 1976.

Chapter Thirteen

STUDYING RELATIONSHIPS BETWEEN VARIABLES
BY LINEAR CORRELATION AND REGRESSION

We now take up a useful method of analyzing the relationship between two variables. We make simultaneous observations on the two variables x and y, say, x_1 and y_1, then later on another pair x_2 and y_2, and so on to x_n and y_n. Each such pair is at the same time or place, or on the same material. Then we seek to study the relationship between the two variables. For example, can we estimate or predict y from x? Are they closely related, loosely related, or unrelated? One very simple way to gain some insight into the relation is to make a "scatter diagram," that is, to plot each pair of x and y on a graph. Thus with a horizontal x-axis and a vertical y-axis, we first plot y_1 against x_1, then y_2 against x_2, and so on. Then a relationship may or may not emerge. Sometimes such a scatter diagram is all that we need in a study.

13.1 TWO GENERAL PROBLEMS

In the first problem, we are especially interested in the estimation or prediction of y from x. For example, we may be using a standard analytical technique and wish to see whether we can accurately estimate its result from that of an alternative less expensive or less time-consuming analysis. Or, we may compare two gauges, or wish to see whether two physical properties such as hardness and tensile strength are closely enough related to predict the latter from the former. Such problems frequently arise in industry.

Another general problem is to study a collection of "input" variables to see which is most closely related to an "output" or quality variable. That variable or those variables most closely

related are the ones to work on in trying to improve the process
and obtain better quality. The degree or strength of relationships
is thus the key to the study.

These two problems respectively emphasize "regression" and
"correlation," although there is not really any hard and fast dis-
tinction between these two concepts.

In this book, we shall only consider quite basic techniques.
For those who wish to go more deeply into the subject, Refs. 1 and
2 are recommended.

13.2 FIRST EXAMPLE--ESTIMATION

In the manufacture of piston rings, C-shaped castings are precisely
ground and machined to specifications. A processing of the completed rings called "ferroxing" was being used. It used high-tem-
perature steam to form a blue oxide coating to protect the rings
against rust. The measurement in question was the force necessary
to close the ring to the specified gap and is called "tension."
This tension must not be too great, or it will cause excessive wear,
nor too little for then the piston may leak pressure and the cylinder lose power. During ferroxing, strains are relaxed and the tension is increased. Three different forms of suspension during the
ferroxing were being studied. The data we shall use are from a T-
bar suspension with the gaps up. We give here 20 rings out of the
300 used in the study. The tension of each ring was measured before and after ferroxing. The correlating feature is that x_1 was
the tension before for the *first* ring and y_1 the tension after for
the *same* ring, and so on. Then the question is as to how we may
predict y from x, and how accurate is the prediction, that is, to
how much error is it subject?

We shall first show a method of calculation in Table 13.1,
which may seem cumbersome, but does help the reader to see what is
happening. Then by coding and other methods analogous to (2.6) and
(2.5), the calculations can be simplified. Now although practically every plant with an electronic computer will have a program

TABLE 13.1. Pounds "Tension" of 20 Piston Rings, before and after "Ferroxing," Respectively, x_i and y_i for the i'th Ring

x	y	$x - \bar{x}$	$y - \bar{y}$	$(x - \bar{x})^2$	$(y - \bar{y})^2$	$(x - \bar{x})(y - \bar{y})$
5.2	5.7	+.185	+.135	.034225	.018225	+.024975
5.8	6.4	+.785	+.835	.616225	.697225	.655475
4.8	5.4	-.215	-.165	.046225	.027225	.035475
6.4	7.0	+1.385	+1.435	1.918225	2.059225	1.987475
5.2	5.8	+.185	+.235	.034225	.055225	.043475
4.6	5.1	-.415	-.465	.172225	.216225	.192975
5.7	6.3	+.685	+.735	.469225	.540225	.503475
5.9	6.6	+.885	+1.035	.783225	1.071225	.915975
4.6	5.2	-.415	-.365	.172225	.133225	.151475
4.9	5.3	-.115	-.265	.013225	.070225	.030475
4.2	4.9	-.815	-.665	.664225	.442225	.541975
4.2	4.8	-.815	-.765	.664225	.585225	.623475
4.3	4.8	-.715	-.765	.511225	.585225	.546975
5.5	6.0	+.485	+.435	.235225	.189225	.210975
4.8	5.3	-.215	-.265	.046225	.070225	.056975
5.6	6.2	+.585	+.635	.342225	.403225	.371475
4.7	5.1	-.315	-.465	.099225	.216225	.146475
4.7	5.1	-.315	-.465	.099225	.216225	.146475
4.7	5.2	-.315	-.365	.099225	.133225	.114975
4.5	5.1	-.515	-.465	.265225	.216225	+.239475
100.3	111.3	.000	.000	7.285500	7.945500	+7.540500

available for linear regression, the reader is strongly advised to carry through some problems by the methods to be shown. In the first two columns of Table 13.1 are listed the before and after ferroxing tension (in pounds), respectively, x and y.

Figure 13.1 shows the y's plotted against the respective x's. The relationship seems to be quite close. That is, given a before ferroxing x, we can predict y with considerable confidence. Thus, say, if x = 5.4, we would estimate by eye that y would be close to

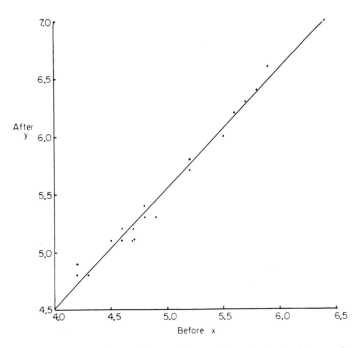

Fig. 13.1. Observations of pounds tension for 20 piston rings of x before versus y after ferroxing. The best-fitting line, y on x, is also shown.

5.9. Moreover, it looks as though a straight line might be made to fit the points quite well, that is, the relation is "linear."

Definition 13.1 (Linear relation). If a plot of y versus x seems to follow a straight line, even if quite loosely, then we call the relation linear.

Definition 13.2 (Nonlinear relation). If a plot of y versus x does not seem to follow a straight line but instead some curve, then we call the relation nonlinear.

Since Fig. 13.1 shows such a linear trend, our object is to fit a straight line to the data given in Table 13.1. In fact, we wish to find the equation of the best fitting straight line. For estimating y from x, we choose to use that line for which the sum of the *squares* of the distances from the line to the points, *measured vertically*, is as small as it can be for the given data. This is called the "least squares fitted line." It can be proved that the following method gives this line.

The reader may well recall that the general equation of a straight line takes the form

$$y = mx + b \qquad\qquad (13.1)$$

where b is the "intercept," that is, the value of y when x is zero, and where m is the "slope," that is, the amount of *change* in y when x *increases* by one unit. Then our problem becomes that of finding what values to use for m and b so as to fit the given data as well as possible (in the least-squares sense).

For this fitting, we may proceed as in Table 13.1. First, we find $\Sigma\, x = 100.3$ and $\Sigma\, y = 111.3$, then

$$\bar{x} = \frac{100.3}{20} = 5.015 \text{ lb}$$

$$\bar{y} = \frac{111.3}{20} = 5.565 \text{ lb}$$

Next we find the respective deviations $x - \bar{x}$ and $y - \bar{y}$ in the next two columns. The totals of these deviations are zero (except possibly for round-off errors). This provides a good check on our deviations. Next we find the squares of the deviations or at least their sums $\Sigma(x - \bar{x})^2$, $\Sigma(y - \bar{y})^2$. Such sums of squares are readily accumulated on many desk calculators, without writing each individual squared deviation. Finally, we obtain in the last column the products of the x deviation $(x - \bar{x})$ by the y deviation $(y - \bar{y})$. For example $(+.185)(+.135) = .024975$. Signs must be carefully watched, but in our problem both deviations carried the *same* sign in all pairs, so that the products were *all* positive.

Now for our "best" slope we have

$$m = \frac{\Sigma(x - \bar{x})(y - \bar{y})}{\Sigma(x - \bar{x})^2} \qquad \text{(slope of "best" line)} \qquad (13.2)$$

For our example, we, therefore, have

$$m = \frac{7.540500}{7.285500} = 1.0350$$

Next we need the intercept b, which is to be found by

$$b = \bar{y} - m\bar{x} \qquad \text{(intercept of "best" line)} \qquad (13.3)$$

This gives here

$$b = 5.565 - 1.0350(5.015) = .374$$

so that our estimation equation of y from x is

$$\hat{y} = 1.035x + .374$$

where $\hat{y}$ designates the estimated y.

The estimating equation $\hat{y} = mx + b$ is always satisfied by the pair $(\bar{x}, \bar{y})$, so that the best fitting straight line of y on x passes through the point of averages $(\bar{x}, \bar{y})$ and, of course, has slope m. Here this point is (5.015, 5.565). In order to draw the best-fitting line, it is desirable to find two more points on the line and plot all three. We find (4.0, 4.514) and (6.4, 6.998). These three lie on a line and supply a check in drawing the line.

Next we would like to know something about how well we are able to estimate y from x. One measure is the "coefficient of linear correlation," called r. It may be defined by

$$r = \frac{\Sigma(x - \bar{x})(y - \bar{y})}{\sqrt{\Sigma(x - \bar{x})^2 \Sigma(y - \bar{y})^2}} \qquad \begin{array}{l}\text{(linear correlation} \\ \text{coefficient)}\end{array} \qquad (13.4)$$

This coefficient measures the relative closeness of fit of the points around the line. The value of r always lies between -1 and +1, the two extremes occurring when, and only when, all points lie exactly on a straight line. As may be seen by comparing (13.2) and (13.4), r and the slope m have the same sign, because the numerators are identical, and both denominators are always positive. The larger *in size* that r is, the relatively better the straight line fits. Perhaps the best way to use r in this connection is through its square. Thus

r^2 = coefficient of determination

= proportion of $\Sigma(y - \bar{y})^2$ which is explained

by the line (13.5)

The total variation of the y's around their average is $\Sigma(y - \bar{y})^2$, and r^2 tells us what part of this can be related to x via the best-fitting straight line.

Now let us use these formulas in our example. We have

$$r = \frac{7.5405}{\sqrt{7.2855(7.9455)}} = .99108$$

which indicates a strong linear relationship (as we see in Fig. 13.1). Also

$$r^2 = .99108^2 = .98224$$

Therefore, over 98% of the total variation in y is linearly related to x via the best-fitting line, for the before and after tensions. This indicates excellent predictability.

Let us consider briefly another way to describe the predictability. Since by (13.5) r^2 is the proportion of $\Sigma(y - \bar{y})^2$ that is *explained* by x via the straight line, then $1 - r^2$ is the *unexplained* proportion or

$$(1 - r^2)\Sigma(y - \bar{y})^2 = \Sigma[y(\text{observed}) - y(\text{estimated})]^2$$

that is, the sum of the squares of the distances from the line to the observed points, *measured vertically*. This sum is best divided by n - 2, after which the square root is taken. We call this result the "standard error of estimate." It gives us an only-to-be-expected distance of the y's from the line, *within the data at hand*. Calling this $s_{y \cdot x}$, we thus have

$$s_{y \cdot x} = \sqrt{(1 - r^2)\Sigma(y - \bar{y})^2/(n - 2)} \quad \text{(standard error} \quad (13.6) \\ \text{of estimate)}$$

In our example this becomes

$$s_{y \cdot x} = \sqrt{(1 - .99108^2)(7.9455)/(20 - 2)} = .0885$$

This means that *within our 20 points*, when we estimate y from x
using $\hat{y}$ = 1.035x + .374, we can expect the observed y to be off
from the estimated y by .0885, on the average. However, when we go
outside our original 20 data points, we can expect to miss hitting
y by a slightly larger amount especially toward the extreme x's as
a result of errors in our obtained m and b. We can make a comparison of $s_{y \cdot x}$ with s_y as follows:

> If we do not know x, our best guess at a random y value is
> $\bar{y}$, whereas if we know x and have a linear relation we use
> the regression equation for an estimate of y. The average
> error in the latter estimate is $s_{y \cdot x}$ = .0885, whereas for
> the former estimate the average error is

$$s_y = \sqrt{\Sigma(y - \bar{y})^2/(n - 1)} = \sqrt{7.9455/19} = .647$$

> Therefore, knowledge of x cuts the estimating error down
> from .647 to .0885.

This $s_{y \cdot x}$ thus gives us a useful picture of the accuracy with which
we may estimate y from x using the best-fitting line. We may sometimes be off by $2s_{y \cdot x}$, but very seldom by $3s_{y \cdot x}$.
   One further use of the best-fitting line is in predicting a
lot mean $\bar{y}$, that is, the average tension *after* ferroxing from a lot
mean $\bar{x}$, the average tension *before* ferroxing. This we do by (using
the carat for the estimate)

$$\hat{\bar{y}} = 1.035\bar{x} + .374$$

Hence, we can tell what average tension to expect after ferroxing,
if we estimate the lot mean tension by $\bar{x}$ before ferroxing.
   This story on piston rings is a good example of a case in which
reliable estimations of y from x may be made because the correlation
coefficient r was high. The results of the full study compared
three forms of ring suspension during ferroxing and proved useful
in choosing one method and in making predictions on lots and controlling the production process to meet specifications.

13.3 SECOND EXAMPLE--CORRELATION

Our second example illustrates the second general type of application mentioned in Sec. 13.1, namely, using correlation to sift through the various "input" variables to find which are most closely related to an important "output" variable. In this case, the output variable was tons rejected for bad surface on an open-hearth heat of steel. Altogether in the original study, there were about 20 or more input variables and 130 heats of steel. (Careful records are made in steel plants on each heat of steel.) Now in such process industries as steel making, plastics, foundries, and chemicals and food production, we must not expect large correlations for at least two reasons: (1) there are so many factors involved, which may influence the quality produced, and (2) our predecessors have probably already discovered the strong relations of really influential input variables. (Another possible trouble is inaccurate measurement of the output variable itself; that is, our measurements are not highly reproducible.)

Therefore, our problem becomes that of trying to decide which of many rather loose or poor relationships are the best, or perhaps we should say least poor, *within current practice specifications*. Which input variables might it be desirable to control more closely than we have in the past? In this way correlation coefficients may be used as a criterion of choice.

In this approach, however, the reader is urged to draw scatter diagrams of many, if not all, pairs of points, because some relationships may be nonlinear, which may tend to make the correlation coefficient r rather meaningless. Sometimes r may be very small, even though there is a fairly distinct curvilinear relationship.

Now see Table 13.2, in which we give a small portion of the original data, namely for only 20 heats of steel and only five of the input or independent variables. (These variables are not "independent" in any mathematical or statistical sense, but they are subject to control.) We have plotted five scatter diagrams in Fig. 13.2 for the five variables versus the quality variable--tons

TABLE 13.2. Data on Five Input and One Output Variables for 20
Heats of Steel, a Portion of the Original Data Comprising 130 Heats
and About 20 Input Variables

Heat number	0.01% ladle manganese	0.01% ladle carbon	Number of bags of coke	Pouring temperature (°F)	Average sealing time (sec)	Tons rejected for bad surface
1	47	28	12	2870	31	38
2	45	25	2	2860	26	0
3	81	20	0	2855	115	2
4	51	21	4	2840	67	12
5	87	19	4	2855	14	67
6	60	22	4	2845	75	12
7	51	22	8	2850	120	55
8	49	27	8	2878	46	33
9	51	25	10	2845	70	26
10	84	16	2	2875	45	16
11	88	21	4	2875	31	12
12	44	17	5	2870	25	11
13	47	28	14	2870	27	8
14	50	25	0	2840	120	3
15	48	22	8	2860	127	6
16	53	18	2	2880	23	3
17	46	20	1	2840	44	1
18	43	19	1	2890	98	7
19	46	26	6	2860	44	0
20	48	14	5	2860	40	0

rejected. None of the graphs show much strength of relation. In
particular, the heat with 67 tons rejected seems to be quite a dis-
turbing factor to the relations. It may well have been largely
caused by some other variable or condition than those given here.

Let us again illustrate the calculations for r by use of deviations of the variables (see Table 13.3). There we see the same
five column headings as in Table 13.1. But in Table 13.1 every

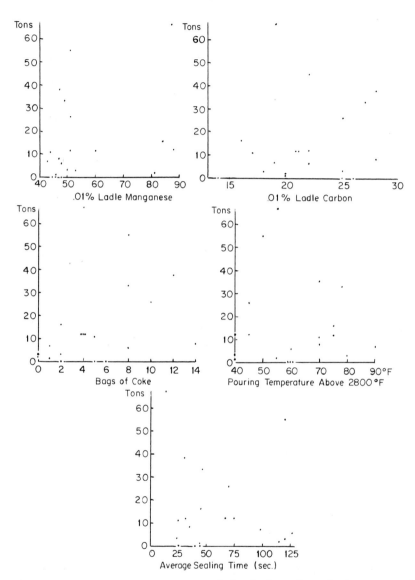

Fig. 13.2. Tons of steel rejected for bad surface in heat versus five input variables. Data from Table 13.2, being a small portion from the original study.

TABLE 13.3. Calculation Table for Correlation Between y = Tons
Rejected for Bad Surface in Heat and x = .01% Ladle Manganese.
Use Made of Deviations of Variables.

x	y	$x - \bar{x}$	$y - \bar{y}$	$(x - \bar{x})^2$	$(y - \bar{y})^2$	$(x - \bar{x})(y - \bar{y})$
47	38	-9	+22.4	81	501.76	-201.6
45	0	-11	-15.6	121	243.36	+171.6
81	2	+25	-13.6	625	184.96	-340.0
51	12	-5	-3.6	25	12.96	+18.0
87	67	+31	+51.4	961	2641.96	+1593.4
60	12	+4	-3.6	16	12.96	-14.4
51	55	-5	+39.4	25	1552.36	-197.0
49	33	-7	+17.4	49	302.76	-121.8
51	26	-5	+10.4	25	108.16	-52.0
84	16	+28	+.4	784	.16	+11.2
88	12	+32	-3.6	1024	12.96	-115.2
44	11	-12	-4.6	144	21.16	+55.2
47	8	-9	-7.6	81	57.76	+68.4
50	3	-6	-12.6	36	158.76	+75.6
48	6	-8	-9.6	64	92.16	+76.8
53	3	-3	-12.6	9	158.76	+37.8
46	1	-10	-14.6	100	213.16	+146.0
43	7	-13	-8.6	169	73.96	+111.8
46	0	-10	-15.6	100	243.36	+156.0
48	0	-8	-15.6	64	243.36	+124.8
1119	312	-1	0	4503	6836.80	+1604.6

product $x - \bar{x}$ by $y - \bar{y}$ was positive, and, moreover, the relatively
large positive deviations in the respective columns fell together,
as did the negatives. That is, no large deviation was "wasted" on
a relatively small deviation of the same or opposite sign. This
gave a relatively large positive sum for $(x - \bar{x})(y - \bar{y})$'s, of about
the same size as $\Sigma(x - \bar{x})^2$ and $\Sigma(y - \bar{y})^2$. Thus (13.4) gave a large
positive r of .99. But in Table 13.3 we see quite a different situ-
ation in the building up of $\Sigma(x - \bar{x})(y - \bar{y})$, with products both plus

and minus occurring. Note how the algebraic sum of this column is much less than that for either sum of squares of deviations. Now we use (13.4), finding

$$r = \frac{+1604.6}{\sqrt{4503(6836.80)}} = \frac{1604.6}{5548.5} = .28919$$

Thus we have a rather low positive correlation. Only $.28919^2$ = $.08363$ of the variation, $\Sigma(y - \bar{y})^2$, among these 20 points is linearly related to percent of ladle manganese. And yet within the 130 heats in the original study, percent of mangances was one of the variables showing at least some relation to tons rejected for bad surface condition. Prediction of y from this x is, of course, relatively poor. But at least

$$m = \frac{\Sigma(x - \bar{x})(y - \bar{y})}{\Sigma(x - \bar{x})^2} = \frac{+1604.6}{4503} = +.356$$

was positive, indicating that higher rejection tended to go with higher manganese.

In the whole study, by means of scatter diagrams and correlation coefficients, four input variables were selected: percent manganese, percent sulfur, bags of coke and sealing time (time delay before capping the ingot mold to kill the action). The actions taken were as follows: Higher rejection tended to go with higher sulfur (m positive), so further attempt was made to hold down the percent of sulfur. It was found from the scatter diagram of rejection versus sealing time for 130 heats that there was a somewhat curved relation. Low sealing times giving fair results; intermediate times, relatively higher rejection; and best results at 2 minutes or more. Sealing times were set at times of about 2 min. Finally, percent manganese and bags of coke seemed somewhat related. A cross-tabulation was made for 356 heats. The average percent rejection was found for all combinations of the two variables in question. Classes of percent manganese were .30 - .39, .40 - .49, ..., .80 and over, and for bags of coke: 0, 1 and 2, 3 and 4, 5 to 7, 8 to 10, and 11 and over. If manganese was .30 - .39, bags of coke

were no problem, but as percent manganese increased, the number of
bags of coke should be held down to three or four at most, if at
all possible.

These practice changes when incorporated led to very considerable improvement in surface quality.

13.4 SIMPLIFYING THE CALCULATIONS

In looking over Table 13.1, the reader may well be feeling unhappy
and saying to himself "This is not for me!" So he should be well
motivated by now for some simplifications. Of course, one approach
is to use an electronic calculator, programmed to complete the whole
set of correlation and regression calculations, once the original
data for x and y are fed in. Nearly all company computing centers
will have available appropriate programs. One can, however, go
overboard in using such programs, accepting the results at full face
value without making the scatter diagrams. The latter may reveal a
nonlinear trend which could render the *linear* regression calculations
inefficient or even downright misleading. Or there may be one or
more really off-trend or isolated points which could well "muddy"
the whole picture, and of whose influence the computer might well
be ignorant. So be sure to make scatter diagrams for variables
whose relations are being studied.

Another point is that if the correlation is high, it is possible for the calculational methods to be discussed next, to lead to
very "heavy cancellation," and thereby to create problems. For this
reason, some programmers prefer to use formulas like (13.2) and
(13.4) instead of the formulas now to be discussed.

As we have seen in Tables 13.1 and 13.3, one first finds $\bar{x}$ and
$\bar{y}$, then subtracts them from each of the x's and y's to find the deviations $x - \bar{x}$, $y - \bar{y}$. Then their sums of squares and the *algebraic*
sum of the cross-products must be found. Instead of all this work,
we can work directly with our x and y values to accumulate five sums
$\Sigma\, x$, $\Sigma\, y$, $\Sigma\, x^2$, $\Sigma\, y^2$, and $\Sigma\, xy$. For example, for Table 13.1, the
last sum would be $\Sigma\, xy = 5.2(5.7) + 5.8(6.4) + \cdots + 4.5(5.1)$.

Now even with motor-driven desk calculators, it is often possible to accumulate all five totals in just one time through the data. (This is done by making use of $(x + y)^2 = x^2 + 2xy + y^2$, keeping x and y well separated so that the accumulating sums Σx^2, $2\Sigma xy$, and Σy^2 do not run together.)

Then we form the following

$$n \Sigma x^2 - (\Sigma x)^2, \quad n \Sigma y^2 - (\Sigma y)^2, \quad n \Sigma xy - (\Sigma x)(\Sigma y)$$

and use these results to find r, m, $s_{y \cdot x}$ and also b.

$$r = \frac{n \Sigma xy - (\Sigma x)(\Sigma y)}{\sqrt{[n \Sigma x^2 - (\Sigma x)^2][n \Sigma y^2 - (\Sigma y)^2]}} \qquad (13.7)$$

$$m = \frac{n \Sigma xy - (\Sigma x)(\Sigma y)}{n \Sigma x^2 - (\Sigma x)^2} \qquad (13.8)$$

$$s_{y \cdot x} = \sqrt{\frac{(1 - r^2)[n \Sigma y^2 - (\Sigma y)^2]}{n(n - 2)}} \qquad (13.9)$$

But note the possibility of very heavy cancellation.

For example, for pouring temperatures in Table 13.2, we have small *relative* variation. We find $\Sigma x = 57,218$ and $\Sigma x^2 = 163,699,134$. But if instead we use v = x - 2800, we find $\Sigma v = 1218$ and $\Sigma v^2 = 78,334$. These give, respectively,

$n \Sigma x^2 = 3,273,982,680$	$n \Sigma v^2 = 1,566,680$
$(\Sigma x)^2 = 3,273,899,524$	$(\Sigma v)^2 = 1,483,524$
83,156	83,156

Note the very heavy cancellation for the x's showing why full precision must be carried.

For the first example of Sec. 13.2, we have the following results:

$$\Sigma x = 100.3, \ \Sigma y = 111.3, \ \Sigma x^2 = 510.29, \ \Sigma y^2 = 627.33,$$
$$\Sigma xy = 565.71, \ n \Sigma x^2 - (\Sigma x)^2 = 145.71,$$
$$n \Sigma y^2 - (\Sigma y)^2 = 158.91, \ n \Sigma xy - (\Sigma x)(\Sigma y) = 150.81$$

$$r = \frac{150.81}{\sqrt{145.71(158.91)}} = .99108$$

$$m = \frac{150.81}{145.71} = 1.0350$$

$$s_{y \cdot x} = \sqrt{\frac{(1 - .99108^2)(158.91)}{20(18)}} = .0885$$

$$b = \bar{y} - m\bar{x} = 5.565 - 1.0350(5.015) = .374$$

All of these results agree with those found before. Note that we save time (and possibly errors) in avoiding all subtractions for $x - \bar{x}$ and $y - \bar{y}$. Moreover, if n is not as simple as 20, we would have to decide as to how much precision to carry in $\bar{x}$ and $\bar{y}$. Formulas (13.7) through (13.9) require no rounding off until practically the end of the calculation.

Another aid to calculation is to use "coding" as in Sec. 2.6. For example, in Table 13.2, the pouring temperatures were all 2840 to 2890PF. We could subtract from each temperature the same amount, 2800 or 2840°F so as to avoid any negative numbers. Then the coded numbers would be much simpler to use. Also we avoid heavy cancellation in $n \Sigma x^2 - (\Sigma x)^2$. We can use the following to code and simplify x and y as in (2.6)

$$v = \frac{x - x_0}{d} \qquad w = \frac{y - y_0}{e} \quad \text{(d, e positive numbers)} \qquad (13.10)$$

Using such coding leaves r unchanged, that is,

$$r_{vw} = r_{xy} \tag{13.11}$$

but we do need to decode the other quantities. Thus in line with (2.9) and (2.10)

$$\bar{x} = x_0 + d\bar{v} \qquad \bar{y} = y_0 + e\bar{w} \qquad m_{y \cdot x} = m_{w \cdot v}\left(\frac{e}{d}\right)$$

$$s_{y \cdot x} = s_{w \cdot v}(e) \tag{13.12}$$

The simplest is when we merely subtract x_0 and y_0, then we need not bother with d and e since they would each be one.

Of course, if a properly programmed computer is used, coding is unnecessary.

13.5 INTERPRETATIONS AND PRECAUTIONS

Again, let us emphasize the desirability of making scatter diagrams when studying relationships, preferably for all pairs of points, but in any case a substantial portion of them. This will help in visualizing the relation. It may bring to light a nonlinear tendency. It is possible for r to be zero, even though there is a strong nonlinear trend with excellent predictability. Or there may be isolated points way off from the general trend, or a point which is along the trend but very far away from the rest of the points. Such are evidences of nonhomogeneity. The author saw another type of nonhomogeneity in a steel-making case. The r was about +.2, but the scatter diagram revealed two patches of points in each of which the correlation coefficient was about -.5, that is, as the value of x increased, y tended to decrease within each patch. It resulted from two different ways of calculating the quality variable.

Another aspect of correlation and regression is the sampling error. Would we obtain the same values for r, m, b, and $s_{y \cdot x}$ if we were to take a larger sample of points (x,y), or another sample of the same number? Quite obviously not. Formulas are available; see, for example, Refs. 1 or 2. Such errors, as is usual in statistics, vary inversely as the square root of the sample size n, so that to cut an error in half we need to make n four times as large.

We shall include one test for the significance of r. Consider first the correlation between two variables x and y which are completely *independent*. Therefore, the true correlation coefficient ρ is *zero* for the *population* of pairs. Nevertheless, if we gather a sample of, say, 10 pairs (x,y), then calculate r, it will not be zero. (The author has only seen one correlation coefficient of zero in his career.) Suppose that the true correlation coefficient ρ is zero, then how far can the observed sample correlation coefficient r be off from zero? It depends upon the sample size n.

A good test as to whether ρ could be zero and still yield an r as large *in size* as we did observe is the following one. Form

$$t = \frac{r\sqrt{n - 2}}{\sqrt{1 - r^2}} \qquad (13.13)$$

Now if we have at least 30 pairs (n = 30) and if ρ = 0, then only once in about 20 times will t be larger in size than 2.0. Thus if ρ = 0 in fact, then the probability of $|t| > 2.0$ is about .05. Therefore, whenever (13.13) does yield a t bigger *in size* than 2.0, we can quite reliably conclude that ρ is *not zero* and that there really is some linear correlation present.

Let us check on the two examples given. In the second r = .28919, n = 20. This is given by (13.13)

$$t = \frac{.28919\sqrt{20 - 2}}{\sqrt{1 - .28919^2}} = 1.28$$

Thus such a correlation as was observed could perfectly well come from an uncorrelated population, and we thus have no reliable evidence that ρ is not zero. However, in the actual application n was 130 and the r coefficient was clearly significant.

For the first example, r = .99108, n = 20. Therefore,

$$t = \frac{.99108\sqrt{20 - 2}}{\sqrt{1 - .99108^2}} = 31.6$$

which is far above 2.0. So that even with only 20 pairs (x,y), we have extremely reliable evidence that ρ is not zero, but rather some strong positive value. Of course, with the 300 pairs in the original study, there was a much higher t, and actually r was high at .968. Little use can nor should be made of nonsignificant correlations, but if significant, an r as low as ±.3 may well prove useful.

Next we must discuss cause and effect versus correlation. A large-sized correlation coefficient does not necessarily mean that either variable "causes" the other to vary as observed. There often is some common cause affecting both variables. For example, if you heat a block of steel, measuring accurately length and width from time to time, a close linear relation will be observed. But neither "causes" the other to vary. Both increase because of the temperature rise.

But if a significant correlation is observed there would seem in general to be some causative factor at work, in practical cases. And we would do well to study carefully those input variables showing the largest r's *in size* versus the quality variable or variables. High r's can also be well used for predictive purposes, even if there is no cause-and-effect relation present. Of course, if there is a causative relation, so much the better. In all of this, engineering knowledge and judgment are invaluable.

Following the preceding points, we also mention that two or more input factors may have a joint effect not suspected by studying the separate relation of either to the dependent variable. Such effects can be studied by "multiple regression and correlation," and by "analysis of variance" in designed experiments. In multiple regression, an objective is to develop an estimation equation involving two or more independent variables (see Refs. 1 or 2).

13.6 SOME APPLICATIONS

The bore hardness of a certain cylinder block was subject to specification limits. Normally the hardness on the Rockwell B scale ran from 88 to 96. The problem was that in order to measure the bore hardness the block had to be cut in half, destroying the block. It was conjectured that hardness measured on the top deck (a nondestructive test) might be sufficiently related for prediction. This hardness was measured by Brinell readings. We call the respective hardnesses y and x, each being actually an average of five readings on a single block. A study was made for 25 blocks, yielding r = .79. This provided an estimation equation which was sufficiently accurate to make bore hardness measurement unnecessary, although an occasional check measurement was still run. Thus, considerable time and money was saved.

In another somewhat similar study in the meat packing industry,
a study was made of percent fat content y versus percent moisture
content x of a type of meat. Both x and y are subject to specification limits, but y is a much messier and more time-consuming measurement. In the scatter diagram the author saw, r was at least
.99, a better correlation than the author might have expected between repeat measurement of y alone! Further analysis was made,
but the scatter diagram was almost sufficient by itself. This enabled the company to very considerably decrease the making of percent fat content measurements and still to meet specifications on
both characteristics.

In an effort to control the thickness of fiberglas after re-
covery y (the quality specification for usage), the machine thickness x was studied as a predictor. Fifty pairs of measurements were
taken, yielding r = .9963, or 99.26% of the variation in y was explained. The estimation equation was $\hat{y}$ = .868x - .072 in., cutting
s_y = .4338 down to $s_{y \cdot x}$ = .0377. The close relation and the estimation equation aided in controlling the machine thickness to give
desired results on recovered thickness.

A certain large-running order of steel was experiencing considerable trouble in meeting the customer's Rockwell specifications.
The latter measurements were made in the customer's plant. By correlating his results with about 20 input variables per heat in the
steel plant, three variables were chosen having the largest (least
small) correlations with the Rockwell values. The three were all
ladle chemistry: carbon, manganese, and residual (tin and other
trace elements). These were studied and a relation developed, involving the three variables. This led to a table for the melt foreman, which gave for various combinations of carbon and manganese
the maximum and minimum limits for residual, so that the heat would
have at least a 90% chance to meet the customer's specification
range. No further trouble was experienced. If the chemistry of a
heat did not have a 90% chance of success with this customer, the
steel would be rolled on some other order.

Failure to meet specifications of $90° \pm .333°$ on the angle of
bend after forming was blamed on *variation* in the thickness of stock.
A correlation study was made on 128 pieces with stock thicknesses
running from .076 to .081 in., giving s_y = .1115°, so $\pm 3s_y$ = $\pm.3345°$,
which is not excessive relative to $90° \pm .333°$. But $\bar{y}$ was *not* $90°$
but instead was 89.798°, that is, about $2s_y$ off from 90°, leading
to many angles out of limits. Were the two variables related? For
the 128 pieces, r = -.7078, which was highly significant. $\bar{x}$ for
thickness was .07818 in. The estimation equation was $\hat{y}$ = -85.43x
+ 6.47774. Substituting the desired $\bar{y}$ = 0 (from 90°), we solve for
$\bar{x}$ = 6.47774/85.43 = .0758 in. Therefore, if the average thickness
of stock is decreased to about .0758 in., *with the same variation
in thickness*, then y will meet $90° \pm .333°$ very well. Note that
this approach assumes that the linearity continues on into lower
x's. Linearity was good in the 128 pieces.

In a study of the diameters of coil condensers for automobiles,
179 condensers showed a correlation of r = +.6591 for diameter versus total foil thickness. Other input factors were total paper
thickness and number of turns. They too provided good correlations.
This study permitted better process control for meeting diameter
specifications.

PROBLEMS

13.1 The data given below are for thicknesses in .00001 in. of
nonmagnetic coatings of galvanized zinc on 11 pieces of iron and
steel. The destructive (stripping) thickness is y; the nondestructive (magnetic) thickness is x.

x	105	120	85	121	115	127	630	155	250	310	443
y	116	132	104	139	114	129	720	174	312	338	465

We find n = 11, Σ x = 2461, Σ y = 2743, Σx^2 = 852,419, Σy^2 =
1,067,143, Σ xy = 952,517.

   (a) Plot a scatter diagram and comment.

   (b) Find r, m, b, and $s_{y \cdot x}$, write the estimation equation,
       and draw it on the scatter diagram.

(c) Is r significant?

(d) Estimate y, if x = 150.

13.2 The following data are for pounds tension of piston rings before and after ferroxing as in Table 13.1, but with an inverted V
bar suspension with the gaps up, first 20 rings out of 300.

x	4.3	5.4	4.9	5.5	4.8	4.4	4.5	4.8	4.9	4.4
y	4.8	5.9	5.5	6.0	5.4	5.1	5.2	5.4	5.5	4.8
x	4.3	4.6	5.0	4.4	4.8	5.9	5.3	5.7	4.8	4.9
y	4.8	5.3	5.6	5.1	5.3	6.6	6.0	6.3	5.6	5.5

(a) Plot a scatter diagram and comment.

(b) Find r, m, b, and $s_{y \cdot x}$, write the estimation equation,
and draw the line on the scatter diagram.

(c) Is r significant?

(d) Estimate y, if x = 4.8.

For Probs. 13.3 and 13.4, for the respective variables of
Table 13.2

(a) Plot a scatter diagram and comment.

(b) Find r, m, b, and $s_{y \cdot x}$, write the estimation equation,
and draw the line on the scatter diagram.

13.3 Tons rejected y versus bags of coke x. Estimate y for x = 2.

13.4 Tons rejected y versus sealing time in seconds x. Estimate
y for x = 25.

13.5 Thirty-five springs were made at each of the following temperatures: 300, 350, 400, ..., 600, so that n = 245. The dependent
variable was initial tension of spring in pounds. Calculations gave
r = -.9346, m = -.004398, b = 7.500, $\Sigma(y - \bar{y})^2$ = 54.26. Test r for
significance. Find $s_{y \cdot x}$ and estimate y for x = 300 and 600.

13.6 In a research study of bituminous road mixes, seven samples
were tested for x = percent water absorbed versus y = percent asphalt
absorbed. We find Σx = 43.18, Σy = 36.23, Σx^2 = 285.7292, Σy^2
= 207.5937, Σxy = 242.6675.

(a) Find r and test it for significance.

(b) Find m and b and write the estimation equation.

(c) Find $s_{y \cdot x}$.

(d) Estimate y from x = 4 and 9. (The range of x's was 3.96 to 8.57.)

13.7 For 49 pieces of 1/6-in. poplar about 110 × 51 in., a study was made of wet, x, versus dry, y, lengths. This was to study the shrinkage so as to control y to specification needs. The following results were found: Σ x = 5523, Σ y = 5162 in., and n Σ x^2 - $(\Sigma$ x$)^2$ = 19,600, n Σ y^2 - $(\Sigma$ y$)^2$ = 23,672, n Σ xy - $(\Sigma$ x$)(\Sigma$ y$)$ = 17,584.

(a) Find x, y, s_y, m, b, r, and $s_{y \cdot x}$.

(b) Test r for significance.

(c) Write the estimation equation, and estimate y for x = 109 and 119.

REFERENCES

1. I. W. Burr, *Applied Statistical Methods*, Academic Press, New York, 1974.

2. N. Draper and H. Smith, *Applied Regression Analysis*, Wiley, New York, 1966.

Chapter Fourteen

A FEW RELIABILITY CONCEPTS

14.1 RELIABILITY IN GENERAL

The subject of reliability of product and its performance have come
very much to the forefront in industry in the last 20 years. Although reliability is an integral part of the total quality program
of a company or corporation, the reliability field is so broad that
it can provide a field of specialization for the individual and is
often serviced by a reliability department or division. Probably
the field of reliability began to emerge as a discipline with the
advent of the air and space age. But the need for reliability is
no less important in the automotive, pharmaceutical, and foods
industries.

Reliability is, of course, intimately tied in with the various
techniques of statistical quality control, which we have been studying in this book. For, in order to obtain reliability, we need to
have a product which is produced by well-controlled processes so
that their output will function as intended and specified. Thus
process control and acceptance sampling are of great importance in
securing reliability.

Since the reader is quite likely to work with specialists in
reliability, we shall here include some techniques and concepts
which are of importance in reliability, and which supplement the
preceding parts of this book. There are many books available on
various aspects of reliability. An excellent book covering many
such aspects is *Reliability: Management, Methods and Mathematics*
[1].

14.2 DEFINITIONS OF RELIABILITY

Nearly everyone has some conception of the meaning of "reliability."
Equipment or product can be called "reliable" if it can be counted
upon to perform satisfactorily its intended function or functions.
Now this involves several ingredients. The product is designed and
manufactured to be used over some field of application. For example,
a chain saw is for cutting wood, green or dry, hard or soft, but not
intended for cutting metal which may occur in the wood. Nevertheless, some safeguards can and are built into a chain saw to protect
the user against moderate misuse should it occur. Some designs are
used to make product not only fool proof but "damn-fool proof."

Another facet of reliability is the time factor or repetitive
usage. A product must be designed and produced so that it will perform its function for at least a minimum length of life, or a minimum number of cycles, such as startings. These guaranteed lengths
of life may be short or long, but are an integral part of the
picture.

Another concept is that of the contents or composition which
is an area much to the forefront today. Products such as foods or
pharmaceuticals *must* contain the prescribed or guaranteed amounts
of the contents desired and must contain none or else not over a
permissible amount of undesirable contents. Such requirements are
controllable by process and testing controls and are a basic part
of the reliability picture. Correct labeling also comes into the
picture, very especially in pharmaceuticals.

Then too there is the probability facet. In this imperfect
world, there is usually no way to guarantee in *absolute terms* the
functioning of product. About all we can do is to make the *probability* of functioning sufficiently high. This was a part of every
space trip. And the unreliability came home on three men, most
unfortunately.

We may summarize the foregoing with a commonly given definition of reliability.

Definition 14.1 (Reliability). The reliability of product is the probability of its successful functioning under prescribed conditions of usage and for the prescribed minimum time or number of cycles.

There are also uses of the word "reliability," such as reliability of design, reliability of production, proving of reliability by tests, such as life testing, environmental testing, and receiving testing. And then too there is the sample or 100% testing of inventory or stockpile for reliability or functionability. Probability is, however, part of the picture in all of these.

14.3 TIME TO FIRST FAILURE, THE GEOMETRIC DISTRIBUTION

As we have seen, it is desired that product should be designed and built so that it will successfully function a certain minimum number of times without a miss. As the appropriate model, we here present the "geometric distribution."

Conditions for the geometric distribution.
1. At each stage or trial the product functions as required, or it does not.
2. The probability of its functioning at each trial is constantly q' of failing to function p' (p' = 1 - q').
3. Functioning or nonfunctioning on trials are independent.
4. Interest is on the number of trials till the first failure or nonfunctioning.

The first three of these conditions are also assumed for the binomial distribution, so the trials are similar. But here interest centers on the number of trials until the first failure, so that the variable whose probability we want is the sample size n, not the number of failures in a fixed, specified sample size n.

Let us, therefore, seek a formula for P(n), the probability that the first failure occurs on the n'th trial or piece tested, where n may be 1, 2, 3, ...

$$P(n = 1) = p' \qquad \text{(geometric distribution)} \qquad (14.1)$$

so that the first trial results in failure. For the *first* failure

to occur on the second trial, we must have a success followed by a
failure. The respective probabilities are q' and p', and since
there is independence among our conditions, we may multiply giving

$$P(n = 2) = q'p' \qquad \text{(geometric distribution)} \qquad (14.2)$$

Next, for the first failure to occur on the third trial, there must
have been two consecutive successes, followed by a failure, with
probabilities q', q', p'. Multiplying gives

$$P(n = 3) = q'^2 p' \qquad \text{(geometric distribution)} \qquad (14.3)$$

In general, if the *first* failure occurs on the n'th trial, then
there had to be n - 1 successes followed by a failure, with probabilities q', ..., q', p', there being n - 1 q' values. Thus

$$P(n) = q'^{n-1} p' \qquad \text{(geometric distribution)} \qquad (14.4)$$

We note that the largest of these is the very first one, n = 1,
and that each successive probability is q' times as large as the
preceding probability (in line with the so-called "geometric pro-
gression" from which the geometric distribution gets its name). Of
interest is the *average* number of trials until the first failure
and also the standard deviation of the number of trials. For the
average, we would need the sum $1P(1) + 2P(2) + 3P(3) + \cdots + nP(n)$
$+ \cdots$. This proves to be

$$E(n) = \frac{1}{p'} \qquad \text{(geometric distribution)} \qquad (14.5)$$

while we also have

$$\sigma_n = \frac{\sqrt{q'}}{p'} \qquad \text{(geometric distribution)} \qquad (14.6)$$

Thus, p' = .01, then the average number of trials until the first
failure is 1/.01 = 100, which seems very natural. The typical departure from this average of 100 is $\sqrt{.99}/.01 = 99.5$, or about as
large as the average itself. (This relatively large σ_n in relation
E(n) sometimes suggests use of the "negative binomial" distribution
which gives the probability of there being n trials until the c'th
failure. See Ref. 1.)

The geometric distribution applies equally to consecutive trials on a *single piece* or to single tests or inspections once to each of a *series of pieces*. The distribution may be used to estimate the failure rate and for qualification tests.

14.4 LOWER CONFIDENCE LIMIT ON RELIABILITY

The reliability of a piece or assembly is its probability of successful operation under the prescribed conditions and time interval. Thus reliability is analogous to q', where 1 - q' = p' is the failure probability or rate.

Now in practice we can never obtain the reliability, q', exactly. The best we can do is to run a series of trials or experiments, and then use the results to obtain an estimate of q'. Thus suppose we find two failures in 1000 trials. We could then estimate the reliability q' to be 998/1000 = .998, or the failure rate p' to be 2/1000 = .002. But if we were to repeat such a series of experiments, we might well obtain different results and estimates. Thus such a "point estimate" is subject to error.

We can determine the amount of error by setting what are commonly called "confidence limits." In reliability testing, however, our chief concern is that the reliability, q', be not less than a certain amount, and thus we wish to set a *lower* confidence limit on the reliability, from the data at hand. This takes the form of finding d failures out of n trials, and then stating with some desired confidence or probability that $q' \geq q'_L$. For example, suppose that we observe three failures in 500 trials. Then we can be 90% confident that the reliability is at least .9867. This is because, if q' were .9867, the probability of as many as or more successes than 497 in 500 is only .10. But we *did* observe 497 successes. Therefore, we can be 90% confident that q' is at least as high as .9867. Note that the *point estimate* of reliability q' is 497/500 = .994, that is much higher.

In Table 14.1, we give a few combinations of confidence and reliability and the required sample size, n, of pieces all to be free from failures in order to give the specified protection.

TABLE 14.1. Sample Size n of Tests with No Failures to Provide at Least the Specified Minimum Reliability with Listed Confidence Level

Minimum reliability	Confidence level			
	.90	.95	.98	.99
.998	1151	1497	1954	2301
.995	460	598	781	919
.99	230	299	390	459
.95	45	59	77	90
.90	22	29	38	44

Thus, for example, if we wish to be 90% confident that an article is 99% reliable, we can achieve this protection by having 230 trials or tests without a single failure. But if we wish to be 99% confident, we must have 459 tests with no failures.

A much larger table is given in Ref. 1, with many more confidence levels and minimum reliabilities.

We also give in Table 14.2 a set of minimum reliabilities with several confidence coefficients when zero to three failures have been observed in the given number of tests. For example, we are satisfied with 90% confidence and decide to run n = 500 tests. We find two failures. Then we are 90% confident that the reliability is at least .989.

A very much larger table is given in a different form in Ref. 2.

The practical usages of these two tables are quite different. For entries of Table 14.1, we decide to run n tests or trials and hope for no failures. We decide on one of the confidence levels given and a desired minimum reliability to be demonstrated with this confidence. The table then tells us how many tests to run which must be with *zero* failures. On the other hand, with Table 14.2, we again decide upon a confidence level given, and now choose one of the available sample sizes. After running n tests and observing failures from zero to three, we enter the table to find

TABLE 14.2. Lower Limits to Reliability with Confidence Coefficients .90, .975, .995 When Zero to Three Failures Have Been Observed in n Tests

Number of tests n	Confidence coefficient	Number of failures in n tests			
		0	1	2	3
100	.90	.977	.962	.948	.934
	.975	.964	.946	.930	.915
	.995	.948	.928	.911	.894
150	.90	.985	.974	.965	.956
	.975	.976	.963	.953	.943
	.995	.965	.951	.940	.929
200	.90	.989	.981	.974	.967
	.975	.982	.972	.964	.957
	.995	.974	.963	.954	.946
300	.90	.992	.987	.982	.978
	.975	.988	.982	.976	.971
	.995	.983	.976	.970	.964
500	.90	.995	.992	.989	.987
	.975	.993	.989	.986	.983
	.995	.989	.985	.982	.978
1000	.90	.998	.996	.995	.993
	.975	.996	.994	.993	.991
	.995	.995	.992	.991	.989

what reliability we have demonstrated as a minimum.

If while using the approach to Table 14.1, we should encounter a failure, *we do not* then decide to run enough more tests to complete a sample size in Table 14.2. This would change the probabilities unpredictably.

14.5 THE EXPONENTIAL DISTRIBUTION FOR LENGTH OF LIFE

The reader is no doubt familiar with the way the length of life of a product varies from a very short life for a few on steadily toward a very long life for others. Electric light bulbs are a good example.

Some may fail almost immediately, whereas others seemingly go on operating "forever." It is, therefore, desirable to have theoretical distribution models for length of life tests. It is hardly necessary to point out the great need for adequate length of life for *all* important components of any assembly or product.

The normal distribution for measurements does not commonly have the desired characteristics for the distribution of length of life of product. Instead, a most commonly used model is the so-called "exponential distribution." It is quite a standard distribution for lengths of life, much like the normal distribution is in other fields of application.

For the exponential distribution we have

$$f(x) = \frac{1}{\mu} e^{-x/\mu} \qquad x, \mu > 0 \qquad \text{(exponential distribution) (14.7)}$$

where e is the natural logarithmic base, namely 2.71828..., and μ is the theoretical average length of life.

Figure 14.1 shows an example of the exponential distribution. There we see that the density function $f(x)$ is at its maximum when x is zero, and that $f(x)$ steadily decreases as x increases, tailing far out to the few very long lives. The x scale is shown for $\mu = 500$ hr, the average length of life, whereas the vertical scale is in one-over-hours units, and is such as to make the total area

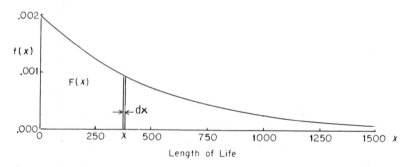

Fig. 14.1. An exponential distribution. The scales are drawn for an average length of life of $\mu = 500$ hr. Total area under curve using these scales is one, representing certainty for some length of life to occur.

under the curve to be one, so as to interpret areas under the curve as probabilities.

It may be shown by calculus that

$$\sigma_x = \mu \quad \text{(exponential distribution)} \tag{14.8}$$

that is, that the standard deviation of the lengths of life is just the same as the mean length of life! Moreover

$$F(x) = 1 - e^{-x/\mu} \tag{14.9}$$
$$= \text{cumulative probability of failure by time x}$$

The value of $F(x)$ tells us what proportion of the pieces put in test or service can be expected to have failed by time x. Note that $F(0) = 0$, and that $F(x)$ increases toward the limit one as x increases without limit.

An important characteristic of distributions of length of life is the "hazard function" $h(x)$. The probability of a unit or piece failing in some short time interval x to x + dx is the area under the curve (such as in Fig. 14.1) between ordinates at these two x values. This may be approximated by $f(x)dx$. So if we start with 1,000,000 units, we may expect that approximately $1,000,000 \cdot f(x)dx$ will fail between times x and x + dx. But how many were there left out of the 1,000,000 at time x? This is the original 1,000,000 minus those already failed, that is, $1,000,000 \cdot F(x)$, giving the expected number left as $1,000,000 - 1,000,000F(x) = 1,000,000[1 - F(x)]$. What then is the failure *rate* over the interval x to x + dx? It is the number failing divided by the number present at time x, that is

$$\frac{1,000,000f(x)dx}{1,000,000[1 - F(x)]} = \frac{f(x)dx}{1 - F(x)}$$

The latter is in general terms for all length-of-life distributions. We define then the "hazard function" in general as

$$h(x) = \frac{f(x)}{1 - F(x)} \tag{14.10}$$

For the exponential distribution (14.7) and (14.9), we have

$$h(x) = \frac{e^{-x/\mu}/\mu}{1 - [1 - e^{-x/\mu}]} = \frac{e^{-x/\mu}/\mu}{e^{-x/\mu}} = \frac{1}{\mu}$$

(exponential distribution)   (14.11)

Therefore, the hazard function or failure rate function is constant and equal to the reciprocal of the mean length of life. This constant failure rate is a unique property of the exponential distribution. This situation is sometimes called "random failure."

There can well be two additional components to a general hazard function $h(x)$. The first is an even higher initial failure rate than the $1/\mu$ of the exponential distribution. This is sometimes called "infant mortality." Then, after a substantial length of time during which the hazard function remained constant, the hazard function may gradually begin to increase. In some models, $h(x)$ may gradually increase or decrease throughout the range of x's, or $h(x)$ may decrease for early x's and then gradually increase.

The properties of the exponential distribution model make it possible to estimate average length of life and to set tests for determining reliability of life to a required minimum of time.

14.6 RELIABILITY OF COMPLEX EQUIPMENT

We now consider very briefly the difficult problem of securing adequate reliability for a complex assembly. Such an assembly may well consist of several thousand component parts. Of these, some are, of course, more crucial than others. In fact, it may well be possible to divide up the list of parts into: (1) those, the failure of any one of which, will probably cause the assembly to fail, and (2) those parts in which failure will very probably *not* cause the assembly to fail. The former are the "lethal" or vital components which must have very high reliability in order for the assembly to have high reliability. Let us see.

Now suppose that there are 500 lethal components in the assembly. Let the reliabilities (under specified conditions and length of time) be R_1, R_2, ..., R_{500}. Now we next make the substantial assumption that failures among these 500 components occur entirely independently. Then the reliability of the system, that is, the probability that *all* 500 components function without failure, is the product of the reliabilities of the separate lethal components, that is

$$R(\text{assembly}) = R_1 R_2 \cdots R_{500} \tag{14.12}$$

by (3.7). Carefully note how very troublesome this product of separate reliabilities is: If just *one* of these lethal components has a low reliability, then the assembly does also, even if the reliabilities of the other 499 components are virtually one! *All* must have high reliabilities. (R cannot exceed the lowest R_i.)

As a further analysis of (14.12), let us make the hypothetical assumption that we can make all component reliabilities equally high. How high must they be to provide a reliability for the *assembly* of, say, .99? We have from (14.12)

$$R = R_i^{500}$$

or substituting .99 for R, we have

$$.99 = R_i^{500} \quad \text{or} \quad R_i = .99^{1/500}$$

We can use logarithms to solve the last equation for R_i, obtaining $R_i = .99998$ (or we could use the binomial expansion).

Now it is possible that many of the component reliabilities are exactly one. But we cannot very well prove this, and to establish a single reliability of .99998 would require a fantastically large sample size. Thus, we have many practical problems.

One approach is to test components under conditions very much more rigorous than the specified conditions for the assembly. If *all* of a reasonable sample size pass the test of greatly "increased severity" then a very high reliability may be assumed for the specified conditions.

Another often-used technique for achieving very high reliability is to use "redundancy." Thus suppose we wish to use a relay which will open a circuit under certain conditions with a reliability

of .995 = q_1'. We may place another design of relay with the same
purpose in series with the first one. Let its reliability be
q_2' = .996. Now under the critical undesirable conditions if either
one or both relays open the circuit their mission is accomplished.
The only way failure of the system of the two can occur is by *both*
failing to open the circuit. If we assume that the failing of the
two are independent events, then the probability of failure of the
redundant system of the two is by (14.12)

$$p' = p_1' p_2' = (1 - q_1')(1 - q_2') = .005(.004) = .000020$$

so that the *reliability* of the two in series is .999980.

One of the facets of our space program, facilitating very high
reliability, has been our great ability at miniaturizing circuits
permitting much use of redundancy without undue weight.

The reader may wonder whether there might be as many as 500
lethal components in any system. But consider a large plane. All
six descendants of a good friend of the author's were killed in a
plane crash which was traced to a single missing cotter pin. Then
there was John Glenn, who after his first space flight, was asked
for his thoughts; and he is reported to have said that he was calling to mind that each of the components was made by the lowest
bidder!

14.7 SUMMARY

Our objective in this chapter has been to give the reader a speak-
ing acquaintance with a few of the techniques with which those in
reliability work. In this way it is hoped that cooperation and com-
munication may be facilitated. Anyone seriously interested in reliability will, of course, wish to study books on the subject, such
as Ref. 1.

But here let us say that one of the absolutely basic building
blocks to reliability is control of processes and incoming product.
The techniques we have discussed in Chapters 1 to 13 are thus of
great use in obtaining high reliability of the finished product, of
whatever type.

PROBLEMS

For Probs. 14.1 and 14.2, find for the respective geometric distributions P(1), P(2), P(3), P(4), μ, and σ.

14.1   p' = .10.

14.2   p' = .05.

14.3   In an audit of missile systems we want 90% confidence of a reliability of at least .90. What size of random sample of missile systems do we test for no failures to secure the desired confidence?

14.4   We want to prove .99 minimum reliability with 98% confidence. What sample size do we require with no failures to establish this degree of protection?

14.5   Suppose that we observe one failure in 200 trials or tests. Find the minimum established limit for reliability with confidence .90, .975, .995, and interpret the results.

14.6   Suppose that we observe one failure in 500 trials or tests. Find the minimum established limit for reliability with confidence .90, .975, .995, and interpret the results. Find also for confidence .95.

14.7   If the average length of life for an exponential distribution is 2000 hr, find h(x). What number of the 1200 components remaining on test, not having failed, may be expected to fail in the next 10 hours? (*Hint*: Let dx = 10 hr.)

14.8   Two relays each expected to open under certain dangerous conditions are placed in series in a circuit. Their reliabilities are .99 and .995, and they act independently. Find the reliability of the system of the two.

14.9   Two relays each expected to close under certain dangerous conditions are placed in parallel in a circuit. Their reliabilities are .98 and .99. Find the reliability of the system of the two. Assume they act independently.

REFERENCES

1. D. K. Lloyd and M. Lipow, *Reliability: Management, Methods and Mathematics,* Prentice-Hall, Englewood Cliffs, N. J., 1962.

2. D. Mainland, L. Herrera, and M. I. Sutcliffe, *Tables for Use with Binomial Samples,* Dept. Medical Statistics, New York University, New York, 1956.

Appendix

TABLES OF STATISTICAL AND
MATHEMATICAL
FUNCTIONS

TABLE A. Cumulative Probability ≤z, to Four Decimal Places, for Standard Normal Distribution; Units and Tenths for z in Left Column, Hundredths in Column Headings

z	.00	.01	.02	.03	.04	.05	.06	.07	.08	.09
−3.5	.0002	.0002	.0002	.0002	.0002	.0002	.0002	.0002	.0002	.0002
−3.4	.0003	.0003	.0003	.0003	.0003	.0003	.0003	.0003	.0003	.0002
−3.3	.0005	.0005	.0005	.0004	.0004	.0004	.0004	.0004	.0004	.0003
−3.2	.0007	.0007	.0006	.0006	.0006	.0006	.0006	.0005	.0005	.0005
−3.1	.0010	.0009	.0009	.0009	.0008	.0008	.0008	.0008	.0007	.0007
−3.0	.0013	.0013	.0013	.0012	.0012	.0011	.0011	.0011	.0010	.0010
−2.9	.0019	.0018	.0018	.0017	.0016	.0016	.0015	.0015	.0014	.0014
−2.8	.0026	.0025	.0024	.0023	.0023	.0022	.0021	.0021	.0020	.0019
−2.7	.0035	.0034	.0033	.0032	.0031	.0030	.0029	.0028	.0027	.0026
−2.6	.0047	.0045	.0044	.0043	.0041	.0040	.0039	.0038	.0037	.0036
−2.5	.0062	.0060	.0059	.0057	.0055	.0054	.0052	.0051	.0049	.0048
−2.4	.0082	.0080	.0078	.0075	.0073	.0071	.0069	.0068	.0066	.0064
−2.3	.0107	.0104	.0102	.0099	.0096	.0094	.0091	.0089	.0087	.0084
−2.2	.0139	.0136	.0132	.0129	.0125	.0122	.0119	.0116	.0113	.0110
−2.1	.0179	.0174	.0170	.0166	.0162	.0158	.0154	.0150	.0146	.0143
−2.0	.0228	.0222	.0217	.0212	.0207	.0202	.0197	.0192	.0188	.0183
−1.9	.0287	.0281	.0274	.0268	.0262	.0256	.0250	.0244	.0239	.0233
−1.8	.0359	.0351	.0344	.0336	.0329	.0322	.0314	.0307	.0301	.0294
−1.7	.0446	.0436	.0427	.0418	.0409	.0401	.0392	.0384	.0375	.0367
−1.6	.0548	.0537	.0526	.0516	.0505	.0495	.0485	.0475	.0465	.0455
−1.5	.0668	.0655	.0643	.0630	.0618	.0606	.0594	.0582	.0571	.0559
−1.4	.0808	.0793	.0778	.0764	.0749	.0735	.0721	.0708	.0694	.0681
−1.3	.0968	.0951	.0934	.0918	.0901	.0885	.0869	.0853	.0838	.0823
−1.2	.1151	.1131	.1112	.1093	.1075	.1056	.1038	.1020	.1003	.0985
−1.1	.1357	.1335	.1314	.1292	.1271	.1251	.1230	.1210	.1190	.1170
−1.0	.1587	.1562	.1539	.1515	.1492	.1469	.1446	.1423	.1401	.1379
−0.9	.1841	.1814	.1788	.1762	.1736	.1711	.1685	.1660	.1635	.1611
−0.8	.2119	.2090	.2061	.2033	.2005	.1977	.1949	.1922	.1894	.1867
−0.7	.2420	.2389	.2358	.2327	.2296	.2266	.2236	.2206	.2177	.2148
−0.6	.2743	.2709	.2676	.2643	.2611	.2578	.2546	.2514	.2483	.2451
−0.5	.3085	.3050	.3015	.2981	.2946	.2912	.2877	.2843	.2810	.2776
−0.4	.3446	.3409	.3372	.3336	.3300	.3264	.3228	.3192	.3156	.3121
−0.3	.3821	.3783	.3745	.3707	.3669	.3632	.3594	.3557	.3520	.3483
−0.2	.4207	.4168	.4129	.4090	.4052	.4013	.3974	.3936	.3897	.3859
−0.1	.4602	.4562	.4522	.4483	.4443	.4404	.4364	.4325	.4286	.4247
−0.0	.5000	.4960	.4920	.4880	.4840	.4801	.4761	.4721	.4681	.4641

Reproduced with permission from I. W. Burr, *Engineering Statistics and Quality Control*, McGraw-Hill, New York, 1953, pp. 404, 405.

TABLE A (*continued*)

z	.00	.01	.02	.03	.04	.05	.06	.07	.08	.09
+0.0	.5000	.5040	.5080	.5120	.5160	.5199	.5239	.5279	.5319	.5359
+0.1	.5398	.5438	.5478	.5517	.5557	.5596	.5636	.5675	.5714	.5753
+0.2	.5793	.5832	.5871	.5910	.5948	.5987	.6026	.6064	.6103	.6141
+0.3	.6179	.6217	.6255	.6293	.6331	.6368	.6406	.6443	.6480	.6517
+0.4	.6554	.6591	.6628	.6664	.6700	.6736	.6772	.6808	.6844	.6879
+0.5	.6915	.6950	.6985	.7019	.7054	.7088	.7123	.7157	.7190	.7224
+0.6	.7257	.7291	.7324	.7357	.7389	.7422	.7454	.7486	.7517	.7549
+0.7	.7580	.7611	.7642	.7673	.7704	.7734	.7764	.7794	.7823	.7852
+0.8	.7881	.7910	.7939	.7967	.7995	.8023	.8051	.8078	.8106	.8133
+0.9	.8159	.8186	.8212	.8238	.8264	.8289	.8315	.8340	.8365	.8389
+1.0	.8413	.8438	.8461	.8485	.8508	.8531	.8554	.8577	.8599	.8621
+1.1	.8643	.8665	.8686	.8708	.8729	.8749	.8770	.8790	.8810	.8830
+1.2	.8849	.8869	.8888	.8907	.8925	.8944	.8962	.8980	.8997	.9015
+1.3	.9032	.9049	.9066	.9082	.9099	.9115	.9131	.9147	.9162	.9177
+1.4	.9192	.9207	.9222	.9236	.9251	.9265	.9279	.9292	.9306	.9319
+1.5	.9332	.9345	.9357	.9370	.9382	.9394	.9406	.9418	.9429	.9441
+1.6	.9452	.9463	.9474	.9484	.9495	.9505	.9515	.9525	.9535	.9545
+1.7	.9554	.9564	.9573	.9582	.9591	.9599	.9608	.9616	.9625	.9633
+1.8	.9641	.9649	.9656	.9664	.9671	.9678	.9686	.9693	.9699	.9706
+1.9	.9713	.9719	.9726	.9732	.9738	.9744	.9750	.9756	.9761	.9767
+2.0	.9772	.9778	.9783	.9788	.9793	.9798	.9803	.9808	.9812	.9817
+2.1	.9821	.9826	.9830	.9834	.9838	.9842	.9846	.9850	.9854	.9857
+2.2	.9861	.9864	.9868	.9871	.9875	.9878	.9881	.9884	.9887	.9890
+2.3	.9893	.9896	.9898	.9901	.9904	.9906	.9909	.9911	.9913	.9916
+2.4	.9918	.9920	.9922	.9925	.9927	.9929	.9931	.9932	.9934	.9936
+2.5	.9938	.9940	.9941	.9943	.9945	.9946	.9948	.9949	.9951	.9952
+2.6	.9953	.9955	.9956	.9957	.9959	.9960	.9961	.9962	.9963	.9964
+2.7	.9965	.9966	.9967	.9968	.9969	.9970	.9971	.9972	.9973	.9974
+2.8	.9974	.9975	.9976	.9977	.9977	.9978	.9979	.9979	.9980	.9981
+2.9	.9981	.9982	.9982	.9983	.9984	.9984	.9985	.9985	.9986	.9986
+3.0	.9987	.9987	.9987	.9988	.9988	.9989	.9989	.9989	.9990	.9990
+3.1	.9990	.9991	.9991	.9991	.9992	.9992	.9992	.9992	.9993	.9993
+3.2	.9993	.9993	.9994	.9994	.9994	.9994	.9994	.9995	.9995	.9995
+3.3	.9995	.9995	.9995	.9996	.9996	.9996	.9996	.9996	.9996	.9997
+3.4	.9997	.9997	.9997	.9997	.9997	.9997	.9997	.9997	.9997	.9998
+3.5	.9998	.9998	.9998	.9998	.9998	.9998	.9998	.9998	.9998	.9998

TABLE B. Poisson Distribution. Probabilities of
c or Less, Given c', Appear in Body of Table Multiplied by 1000.

c' or np'	0	1	2	3	4	5	6	7	8	9
0.02	980	1,000								
0.04	961	999	1,000							
0.06	942	998	1,000							
0.08	923	997	1,000							
0.10	905	995	1,000							
0.15	861	990	999	1,000						
0.20	819	982	999	1,000						
0.25	779	974	998	1,000						
0.30	741	963	996	1,000						
0.35	705	951	994	1,000						
0.40	670	938	992	999	1,000					
0.45	638	925	989	999	1,000					
0.50	607	910	986	998	1,000					
0.55	577	894	982	998	1,000					
0.60	549	878	977	997	1,000					
0.65	522	861	972	996	999	1,000				
0.70	497	844	966	994	999	1,000				
0.75	472	827	959	993	999	1,000				
0.80	449	809	953	991	999	1,000				
0.85	427	791	945	989	998	1,000				
0.90	407	772	937	987	998	1,000				
0.95	387	754	929	984	997	1,000				
1.00	368	736	920	981	996	999	1,000			
1.1	333	699	900	974	995	999	1,000			
1.2	301	663	879	966	992	998	1,000			
1.3	273	627	857	957	989	998	1,000			
1.4	247	592	833	946	986	997	999	1,000		
1.5	223	558	809	934	981	996	999	1,000		
1.6	202	525	783	921	976	994	999	1,000		
1.7	183	493	757	907	970	992	998	1,000		
1.8	165	463	731	891	964	990	997	999	1,000	
1.9	150	434	704	875	956	987	997	999	1,000	
2.0	135	406	677	857	947	983	995	999	1,000	

Reproduced with permission from I. W. Burr, *Engineering Statistics and Quality Control*, McGraw-
Hill, New York, 1953, pp. 417-421.

TABLE B *(continued)*

c $\backslash$ c' or np'	0	1	2	3	4	5	6	7	8	9
2.2	111	355	623	819	928	975	993	998	1,000	
2.4	091	308	570	779	904	964	988	997	999	1,000
2.6	074	267	518	736	877	951	983	995	999	1,000
2.8	061	231	469	692	848	935	976	992	998	999
3.0	050	199	423	647	815	916	966	988	996	999
3.2	041	171	380	603	781	895	955	983	994	998
3.4	033	147	340	558	744	871	942	977	992	997
3.6	027	126	303	515	706	844	927	969	988	996
3.8	022	107	269	473	668	816	909	960	984	994
4.0	018	092	238	433	629	785	889	949	979	992
4.2	015	078	210	395	590	753	867	936	972	989
4.4	012	066	185	359	551	720	844	921	964	985
4.6	010	056	163	326	513	686	818	905	955	980
4.8	008	048	143	294	476	651	791	887	944	975
5.0	007	040	125	265	440	616	762	867	932	968
5.2	006	034	109	238	406	581	732	845	918	960
5.4	005	029	095	213	373	546	702	822	903	951
5.6	004	024	082	191	342	512	670	797	886	941
5.8	003	021	072	170	313	478	638	771	867	929
6.0	002	017	062	151	285	446	606	744	847	916

	10	11	12	13	14	15	16
2.8	1,000						
3.0	1,000						
3.2	1,000						
3.4	999	1,000					
3.6	999	1,000					
3.8	998	999	1,000				
4.0	997	999	1,000				
4.2	996	999	1,000				
4.4	994	998	999	1,000			
4.6	992	997	999	1,000			
4.8	990	996	999	1,000			
5.0	986	995	998	999	1,000		
5.2	982	993	997	999	1,000		
5.4	977	990	996	999	1,000		
5.6	972	988	995	998	999	1,000	
5.8	965	984	993	997	999	1,000	
6.0	957	980	991	996	999	999	1,000

TABLE B   *(continued)*

c' or np' \ c	0	1	2	3	4	5	6	7	8	9
6.2	002	015	054	134	259	414	574	716	826	902
6.4	002	012	046	119	235	384	542	687	803	886
6.6	001	010	040	105	213	355	511	658	780	869
6.8	001	009	034	093	192	327	480	628	755	850
7.0	001	007	030	082	173	301	450	599	729	830
7.2	001	006	025	072	156	276	420	569	703	810
7.4	001	005	022	063	140	253	392	539	676	788
7.6	001	004	019	055	125	231	365	510	648	765
7.8	000	004	016	048	112	210	338	481	620	741
8.0	000	003	014	042	100	191	313	453	593	717
8.5	000	002	009	030	074	150	256	386	523	653
9.0	000	001	006	021	055	116	207	324	456	587
9.5	000	001	004	015	040	089	165	269	392	522
10.0	000	000	003	010	029	067	130	220	333	458

c' or np' \ c	10	11	12	13	14	15	16	17	18	19
6.2	949	975	989	995	998	999	1,000			
6.4	939	969	986	994	997	999	1,000			
6.6	927	963	982	992	997	999	999	1,000		
6.8	915	955	978	990	996	998	999	1,000		
7.0	901	947	973	987	994	998	999	1,000		
7.2	887	937	967	984	993	997	999	999	1,000	
7.4	871	926	961	980	991	996	998	999	1,000	
7.6	854	915	954	976	989	995	998	999	1,000	
7.8	835	902	945	971	986	993	997	999	1,000	
8.0	816	888	936	966	983	992	996	998	999	1,000
8.5	763	849	909	949	973	986	993	997	999	999
9.0	706	803	876	926	959	978	989	995	998	999
9.5	645	752	836	898	940	967	982	991	996	998
10.0	583	697	792	864	917	951	973	986	993	997

c' or np' \ c	20	21	22
8.5	1,000		
9.0	1,000		
9.5	999	1,000	
10.0	998	999	1,000

TABLE B (continued)

c' or np' \ c	0	1	2	3	4	5	6	7	8	9
10.5	000	000	002	007	021	050	102	179	279	397
11.0	000	000	001	005	015	038	079	143	232	341
11.5	000	000	001	003	011	028	060	114	191	289
12.0	000	000	001	002	008	020	046	090	155	242
12.5	000	000	000	002	005	015	035	070	125	201
13.0	000	000	000	001	004	011	026	054	100	166
13.5	000	000	000	001	003	008	019	041	079	135
14.0	000	000	000	000	002	006	014	032	062	109
14.5	000	000	000	000	001	004	010	024	048	088
15.0	000	000	000	000	001	003	008	018	037	070

	10	11	12	13	14	15	16	17	18	19
10.5	521	639	742	825	888	932	960	978	988	994
11.0	460	579	689	781	854	907	944	968	982	991
11.5	402	520	633	733	815	878	924	954	974	986
12.0	347	462	576	682	772	844	899	937	963	979
12.5	297	406	519	628	725	806	869	916	948	969
13.0	252	353	463	573	675	764	835	890	930	957
13.5	211	304	409	518	623	718	798	861	908	942
14.0	176	260	358	464	570	669	756	827	883	923
14.5	145	220	311	413	518	619	711	790	853	901
15.0	118	185	268	363	466	568	664	749	819	875

	20	21	22	23	24	25	26	27	28	29
10.5	997	999	999	1,000						
11.0	995	998	999	1,000						
11.5	992	996	998	999	1,000					
12.0	988	994	997	999	999	1,000				
12.5	983	991	995	998	999	999	1,000			
13.0	975	986	992	996	998	999	1,000			
13.5	965	980	989	994	997	998	999	1,000		
14.0	952	971	983	991	995	997	999	999	1,000	
14.5	936	960	976	986	992	996	998	999	999	1,000
15.0	917	947	967	981	989	994	997	998	999	1,000

TABLE B   (continued)

c / c' or np'	4	5	6	7	8	9	10	11	12	13
16	000	001	004	010	022	043	077	127	193	275
17	000	001	002	005	013	026	049	085	135	201
18	000	000	001	003	007	015	030	055	092	143
19	000	000	001	002	004	009	018	035	061	098
20	000	000	000	001	002	005	011	021	039	066
21	000	000	000	000	001	003	006	013	025	043
22	000	000	000	000	001	002	004	008	015	028
23	000	000	000	000	000	001	002	004	009	017
24	000	000	000	000	000	000	001	003	005	011
25	000	000	000	000	000	000	001	001	003	006

	14	15	16	17	18	19	20	21	22	23
16	368	467	566	659	742	812	868	911	942	963
17	281	371	468	564	655	736	805	861	905	937
18	208	287	375	469	562	651	731	799	855	899
19	150	215	292	378	469	561	647	725	793	849
20	105	157	221	297	381	470	559	644	721	787
21	072	111	163	227	302	384	471	558	640	716
22	048	077	117	169	232	306	387	472	556	637
23	031	052	082	123	175	238	310	389	472	555
24	020	034	056	087	128	180	243	314	392	473
25	012	022	038	060	092	134	185	247	318	394

	24	25	26	27	28	29	30	31	32	33
16	978	987	993	996	998	999	999	1,000		
17	959	975	985	991	995	997	999	999	1,000	
18	932	955	972	983	990	994	997	998	999	1,000
19	893	927	951	969	980	988	993	996	998	999
20	843	888	922	948	966	978	987	992	995	997
21	782	838	883	917	944	963	976	985	991	994
22	712	777	832	877	913	940	959	973	983	989
23	635	708	772	827	873	908	936	956	971	981
24	554	632	704	768	823	868	904	932	953	969
25	473	553	629	700	763	818	863	900	929	950

	34	35	36	37	38	39	40	41	42	43
19	999	1,000								
20	999	999	1,000							
21	997	998	999	999	1,000					
22	994	996	998	999	999	1,000				
23	988	993	996	997	999	999	1,000			
24	979	987	992	995	997	998	999	999	1,000	
25	966	978	985	991	994	997	998	999	999	1,000

TABLE C. Control Chart Constants for Averages x̄, Standard Deviations s, and Ranges R, from Normal Populations. Factors for Computing Central Lines and Three Sigma Control Limits.

Sample Size n	Factors for Control limits for x̄		Factors for Standard deviations, s		Factors for control limits for s				Factors for ranges R			Factors for control limits for R		
n	A	A_2	A_3	c_4	c_5	B_3	B_4	B_5	B_6	d_2	d_3	D_1	D_2	D_3	D_4
2	2.121	1.880	2.659	.798	.603	0	3.267	0	2.606	1.128	.853	0	3.686	0	3.267
3	1.732	1.023	1.954	.886	.463	0	2.568	0	2.276	1.693	.888	0	4.358	0	2.575
4	1.500	.729	1.628	.921	.389	0	2.266	0	2.088	2.059	.880	0	4.698	0	2.282
5	1.342	.577	1.427	.940	.341	0	2.089	0	1.964	2.326	.864	0	4.918	0	2.115
6	1.225	.483	1.287	.952	.308	.030	1.970	.029	1.874	2.534	.848	0	5.078	0	2.004
7	1.134	.419	1.182	.959	.282	.118	1.882	.113	1.806	2.704	.833	.205	5.203	.076	1.924
8	1.061	.373	1.099	.965	.262	.185	1.815	.179	1.751	2.847	.820	.387	5.307	.136	1.864
9	1.000	.337	1.032	.969	.246	.239	1.761	.232	1.707	2.970	.808	.546	5.394	.184	1.816
10	.949	.308	.975	.973	.232	.284	1.716	.276	1.669	3.078	.797	.687	5.469	.223	1.777
11	.905	.285	.927	.975	.221	.321	1.679	.313	1.637	3.173	.787	.812	5.534	.256	1.744
12	.866	.266	.886	.978	.211	.354	1.646	.346	1.610	3.258	.778	.924	5.592	.284	1.716
13	.832	.249	.850	.979	.202	.382	1.618	.374	1.585	3.336	.770	1.026	5.646	.308	1.692
14	.802	.235	.817	.981	.194	.406	1.594	.399	1.563	3.407	.762	1.121	5.693	.329	1.671
15	.775	.223	.789	.982	.187	.428	1.572	.421	1.544	3.472	.755	1.207	5.737	.348	1.652
16	.750	.212	.763	.983	.181	.448	1.552	.440	1.526	3.532	.749	1.285	5.779	.364	1.636
17	.728	.203	.739	.985	.175	.466	1.534	.458	1.511	3.588	.743	1.359	5.817	.379	1.621
18	.707	.194	.718	.985	.170	.482	1.518	.475	1.496	3.640	.738	1.426	5.854	.392	1.608
19	.688	.187	.698	.986	.165	.497	1.503	.490	1.483	3.689	.733	1.490	5.888	.404	1.596
20	.671	.180	.680	.987	.161	.510	1.490	.504	1.470	3.735	.729	1.548	5.922	.414	1.586
21	.655	.173	.663	.988	.157	.523	1.477	.516	1.459	3.778	.724	1.606	5.950	.425	1.575
22	.640	.167	.647	.988	.153	.534	1.466	.528	1.448	3.819	.720	1.659	5.979	.434	1.566
23	.626	.162	.633	.989	.150	.545	1.455	.539	1.438	3.858	.716	1.710	6.006	.443	1.557
24	.612	.157	.619	.989	.147	.555	1.445	.549	1.429	3.895	.712	1.759	6.031	.452	1.548
25	.600	.153	.606	.990	.144	.565	1.435	.559	1.420	3.931	.709	1.804	6.058	.459	1.541

Formulas for Control Charts for Variables, x̄, s, R

Purpose of Chart	Chart for	Central Line	3-sigma Control Limits
No Standard Given - used for analyzing past data for control. (x̄, R̄, s̄ are average values for data being analyzed.)	Averages, x̄	x̄	$\bar{\bar{x}} \pm A_2\bar{R}$, or $\bar{\bar{x}} \pm A_3\bar{s}$
	Ranges, R	R̄	$D_3\bar{R}$, $D_4\bar{R}$
	Std. Devs., s	s̄	$B_3\bar{s}$, $B_4\bar{s}$
Standards Given - Used for controlling quality with respect to standards given μ, σ. $R'_n = d_2\sigma$ for n.	Averages, x̄	μ	$\mu \pm A\sigma$
	Ranges, R	$d_2\sigma$ or R'_n	$D_1\sigma$, $D_2\sigma$ $D_3R'_n$, $D_4R'_n$
	Std. devs., s	$c_4\sigma$	$B_5\sigma$, $B_6\sigma$

$E(s) = c_4\sigma$, $\sigma_s = c_5\sigma$, $E(R) = d_2\sigma$, $\sigma_R = d_3\sigma$.

TABLE D. Random Numbers

1368	9621	9151	2066	1208	2664	9822	6599	6911	5112
5953	5936	2541	4011	0408	3593	3679	1378	5936	2651
7226	9466	9553	7671	8599	2119	5337	5953	6355	6889
8883	3454	6773	8207	5576	6386	7487	0190	0867	1298
7022	5281	1168	4099	8069	8721	8353	9952	8006	9045
4576	1853	7884	2451	3488	1286	4842	7719	5795	3953
8715	1416	7028	4616	3470	9938	5703	0196	3465	0034
4011	0408	2224	7626	0643	1149	8834	6429	8691	0143
1400	3694	4482	3608	1238	8221	5129	6105	5314	8385
6370	1884	0820	4854	9161	6509	7123	4070	6759	6113
4522	5749	8084	3932	7678	3549	0051	6761	6952	7041
7195	6234	6426	7148	9945	0358	3242	0519	6550	1327
0054	0810	2937	2040	2299	4198	0846	3937	3986	1019
5166	5433	0381	9686	5670	5129	2103	1125	3404	8785
1247	3793	7415	7819	1783	0506	4878	7673	9840	6629
8529	7842	7203	1844	8619	7404	4215	9969	6948	5643
8973	3440	4366	9242	2151	0244	0922	5887	4883	1177
9307	2959	5904	9012	4951	3695	4529	7197	7179	3239
2923	4276	9467	9868	2257	1925	3382	7244	1781	8037
6372	2808	1238	8098	5509	4617	4099	6705	2386	2830
6922	1807	4900	5306	0411	1828	8634	2331	7247	3230
9862	8336	6453	0545	6127	2741	5967	8447	3017	5709
3371	1530	5104	3076	5506	3101	4143	5845	2095	6127
6712	9402	9588	7019	9248	9192	4223	6555	7947	2474
3071	8782	7157	5941	8830	8563	2252	8109	5880	9912
4022	9734	7852	9096	0051	7387	7056	9331	1317	7833
9682	8892	3577	0326	5306	0050	8517	4376	0788	5443
6705	2175	9904	3743	1902	5393	3032	8432	0612	7972
1872	8292	2366	8603	4288	6809	4357	1072	6822	5611
2559	7534	2281	7351	2064	0611	9613	2000	0327	6145
4399	3751	9783	5399	5175	8894	0296	9483	0400	2272
6074	8827	2195	2532	7680	4288	6807	3101	6850	6410
5155	7186	4722	6721	0838	3632	5355	9369	2006	7681
3193	2800	6184	7891	9838	6123	9397	4019	8389	9508
8610	1880	7423	3384	4625	6653	2900	6290	9286	2396
4778	8818	2992	6300	4239	9595	4384	0611	7687	2088
3987	1619	4164	2542	4042	7799	9084	0278	8422	4330
2977	0248	2793	3351	4922	8878	5703	7421	2054	4391
1312	2919	8220	7285	5902	7882	1403	5354	9913	7109
3890	7193	7799	9190	3275	7840	1872	6232	5295	3148
0793	3468	8762	2492	5854	8430	8472	2264	9279	2128
2139	4552	3444	6462	2524	8601	3372	1848	1472	9667
8277	9153	2880	9053	6880	4284	5044	8931	0861	1517
2236	4778	6639	0862	9509	2141	0208	1450	1222	5281
8837	7686	1771	3374	2894	7314	6856	0440	3766	6047
6605	6380	4599	3333	0713	8401	7146	8940	2629	2006
8399	8175	3525	1646	4019	8390	4344	8975	4489	3423
8053	3046	9102	4515	2944	9763	3003	3408	1199	2791
9837	9378	3237	7016	7593	5958	0068	3114	0456	6840
2557	6395	9496	1884	0612	8102	4402	5498	0422	3335

Reproduced with permission from D. B. Owen, *Handbook of Statistical Tables,* Addison-Wesley, Reading, Mass., 1962, pp. 519, 520.

TABLE D (*continued*)

2671	4690	1550	2262	2597	8034	0785	2978	4409	0237
9111	0250	3275	7519	9740	4577	2064	0286	3398	1348
0391	6035	9230	4999	3332	0608	6113	0391	5789	9926
2475	2144	1886	2079	3004	9686	5669	4367	9306	2595
5336	5845	2095	6446	5694	3641	1085	8705	5416	9066
6808	0423	0155	1652	7897	4335	3567	7109	9690	3739
8525	0577	8940	9451	6726	0876	3818	7607	8854	3566
0398	0741	8787	3043	5063	0617	1770	5048	7721	7032
3623	9636	3638	1406	5731	3978	8068	7238	9715	3363
0739	2644	4917	8866	3632	5399	5175	7422	2476	2607
6713	3041	8133	8749	8835	6745	3597	3476	3816	3455
7775	9315	0432	8327	0861	1515	2297	3375	3713	9174
8599	2122	6842	9202	0810	2936	1514	2090	3067	3574
7955	3759	5254	1126	5553	4713	9605	7909	1658	5490
4766	0070	7260	6033	7997	0109	5993	7592	5436	1727
5165	1670	2534	8811	8231	3721	7947	5719	2640	1394
9111	0513	2751	8256	2931	7783	1281	6531	7259	6993
1667	1084	7889	8963	7018	8617	6381	0723	4926	4551
2145	4587	8585	2412	5431	4667	1942	7238	9613	2212
2739	5528	1481	7528	9368	1823	6979	2547	7268	2467
8769	5480	9160	5354	9700	1362	2774	7980	9157	8788
6531	9435	3422	2474	1475	0159	3414	5224	8399	5820
2937	4134	7120	2206	5084	9473	3958	7320	9878	8609
1581	3285	3727	8924	6204	0797	0882	5945	9375	9153
6268	1045	7076	1436	4165	0143	0293	4190	7171	7932
4293	0523	8625	1961	1039	2856	4889	4358	1492	3804
6936	4213	3212	7229	1230	0019	5998	9206	6753	3762
5334	7641	3258	3769	1362	2771	6124	9813	7915	8960
9373	1158	4418	8826	5665	5896	0358	4717	8232	4859
6968	9428	8950	5346	1741	2348	8143	5377	7695	0685
4229	0587	8794	4009	9691	4579	3302	7673	9629	5246
3807	7785	7097	5701	6639	0723	4819	0900	2713	7650
4891	8829	1642	2155	0796	0466	2946	2970	9143	6590
1055	2968	7911	7479	8199	9735	8271	5339	7058	2964
2983	2345	0568	4125	0894	8302	0506	6761	7706	4310
4026	3129	2968	8053	2797	4022	9838	9611	0975	2437
4075	0260	4256	0337	2355	9371	2954	6021	5783	2827
8488	5450	1327	7358	2034	8060	1788	6913	6123	9405
1976	1749	5742	4098	5887	4567	6064	2777	7830	5668
2793	4701	9466	9554	8294	2160	7486	1557	4769	2781
0916	6272	6825	7188	9611	1181	2301	5516	5451	6832
5961	1149	7946	1950	2010	0600	5655	0796	0569	4365
3222	4189	1891	8172	8731	4769	2782	1325	4238	9279
1176	7834	4600	9992	9449	5824	5344	1008	6678	1921
2369	8971	2314	4806	5071	8908	8274	4936	3357	4441
0041	4329	9265	0352	4764	9070	7527	7791	1094	2008
0803	8302	6814	2422	6351	0637	0514	0246	1845	8594
9965	7804	3930	8803	0268	1426	3130	3613	3947	8086
0011	2387	3148	7559	4216	2946	2865	6333	1916	2259
1767	9871	3914	5790	5287	7915	8959	1346	5482	9251

TABLE E. Logarithms of Factorials; Tens to Left, Units in Columns

	0	1	2	3	4	5	6	7	8	9
00	0.0000	0.0000	0.3010	0.7782	1.3802	2.0792	2.8573	3.7024	4.6055	5.5598
10	6.5598	7.6012	8.6803	9.7943	10.9404	12.1165	13.3206	14.5511	15.8063	17.0851
20	18.3861	19.7083	21.0508	22.4125	23.7927	25.1906	26.6056	28.0370	29.4841	30.9465
30	32.4237	33.9150	35.4202	36.9387	38.4702	40.0142	41.5705	43.1387	44.7185	46.3096
40	47.9116	49.5244	51.1477	52.7811	54.4246	56.0778	57.7406	59.4127	61.0939	62.7841
50	64.4831	66.1906	67.9066	69.6309	71.3633	73.1037	74.8519	76.6077	78.3712	80.1420
60	81.9202	83.7055	85.4979	87.2972	89.1034	90.9163	92.7359	94.5619	96.3945	98.2333
70	100.0784	101.9297	103.7870	105.6503	107.5196	109.3946	111.2754	113.1619	115.0540	116.9516
80	118.8547	120.7632	122.6770	124.5961	126.5204	128.4498	130.3843	132.3238	134.2683	136.2177
90	138.1719	140.1310	142.0948	144.0632	146.0364	148.0141	149.9964	151.9831	153.9744	155.9700
100	157.9700	159.9743	161.9829	163.9958	166.0128	168.0340	170.0593	172.0887	174.1221	176.1595
110	178.2009	180.2462	182.2955	184.3485	186.4054	188.4661	190.5306	192.5988	194.6707	196.7462
120	198.8254	200.9082	202.9945	205.0844	207.1779	209.2748	211.3751	213.4790	215.5862	217.6967
130	219.8107	221.9280	224.0485	226.1724	228.2995	230.4298	232.5634	234.7001	236.8400	238.9830
140	241.1291	243.2783	245.4306	247.5860	249.7443	251.9057	254.0700	256.2374	258.4076	260.5808
150	262.7569	264.9359	267.1177	269.3024	271.4899	273.6803	275.8734	278.0693	280.2679	282.4693
160	284.6735	286.8803	289.0898	291.3020	293.5168	295.7343	297.9544	300.1771	302.4024	304.6303
170	306.8608	309.0938	311.3293	313.5674	315.8079	318.0509	320.2965	322.5444	324.7948	327.0477
180	329.3030	331.5606	333.8207	336.0832	338.3480	340.6152	342.8847	345.1565	347.4307	349.7071
190	351.9859	354.2669	356.5502	358.8358	361.1236	363.4136	365.7059	368.0003	370.2970	372.5959

	0	1	2	3	4	5	6	7	8	9
200	374.8969	377.2001	379.5054	381.8129	384.1226	386.4343	388.7482	391.0642	393.3822	395.7024
210	398.0246	400.3489	402.6752	405.0036	407.3340	409.6664	412.0009	414.3373	416.6758	419.0162
220	421.3587	423.7031	426.0494	428.3977	430.7480	433.1002	435.4543	437.8103	440.1682	442.5281
230	444.8898	447.2534	449.6189	451.9862	454.3555	456.7265	459.0994	461.4742	463.8508	466.2292
240	468.6094	470.9914	473.3752	475.7608	478.1482	480.5374	482.9283	485.3210	487.7154	490.1116
250	492.5096	494.9093	497.3107	499.7138	502.1186	504.5252	506.9334	509.3433	511.7549	514.1682
260	516.5832	518.9999	521.4182	523.8381	526.2597	528.6830	531.1078	533.5344	535.9625	538.3922
270	540.8236	543.2566	545.6912	548.1273	550.5651	553.0044	555.4453	557.8878	560.3318	562.7774
280	565.2246	567.6733	570.1235	572.5753	575.0287	577.4835	579.9399	582.3977	584.8571	587.3180
290	589.7804	592.2443	594.7097	597.1766	599.6449	602.1147	604.5860	607.0588	609.5330	612.0087
300	614.4858	616.9644	619.4444	621.9258	624.4087	626.8930	629.3787	631.8659	634.3544	636.8444
310	639.3357	641.8285	644.3226	646.8182	649.3151	651.8134	654.3131	656.8142	659.3166	661.8204
320	664.3255	666.8320	669.3399	671.8491	674.3596	676.8715	679.3847	681.8993	684.4152	686.9324
330	689.4509	691.9707	694.4918	697.0143	699.5380	702.0631	704.5894	707.1170	709.6460	712.1762
340	714.7076	717.2404	719.7744	722.3097	724.8463	727.3841	729.9232	732.4635	735.0051	737.5479
350	740.0920	742.6373	745.1838	747.7316	750.2806	752.8308	755.3823	757.9349	760.4888	763.0439
360	765.6002	768.1577	770.7164	773.2764	775.8375	778.3997	780.9632	783.5279	786.0937	788.6608
370	791.2290	793.7983	796.3689	798.9406	801.5135	804.0875	806.6627	809.2390	811.8165	814.3952
380	816.9749	819.5559	822.1379	824.7211	827.3055	829.8909	832.4775	835.0652	837.6540	840.2440
390	842.8351	845.4272	848.0205	850.6149	853.2104	855.8070	858.4047	861.0035	863.6034	866.2044

TABLE E (*continued*)

	0	1	2	3	4	5	6	7	8	9
400	868.8064	871.4096	874.0138	876.6191	879.2255	881.8329	884.4415	887.0510	889.6617	892.2734
410	894.8862	897.5001	900.1150	902.7309	905.3479	907.9660	910.5850	913.2052	915.8264	918.4486
420	921.0718	923.6961	926.3214	928.9478	931.5751	934.2035	936.8329	939.4633	942.0948	944.7272
430	947.3607	949.9952	952.6307	955.2672	957.9047	960.5431	963.1826	965.8231	968.4646	971.1071
440	973.7505	976.3949	979.0404	981.6868	984.3342	986.9825	989.6318	992.2822	994.9334	997.5857
450	1000.2389	1002.8931	1005.5482	1008.2043	1010.8614	1013.5194	1016.1783	1018.8383	1021.4991	1024.1609
460	1026.8237	1029.4874	1032.1520	1034.8176	1037.4841	1040.1516	1042.8200	1045.4893	1048.1595	1050.8307
470	1053.5028	1056.1758	1058.8498	1061.5246	1064.2004	1066.8771	1069.5547	1072.2332	1074.9127	1077.5930
480	1080.2742	1082.9564	1085.6394	1088.3234	1091.0082	1093.6940	1096.3806	1099.0681	1101.7565	1104.4458
490	1107.1360	1109.8271	1112.5191	1115.2119	1117.9057	1120.6003	1123.2958	1125.9921	1128.6893	1131.3874
500	1134.0864	1136.7862	1139.4869	1142.1885	1144.8909	1147.5942	1150.2984	1153.0034	1155.7093	1158.4160
510	1161.1236	1163.8320	1166.5412	1169.2514	1171.9623	1174.6741	1177.3868	1180.1003	1182.8146	1185.5298
520	1188.2458	1190.9626	1193.6803	1196.3988	1199.1181	1201.8383	1204.5593	1207.2811	1210.0037	1212.7272
530	1215.4514	1218.1765	1220.9024	1223.6292	1226.3567	1229.0851	1231.8142	1234.5442	1237.2750	1240.0066
540	1242.7390	1245.4722	1248.2062	1250.9410	1253.6766	1256.4130	1259.1501	1261.8881	1264.6269	1267.3665
550	1270.1069	1272.8480	1275.5899	1278.3327	1281.0762	1283.8205	1286.5655	1289.3114	1292.0580	1294.8054
560	1297.5536	1300.3026	1303.0523	1305.8028	1308.5541	1311.3062	1314.0590	1316.8126	1319.5669	1322.3220
570	1325.0779	1327.8345	1330.5919	1333.3501	1336.1090	1338.8687	1341.6291	1344.3903	1347.1522	1349.9149
580	1352.6783	1355.4425	1358.2074	1360.9731	1363.7395	1366.5066	1369.2745	1372.0432	1374.8126	1377.5827
590	1380.3535	1383.1251	1385.8974	1388.6705	1391.4443	1394.2188	1396.9940	1399.7700	1402.5467	1405.3241

	0	1	2	3	4	5	6	7	8	9
600	1408.1023	1410.8812	1413.6608	1416.4411	1419.2221	1422.0039	1424.7863	1427.5695	1430.3534	1433.1380
610	1435.9234	1438.7094	1441.4962	1444.2836	1447.0718	1449.8607	1452.6503	1455.4405	1458.2315	1461.0232
620	1463.8156	1466.6087	1469.4025	1472.1970	1474.9922	1477.7880	1480.5846	1483.3819	1486.1798	1488.9785
630	1491.7778	1494.5779	1497.3786	1500.1800	1502.9821	1505.7849	1508.5883	1511.3924	1514.1973	1517.0028
640	1519.8090	1522.6158	1525.4233	1528.2316	1531.0404	1533.8500	1536.6602	1539.4711	1542.2827	1545.0950
650	1547.9079	1550.7215	1553.5357	1556.3506	1559.1662	1561.9824	1564.7993	1567.6169	1570.4351	1573.2540
660	1576.0736	1578.8938	1581.7146	1584.5361	1587.3583	1590.1811	1593.0046	1595.8287	1598.6535	1601.4789
670	1604.3050	1607.1317	1609.9591	1612.7871	1615.6158	1618.4451	1621.2750	1624.1056	1626.9368	1629.7687
680	1632.6012	1635.4344	1638.2681	1641.1026	1643.9376	1646.7733	1649.6096	1652.4466	1655.2842	1658.1224
690	1660.9612	1663.8007	1666.6408	1669.4816	1672.3229	1675.1649	1678.0075	1680.8508	1683.6946	1686.5391
700	1689.3842	1692.2299	1695.0762	1697.9232	1700.7708	1703.6190	1706.4678	1709.3172	1712.1672	1715.0179
710	1717.8691	1720.7210	1723.5735	1726.4266	1729.2803	1732.1346	1734.9895	1737.8450	1740.7011	1743.5578
720	1746.4152	1749.2731	1752.1316	1754.9908	1757.8505	1760.7109	1763.5718	1766.4333	1769.2955	1772.1582
730	1775.0215	1777.8854	1780.7499	1783.6150	1786.4807	1789.3470	1792.2139	1795.0814	1797.9494	1800.8181
740	1803.6873	1806.5571	1809.4275	1812.2985	1815.1701	1818.0423	1820.9150	1823.7883	1826.6622	1829.5367
750	1832.4118	1835.2874	1838.1636	1841.0404	1843.9178	1846.7957	1849.6742	1852.5533	1855.4330	1858.3133
760	1861.1941	1864.0755	1866.9574	1869.8399	1872.7230	1875.6067	1878.4909	1881.3757	1884.2611	1887.1470
770	1890.0335	1892.9205	1895.8082	1898.6963	1901.5851	1904.4744	1907.3642	1910.2547	1913.1456	1916.0372
780	1918.9293	1921.8219	1924.7151	1927.6089	1930.5032	1933.3981	1936.2935	1939.1895	1942.0860	1944.9831
790	1947.8807	1950.7789	1953.6776	1956.5769	1959.4767	1962.3771	1965.2780	1968.1794	1971.0814	1973.9840

TABLE E (continued)

	0	1	2	3	4	5	6	7	8	9
800	1976.8871	1979.7907	1982.6949	1985.5996	1988.5049	1991.4107	1994.3170	1997.2239	2000.1313	2003.0392
810	2005.9477	2008.8567	2011.7663	2014.6764	2017.5870	2020.4982	2023.4099	2026.3221	2029.2348	2032.1481
820	2035.0619	2037.9763	2040.8911	2043.8065	2046.7225	2049.6389	2052.5559	2055.4734	2058.3914	2061.3100
830	2064.2291	2067.1487	2070.0688	2072.9894	2075.9106	2078.8323	2081.7545	2084.6772	2087.6005	2090.5242
840	2093.4485	2096.3733	2099.2986	2102.2244	2105.1508	2108.0776	2111.0050	2113.9329	2116.8613	2119.7902
850	2122.7196	2125.6495	2128.5800	2131.5109	2134.4424	2137.3744	2140.3068	2143.2398	2146.1733	2149.1073
860	2152.0418	2154.9768	2157.9123	2160.8483	2163.7848	2166.7218	2169.6594	2172.5974	2175.5359	2178.4749
870	2181.4144	2184.3545	2187.2950	2190.2360	2193.1775	2196.1195	2199.0620	2202.0050	2204.9485	2207.8925
880	2210.8370	2213.7820	2216.7274	2219.6734	2222.6198	2225.5668	2228.5142	2231.4621	2234.4106	2237.3595
890	2240.3088	2243.2587	2246.2091	2249.1599	2252.1113	2255.0631	2258.0154	2260.9682	2263.9215	2266.8752
900	2269.8295	2272.7842	2275.7394	2278.6951	2281.6513	2284.6079	2287.5650	2290.5226	2293.4807	2296.4393
910	2299.3983	2302.3579	2305.3179	2308.2783	2311.2393	2314.2007	2317.1626	2320.1250	2323.0878	2326.0511
920	2329.0149	2331.9792	2334.9439	2337.9091	2340.8748	2343.8409	2346.8075	2349.7746	2352.7421	2355.7102
930	2358.6786	2361.6476	2364.6170	2367.5869	2370.5572	2373.5281	2376.4993	2379.4711	2382.4433	2385.4159
940	2388.3891	2391.3627	2394.3367	2397.3112	2400.2862	2403.2616	2406.2375	2409.2139	2412.1907	2415.1679
950	2418.1457	2421.1238	2424.1025	2427.0816	2430.0611	2433.0411	2436.0216	2439.0025	2441.9839	2444.9657
960	2447.9479	2450.9307	2453.9138	2456.8975	2459.8815	2462.8661	2465.8511	2468.8365	2471.8224	2474.8087
970	2477.7954	2480.7827	2483.7703	2486.7584	2489.7470	2492.7360	2495.7255	2498.7154	2501.7057	2504.6965
980	2507.6877	2510.6794	2513.6715	2516.6640	2519.6570	2522.6505	2525.6443	2528.6387	2531.6334	2534.6286
990	2537.6242	2540.6203	2543.6168	2546.6138	2549.6112	2552.6090	2555.6073	2558.6059	2561.6051	2564.6046
1,000	2567.6046	2570.6051	2573.6059	2576.6072	2579.6090	2582.6111	2585.6137	2588.6168	2591.6202	2594.6241

TABLE F.   Single Sample Tests for $\sigma = \sigma_1$ versus $\sigma = \sigma_2 > \sigma_1$, with Risks $\alpha = \beta$

	Ratio of σ_2/σ_1 for $\alpha = \beta$				Multiplier for σ_1^2 to get K. $\alpha = \beta$			
(1)	(2)	(3)	(4)	(5)	(6)	(7)	(8)	(9)
Sample size	.10	.05	.02	.01	.10	.05	.02	.01
2	13.1	31.3	92.6	206.	2.71	3.84	5.41	6.64
3	4.67	7.63	13.9	21.4	2.30	3.00	3.91	4.60
4	3.27	4.71	7.29	9.93	2.08	2.60	3.28	3.78
5	2.70	3.65	5.22	6.69	1.94	2.37	2.92	3.32
6	2.40	3.11	4.22	5.22	1.85	2.21	2.68	3.02
7	2.20	2.76	3.64	4.39	1.77	2.10	2.51	2.80
8	2.06	2.55	3.26	3.86	1.72	2.01	2.37	2.64
9	1.96	2.38	2.99	3.49	1.67	1.94	2.27	2.51
10	1.88	2.26	2.79	3.22	1.63	1.88	2.19	2.41
11	1.81	2.16	2.63	3.01	1.60	1.83	2.12	2.32
12	1.76	2.07	2.50	2.85	1.57	1.79	2.06	2.25
13	1.72	2.01	2.40	2.71	1.55	1.75	2.00	2.18
14	1.68	1.95	2.31	2.60	1.52	1.72	1.96	2.13
15	1.64	1.90	2.24	2.50	1.50	1.69	1.92	2.08
16	1.61	1.86	2.17	2.42	1.49	1.67	1.88	2.04
17	1.59	1.82	2.12	2.35	1.47	1.64	1.85	2.00
18	1.57	1.78	2.07	2.28	1.46	1.62	1.82	1.97
19	1.55	1.75	2.02	2.23	1.44	1.60	1.80	1.93
20	1.53	1.73	1.98	2.18	1.43	1.59	1.77	1.90
21	1.51	1.70	1.95	2.13	1.42	1.57	1.75	1.88
22	1.50	1.68	1.91	2.09	1.41	1.56	1.73	1.85
23	1.48	1.66	1.88	2.05	1.40	1.54	1.71	1.83
24	1.47	1.64	1.86	2.02	1.39	1.53	1.69	1.81
25	1.46	1.62	1.83	1.99	1.38	1.52	1.68	1.79
26	1.44	1.61	1.81	1.96	1.38	1.51	1.66	1.77
27	1.43	1.59	1.79	1.93	1.37	1.50	1.65	1.76
28	1.42	1.58	1.77	1.91	1.36	1.49	1.63	1.74
29	1.41	1.56	1.75	1.89	1.35	1.48	1.62	1.72
30	1.41	1.55	1.73	1.87	1.35	1.47	1.61	1.71
31	1.40	1.54	1.72	1.84	1.34	1.46	1.60	1.70
40	1.34	1.46	1.60	1.71	1.30	1.40	1.52	1.60
50	1.30	1.40	1.52	1.61	1.27	1.35	1.46	1.53
60	1.27	1.36	1.46	1.54	1.24	1.32	1.41	1.48
70	1.25	1.33	1.42	1.49	1.22	1.30	1.38	1.44
80	1.23	1.30	1.39	1.45	1.21	1.28	1.36	1.41
90	1.21	1.28	1.36	1.42	1.20	1.26	1.33	1.38
100	1.20	1.26	1.34	1.39	1.19	1.24	1.31	1.36

For n, seek entry in columns (2) - (5), $\leq \sigma_2/\sigma_1$, giving n. Then for this n and $\alpha = \beta$, find in columns (6) - (9), the multiplier for σ_1^2 to give K. Then $s^2 \leq K$, accept $\sigma = \sigma_1$, $s^2 > K$, reject $\sigma = \sigma_1$, and conclude $\sigma > \sigma_1$.

Reproduced with permission from I. W. Burr, *Applied Statistical Methods*, Academic Press, New York, 1974, p. 448.

TABLE G. Common Logarithms of Numbers to Four Decimal Places

N	L. 0	1	2	3	4	5	6	7	8	9	Proportional Parts 1	2	3	4	5
10	0000	0043	0086	0128	0170	0212	0253	0294	0334	0374	4	8	12	17	21
11	0414	0453	0492	0531	0569	0607	0645	0682	0719	0755	4	8	11	15	19
12	0792	0828	0864	0899	0934	0969	1004	1038	1072	1106	3	7	10	14	17
13	1139	1173	1206	1239	1271	1303	1335	1367	1399	1430	3	6	10	13	16
14	1461	1492	1523	1553	1584	1614	1644	1673	1703	1732	3	6	9	12	15
15	1761	1790	1818	1847	1875	1903	1931	1959	1987	2014	3	6	8	11	14
16	2041	2068	2095	2122	2148	2175	2201	2227	2253	2279	3	5	8	11	13
17	2304	2330	2355	2380	2405	2430	2455	2480	2504	2529	2	5	7	10	12
18	2553	2577	2601	2625	2648	2672	2695	2718	2742	2765	2	5	7	9	12
19	2788	2810	2833	2856	2878	2900	2923	2945	2967	2989	2	4	7	9	11
20	3010	3032	3054	3075	3096	3118	3139	3160	3181	3201	2	4	6	8	11
21	3222	3243	3263	3284	3304	3324	3345	3365	3385	3404	2	4	6	8	10
22	3424	3444	3464	3483	3502	3522	3541	3560	3579	3598	2	4	6	8	10
23	3617	3636	3655	3674	3692	3711	3729	3747	3766	3784	2	4	6	7	9
24	3802	3820	3838	3856	3874	3892	3909	3927	3945	3962	2	4	5	7	9
25	3979	3997	4014	4031	4048	4065	4082	4099	4116	4133	2	4	5	7	9
26	4150	4166	4183	4200	4216	4232	4249	4265	4281	4298	2	3	5	7	8
27	4314	4330	4346	4362	4378	4393	4409	4425	4440	4456	2	3	5	6	8
28	4472	4487	4502	4518	4533	4548	4564	4579	4594	4609	2	3	5	6	8
29	4624	4639	4654	4669	4683	4698	4713	4728	4742	4757	1	3	4	6	7
30	4771	4786	4800	4814	4829	4843	4857	4871	4886	4900	1	3	4	6	7
31	4914	4928	4942	4955	4969	4983	4997	5011	5024	5038	1	3	4	5	7
32	5051	5065	5079	5092	5105	5119	5132	5145	5159	5172	1	3	4	5	7
33	5185	5198	5211	5224	5237	5250	5263	5276	5289	5302	1	3	4	5	7
34	5315	5328	5340	5353	5366	5378	5391	5403	5416	5428	1	2	4	5	6
35	5441	5453	5465	5478	5490	5502	5514	5527	5539	5551	1	2	4	5	6
36	5563	5575	5587	5599	5611	5623	5635	5647	5658	5670	1	2	4	5	6
37	5682	5694	5705	5717	5729	5740	5752	5763	5775	5786	1	2	4	5	6
38	5798	5809	5821	5832	5843	5855	5866	5877	5888	5899	1	2	3	5	6
39	5911	5922	5933	5944	5955	5966	5977	5988	5999	6010	1	2	3	4	5
40	6021	6031	6042	6053	6064	6075	6085	6096	6107	6117	1	2	3	4	5
41	6128	6138	6149	6160	6170	6180	6191	6201	6212	6222	1	2	3	4	5
42	6232	6243	6253	6263	6274	6284	6294	6304	6314	6325	1	2	3	4	5
43	6335	6345	6355	6365	6375	6385	6395	6405	6415	6425	1	2	3	4	5
44	6435	6444	6454	6464	6474	6484	6493	6503	6513	6522	1	2	3	4	5
45	6532	6542	6551	6561	6571	6580	6590	6599	6609	6618	1	2	3	4	5
46	6628	6637	6646	6656	6665	6675	6684	6693	6702	6712	1	2	3	4	5
47	6721	6730	6739	6749	6758	6767	6776	6785	6794	6803	1	2	3	4	5
48	6812	6821	6830	6839	6848	6857	6866	6875	6884	6893	1	2	3	4	5
49	6902	6911	6920	6928	6937	6946	6955	6964	6972	6981	1	2	3	4	4
50	6990	6998	7007	7016	7024	7033	7042	7050	7059	7067	1	2	3	3	4
51	7076	7084	7093	7101	7110	7118	7126	7135	7143	7152	1	2	3	3	4
52	7160	7168	7177	7185	7193	7202	7210	7218	7226	7235	1	2	3	3	4
53	7243	7251	7259	7267	7275	7284	7292	7300	7308	7316	1	2	2	3	4
54	7324	7332	7340	7348	7356	7364	7372	7380	7388	7396	1	2	2	3	4
N	L. 0	1	2	3	4	5	6	7	8	9	1	2	3	4	5

TABLE G *(continued)*

N	L. 0	1	2	3	4	5	6	7	8	9	1	2	3	4	5
55	7404	7412	7419	7427	7435	7443	7451	7459	7466	7474	1	2	2	3	4
56	7482	7490	7497	7505	7513	7520	7528	7536	7543	7551	1	2	2	3	4
57	7559	7566	7574	7582	7589	7597	7604	7612	7619	7627	1	1	2	3	4
58	7634	7642	7649	7657	7664	7672	7679	7686	7694	7701	1	1	2	3	4
59	7709	7716	7723	7731	7738	7745	7752	7760	7767	7774	1	1	2	3	4
60	7782	7789	7796	7803	7810	7818	7825	7832	7839	7846	1	1	2	3	4
61	7853	7860	7868	7875	7882	7889	7896	7903	7910	7917	1	1	2	3	3
62	7924	7931	7938	7945	7952	7959	7966	7973	7980	7987	1	1	2	3	3
63	7993	8000	8007	8014	8021	8028	8035	8041	8048	8055	1	1	2	3	3
64	8062	8069	8075	8082	8089	8096	8102	8109	8116	8122	1	1	2	3	3
65	8129	8136	8142	8149	8156	8162	8169	8176	8182	8189	1	1	2	3	3
66	8195	8202	8209	8215	8222	8228	8235	8241	8248	8254	1	1	2	3	3
67	8261	8267	8274	8280	8287	8293	8299	8306	8312	8319	1	1	2	3	3
68	8325	8331	8338	8344	8351	8357	8363	8370	8376	8382	1	1	2	3	3
69	8388	8395	8401	8407	8414	8420	8426	8432	8439	8445	1	1	2	3	3
70	8451	8457	8463	8470	8476	8482	8488	8494	8500	8506	1	1	2	3	3
71	8513	8519	8525	8531	8537	8543	8549	8555	8561	8567	1	1	2	3	3
72	8573	8579	8585	8591	8597	8603	8609	8615	8621	8627	1	1	2	3	3
73	8633	8639	8645	8651	8657	8663	8669	8675	8681	8686	1	1	2	2	3
74	8692	8698	8704	8710	8716	8722	8727	8733	8739	8745	1	1	2	2	3
75	8751	8756	8762	8768	8774	8779	8785	8791	8797	8802	1	1	2	2	3
76	8808	8814	8820	8825	8831	8837	8842	8848	8854	8859	1	1	2	2	3
77	8865	8871	8876	8882	8887	8893	8899	8904	8910	8915	1	1	2	2	3
78	8921	8927	8932	8938	8943	8949	8954	8960	8965	8971	1	1	2	2	3
79	8976	8982	8987	8993	8998	9004	9009	9015	9020	9025	1	1	2	2	3
80	9031	9036	9042	9047	9053	9058	9063	9069	9074	9079	1	1	2	2	3
81	9085	9090	9096	9101	9106	9112	9117	9122	9128	9133	1	1	2	2	3
82	9138	9143	9149	9154	9159	9165	9170	9175	9180	9186	1	1	2	2	3
83	9191	9196	9201	9206	9212	9217	9222	9227	9232	9238	1	1	2	2	3
84	9243	9248	9253	9258	9263	9269	9274	9279	9284	9289	1	1	2	2	3
85	9294	9299	9304	9309	9315	9320	9325	9330	9335	9340	1	1	2	2	3
86	9345	9350	9355	9360	9365	9370	9375	9380	9385	9390	1	1	2	2	3
87	9395	9400	9405	9410	9415	9420	9425	9430	9435	9440	1	1	2	2	3
88	9445	9450	9455	9460	9465	9469	9474	9479	9484	9489	0	1	1	2	2
89	9494	9499	9504	9509	9513	9518	9523	9528	9533	9538	0	1	1	2	2
90	9542	9547	9552	9557	9562	9566	9571	9576	9581	9586	0	1	1	2	2
91	9590	9595	9600	9605	9609	9614	9619	9624	9628	9633	0	1	1	2	2
92	9638	9643	9647	9652	9657	9661	9666	9671	9675	9680	0	1	1	2	2
93	9685	9689	9694	9699	9703	9708	9713	9717	9722	9727	0	1	1	2	2
94	9731	9736	9741	9745	9750	9754	9759	9763	9768	9773	0	1	1	2	2
95	9777	9782	9786	9791	9795	9800	9805	9809	9814	9818	0	1	1	2	2
96	9823	9827	9832	9836	9841	9845	9850	9854	9859	9863	0	1	1	2	2
97	9868	9872	9877	9881	9886	9890	9894	9899	9903	9908	0	1	1	2	2
98	9912	9917	9921	9926	9930	9934	9939	9943	9948	9952	0	1	1	2	2
99	9956	9961	9965	9969	9974	9978	9983	9987	9991	9996	0	1	1	2	2
N	L. 0	1	2	3	4	5	6	7	8	9	1	2	3	4	5

ANSWERS TO ODD-NUMBERED PROBLEMS

Numerical answers rounded off at end of calculation after having
carried more precision during calculation. Discussions mostly
omitted here.

Chapter 2

2.1 $\bar{x}$ = 69.3, s = 2.66, R = 6.

2.3 $\bar{x}$ = 140.4, s = 3.21, R = 7 .001 in.

2.5 $\bar{x}$ = 40.3, s = 8.39, R = 15 .0001 in.

2.7 $\bar{x}$ = 2.503825, s = .000479, R = .0010 g/cm^3.

2.9   Class 0-9 10-19 20-29 30-39 40-49
             4   19   10   23   11
      Class 50-59 60-69 70-79 80-89 90-99   .0001 in.
             14   3   3   2   1

2.11 $\bar{x}$ = 36.28, s = 19.75   .0001 in.

2.13 $\bar{x}$ = 2.50428, s = .001283 g/cm^3.

2.15 $\bar{x}$ = 52.86, s = 1.003 1b.

Chapter 3

3.1 σ_d = 5.

3.3   P(2g) = .9604, P(1g) = .0392, P(0g) = .0004.

3.5   P(3g) = .778,688, P(2g) = .203,136, P(1g) = .017,664,
      P(0g) = .000,512.

3.7   P(d = 0) = .625, P(d = 1) = .375.

3.9 $P(d = 0) = 10/28$, $P(d = 1) = 15/28$, $P(d = 2) = 3/28$.

3.11   210, 210, 5040.

Chapter 4

4.1   .663, .121.

4.3   .677, .271, .594.

4.5   (a) .590, (b) hypergeometric, (c) binomial.

4.7   .819, .163, .017.

4.9 $\sigma_c = 1.10$, $P(c \geq 4.5) = P(5 \text{ or more}) = .008$.

4.11 $\sigma_c = 1.41$, $P(c \geq 6.23) = P(7 \text{ or more}) = .005$.

4.13   .4096, .4096, .1536, .0256, .0016.

4.15   10/21, 10/21, 1/21.

Chapter 6

6.1 $n\bar{p} = 5.5$, UCL = 12.1, control perfect. $np' = 5$, UCL = 11.4,
         compatible.

6.3 $\bar{p} = .113$, UCL = .247. Three in succession out, one other
         close. $\bar{p} = .0289$, UCL = .100. Two points out of control.

6.5 $np' = 2$, UCL = 6.2. In control relative to $p' = .02$. Yes.

6.7   Use $\bar{p} = .1064$, $\text{limits}_p = .1064 \pm 3\sqrt{.1064(.8936)/n}$ for n's
         used. Cracks might well be related from iron. Still inves-
         tigate points out.

6.9   Prefer np chart to avoid division by 39. $n\bar{p} = 1.72$,
         UCL = 5.6. Probably to insure delivery of 36 unbroken
         articles.

6.11 $\bar{c} = 405.7$, $\text{limits}_c = 345.3$, 466.1. Many points out.
         Probably repetitive defects.

6.13 $\bar{c} = 7.6$, $\text{UCL}_c = 15.9$, homogeneous. $c' = 8$, $\text{UCL}_c = 16.5$.
         In control relative to $c' = 8$.

6.15   Alignment: $\bar{c} = 9.46$, $\text{limits}_c = .2$, 18.7. In control.

6.17 $\bar{u}$ = 1.97, limits$_u$ = 1.59, 2.35. One high point, 12 low in 18.
For present data $\bar{u}$ = 1.6, limits$_u$ 1.26, 1.94. Several indications of assignable causes. Unless causes found and eliminated, might use 1.6.

Chapter 7

7.1 $\bar{\bar{x}}$ = 3.694, limits$_{\bar{x}}$ = 3.25, 4.14, in control. $\bar{R}$ = .77,
UCL$_R$ = 1.63, in control. Limits$_x$ = 2.70, 4.69, justified by
control. Must lower process average. Distribution runs above
specification limit with one very high x.

7.3 $\bar{\bar{x}}$ = .2818, limits$_{\bar{x}}$ = .222, .341. $\bar{R}$ = .0317, UCL$_R$ = .104.
Evidence of assignable causes on both charts. Not meeting
specification limit. Seek better control, but raise $\bar{\bar{x}}$ at
once. (d) Set up two equations, $\bar{x}$ and R in terms of x_1, x_2:
.26, .25.

7.5

$\bar{\bar{x}}$	Limits$_{\bar{x}}$	$\bar{R}$	UCL$_R$	$\bar{x}$ chart	R chart	x's
2.37	-.4, +5.1	4.8	10.2	two out	3 near limit	many out
2.65	+.2, +5.1	4.25	9.0	in control	two high	many out
2.68	+2.0, +3.4	1.20	2.5	one high	two high	OK

7.7 $\bar{\bar{x}}$ = -2.10, limits$_{\bar{x}}$ = -5.5, +1.3. $\bar{R}$ = 5.85, UCL$_R$ = 12.4.
$\bar{\bar{x}}$ = -.17, limits$_{\bar{x}}$ = -1.6, +1.2. $\bar{R}$ = 2.45, UCL$_R$ = 5.2.
All four charts show lack of control. For second period a
preliminary estimate of σ is 1.05; x limits -3.3, +3.0.
All right if control is improved.

7.9 $\bar{\bar{x}}$ = 78.28, limits$_{\bar{x}}$ = 67.2, 89.3, in control. $\bar{R}$ = 10.81,
UCL$_R$ = 27.8, in control. Can set x limits: 59.1, 97.4,
which are far outside of specifications. Need a fundamental
change in process.

7.11 Must emphasize that it is just as easy to put $\bar{x}$'s between
limits for $\bar{x}$ as it is to run x's between x limits. The
natural spread of x's is much greater than for $\bar{x}$'s, hence $\bar{x}$
limits must be *well inside* of x specifications.

Chapter 8

8.3 (a) $\bar{R} = 1.7$, $\text{UCL}_R = 3.6$, in control. $\hat{\sigma} = .73$ $A_2\bar{R} = .98$
     $3\hat{\sigma} = 2.19$ used for safe process averages. (c) Tool wearing
     in control. First run may have been let run too long.

8.5 (a) $\bar{x} = 484.8$, $\bar{R} = 28.2$, $\text{UCL}_R = 92.0$, R's in control.
     2σ limits for x's: 434.8, 534.8, x's in control. (b) Because
     of control can estimate proportion of heats outside (35%).

8.9 U = 9.2, control lines: 4.11, 6.35, 8.59. Out of control.

Chapter 9

9.1 $p'_{95} = .017$, $p'_{10} = .084$.

9.3 Check points:
     (a) p' = .02, Pa = .783, AOQ = .0157, ASN = 80.0, ATI = 280
         p' = .05, Pa = .238, AOQ = .0119, ASN = 80.0, ATI = 781
     (b) p' = .02, Pa = .842, AOQ = .0168, ASN = 77.6, ATI = 224
         p' = .05, Pa = .268, AOQ = .0134, ASN = 73.1, ATI = 755

9.5 Check points:
         p' = .02, Pa = .819, AOQ = .0164, ASN = 74.5, ATI = 411
         p' = .05, Pa = .299, AOQ = .0150, ASN = 97.0, ATI = 1418

9.7 Ac = 7; n = 392 and n = 3920.

Chapter 10

10.1 (a) n = 32, Ac = 0; (b) n = 50, Ac = 0; (c) n = 13, Ac = 0;
     (d) n = 32, Ac = 0; (e) $n_1 = n_2 = 80$, $\text{Ac}_1 = 0$, $\text{Re}_1 = 3$,
     $\text{Ac}_2 = 3$, $\text{Re}_2 = 4$; (f) n = 200, Ac = 0; (g) $n_1 = n_2 = 50$,
     $\text{Ac}_1 = 11$, $\text{Re}_1 = 16$, $\text{Ac}_2 = 26$, $\text{Re}_2 = 27$; (h) $n_1 = n_2 = 200$,
     $\text{Ac}_1 = 0$, $\text{Re}_1 = 3$, $\text{Ac}_2 = 3$, $\text{Re}_2 = 4$;
     (i) $n_i = 80$, Ac * 0 0 1 2 3 4
                     Re 2 3 3 4 4 5 5
     (j) $n_i = 80$, Ac * * 0 0 1 1 2;
                     Re 2 2 2 3 3 3 3
     (k) $n_1 = n_2 = 20$, $\text{Ac}_1 = 0$, $\text{Re}_1 = 4$, $\text{Ac}_2 = 3$, $\text{Re}_2 = 6$;
     (1) n = 32, Ac = 2, Re = 5.

10.3 Pa = .062 at p' = .04. Compatible.

10.5 CSP-1: i = 96, CSP-2: i = 128. Explain i and f.

10.7 i = 14 or 15. Discussion.

Chapter 11

11.1 Test seven fuses, finding $\bar{x}$. $\bar{x} \le K$ accept, otherwise reject.
K = 112.1, 112.5 or 112.9, by choices made. Sketch.

11.3 Weigh contents of four packages and find $\bar{x}$. $\bar{x} \ge K$ accept,
otherwise reject. K about 501.5. Sketch.

11.5 Measure 11 gaskets and find $\bar{x}$. $\bar{x}$ inside .1044, .1076 in.
accept, otherwise reject. Sketch.

11.7 Use n = 9 and find s^2. $s^2 \le 4.91$ accept. Sketch and
experiment.

11.9 Use n = 22, and find s^2 (in $(.00001 \text{ in.})^2$). $s^2 \le 5.64$
accept, otherwise reject.

11.11 Use n = 147, k = 2.31. $\bar{x} + ks \le 150$ sec accept, otherwise
reject.

11.15 Code K. n = 35, M = 1.87%. Estimate percent p_U above U by
$Q_U = (U - \bar{x})/s$. $p_U \le M$ accept.

Chapter 12

12.1
Three dice total	3	4	5	6	7	8	9	10
Expected d, 108	.5	1.5	3.0	5.0	7.5	10.5	12.5	13.5
Three-dice total	11	12	13	14	15	16	17	18
Expected d, 108	13.5	12.5	10.5	7.5	5.0	3.0	1.5	.5

108

12.3 $\hat{\sigma}_{OD} = .0001376$, $\hat{\sigma}_{ID} = .0001720$. $\hat{\mu}_w = .00063$, $\hat{\sigma}_w = .0002202$
(all in inches), z = -2.41, p' = .0080.

12.5 $\mu = .1987$, $\sigma = .001,855$ in. Needed independence. 3σ limits
.19314, .20426 in.

12.7 For z = x + y, $\mu = 200$, $\sigma = .467$, $3\sigma = 1.40$ ohms. Will meet

200 ± 1.5 ohms.

12.9   For w, μ = 175.7, σ = 6.08 lb.   3.9% low.

Chapter 13

13.1   (b) r = .9964, m = 1.12261, b = 1.794, $s_{y \cdot x}$ = 17.52.
(c) t = 35.2, significant.   (d) $\hat{y}$ = 166.6.

13.3   (b) r = .39745, m = 1.9101, b = 6.05, $s_{y \cdot x}$ = 17.88.
$\hat{y}$ = 9.87 at x = 2.

13.5   t = -41.0, significant. $s_{y \cdot x}$ = .168. $\hat{y}$ = 6.18, 4.86.

13.7   (a) $\bar{x}$ = 112.714, $\bar{y}$ = 105.347, s_y = 3.172, m = .89714, b = 4.2,
r = .81634, $s_{y \cdot x}$ = 1.852.   (b) t = 9.69 significant.
(c) $\hat{y}$ = 102.0, 111.0.

Chapter 14

14.1   P(1) = .1, P(2) = .09, P(3) = .081, P(4) = .0729, μ_n = 10,
σ_n = 9.5.

14.3   Table 14.1 gives 22 tested with zero failures.

14.5   .981, .972, .963.

14.7   h(x) = 1/2000, E(d failures) = 10h(x)1200 = 6.

14.9   .9998 .